Adaptive Nussbaum Design for Nonlinear Dynamical Systems

Yongliang Yang · Guilong Liu · Qing Li

Adaptive Nussbaum Design for Nonlinear Dynamical Systems

Springer

Yongliang Yang
School of Automation and Electrical Engineering
University of Science and Technology Beijing
Beijing, China

Guilong Liu
School of Automation and Electrical Engineering
University of Science and Technology Beijing
Beijing, China

Qing Li
School of Automation and Electrical Engineering
University of Science and Technology Beijing
Beijing, China

ISBN 978-981-95-4996-2 ISBN 978-981-95-4997-9 (eBook)
https://doi.org/10.1007/978-981-95-4997-9

This work was supported by National Nature Science Foundation of China (62373039, 62273033), Beijing Natural Science Foundation (J210005) and Interdisciplinary Research Project for Young Teachers of USTB (Fundamental Research Funds for the Central Universities) (FRF-IDRY-22-017).

This Springer imprint is published by the registered company Springer Nature Singapore Pte Ltd.
The registered company address is: 152 Beach Road, #21-01/04 Gateway East, Singapore 189721, Singapore

Preface

Complex nonlinear dynamical systems present inherent challenges in both stability analysis and control design, particularly when dealing with uncertainties in control directions. To address these issues, advanced control strategies are essential. The Nussbaum-type function has proven to be an effective tool, initially introduced for first-order systems to compensate for unknown control gains. Its scope has since expanded to include time-delay systems, switching systems, and other complex scenarios. Despite these advancements, traditional single Nussbaum methods face significant limitations when applied to systems with multiple unknown control directions, often suffering from gain accumulation and mutual cancellation effects. This monograph focuses on developing enhanced Nussbaum-type function designs that overcome these shortcomings. The proposed methods address key performance aspects of complex nonlinear systems, including steady-state accuracy, transient performance, and complex structural characteristics. By proposing a unified framework, this work advances both the theoretical foundations and practical implementations of robust control strategies under multiple uncertainties.

The main body of this monograph is organized into three parts, each addressing a key performance aspect of complex nonlinear systems. Part I focuses on improving steady-state accuracy through the introduction of asymptotic stability concepts, emphasizing their critical role in achieving precise trajectory tracking under uncertain control directions. An adaptive Nussbaum control framework is developed to ensure that system outputs converge to desired trajectories over time without steady-state errors. Additionally, practical considerations such as physical state constraints are incorporated to enhance robustness and applicability in engineering systems. Part II addresses transient performance by introducing fixed-time stability techniques aimed at significantly improving system responsiveness and reliability. Fixed-time Nussbaum method guarantees convergence of nonlinear systems within a predefined time, independent of initial conditions. The proposed control strategies are constructed to mitigate the effects of modeling inaccuracies and external disturbances, thereby ensuring stability. The final Part III turns to advanced topics with Nussbaum design for complex systems, focusing on the coordination of multiagent systems and underactuated structure of nonholonomic systems. For multi-agent systems, novel coordination schemes based on consensus and bipartite consensus are proposed, effectively resolving challenges arising from interactions and unknown control directions. For nonholonomic systems, resilient adaptive Nussbaum control designs are developed to overcome motion constraints and ensure stability, even under adverse conditions such as false data injection attacks.

Through rigorous theoretical analysis and extensive simulation validations, the proposed Nussbaum methods offer new insights and practical tools for the control of complex nonlinear systems. This monograph aims to provide new insights and tools for researchers and engineers working on advanced nonlinear system control.

Beijing, China
June 2025

Yongliang Yang
Guilong Liu
Qing Li

Acknowledgements Our acknowledgments also go to our fellow colleagues who have offered invaluable support and encouragement throughout this research effort. Thanks go to Zhenzhuo Shan, Tianhao Fei, Liqiang Tang, and Hongyu Han for their commentary. The authors are especially grateful to their families for their encouragement and never ending support when it was most required. Finally, we would like to thank the editors at Springer for their professional and efficient handling of this project.

The writing of this book was partially supported in part by the National Nature Science Foundation of China under Grants 62373039 and 62273033.

Competing Interests The authors have no competing interests to declare that are relevant to the content of this manuscript.

Contents

List of Figures

Chapter 1
Introduction

1.1 Nussbaum Design

Due to the challenge of complex nonlinear dynamical systems, neural networks (NNs) and fuzzy logic systems (FLSs)-based intelligent control techniques have been extensively studied for stability analysis and control synthesis for typical nonlinear systems. However, perfect modeling of engineering systems is impossible, resulting in limited available knowledge and increasing the difficulty of control design. Specifically, classic adaptive control design only compensates uncertain drift dynamics, with the control direction being a known priori. Therefore, additional attention should be paid to the uncertainty with unknown control direction [1]. The conventional solution to the issue of a single unknown control direction is the Nussbaum-based method. The Nussbaum-type function was originally proposed to compensate for the unknown sign of control signals in a first-order nonlinear system [2]. With the development of conventional Nussbaum-type gain design, the Nussbaum-type gain design is efficient in dealing with the unknown control direction, which has been successfully developed for time-delay systems [3], robotic manipulators [4, 5], and autonomous underwater vehicles [6]. Zhou et al. employ the Nussbaum type gain for the containment control of stochastic nonlinear multiagent systems [7]. Dong et al. adopt the Nussbaum method to compensate for the actuator fault in automatic aerial vehicle systems [8]. Li et al. extend the Nussbaum design to take the state-dependent constraints in [9]. Ryan then expanded the applicability of the Nussbaum-type function to encompass a broader class of nonlinear systems, including systems with nonlinear actuator characteristics [10]. However, many engineering systems, such as aircraft systems [11], can be modeled with multiple control directions, which is challenging and cannot be neglected [12, 13].

When trying to apply the above method to a system with multiple unknown control directions, only indirect methods can be used. To be specific, the stability in the Lyapunov sense could be obtained indirectly by recursively applying the Lyapunov-Nussbaum theorem [14, 15]. We aim to use an entire approach to discussing system stability and treating the entire system as a whole. For stability analysis of the entire system, we need to introduce multiple Nussbaum-type gains. However, the aforementioned methods are no longer applicable [16], as the accumulation of multiple

Y. Yang et al., *Adaptive Nussbaum Design for Nonlinear Dynamical Systems*,
https://doi.org/10.1007/978-981-95-4997-9_1

Nussbaum-type functions may offset their ability to counteract multiple unknown control directions [17]. Therefore, multiple Nussbaum-type function designs are desired to simplify stability analysis and have been successfully applied to engineering applications. Multiple Nussbaum-type functions for hypersonic flight vehicles have been designed to ensure the zero-error-constrained tracking and the boundedness of multiple Nussbaum-type gains [18]. Hybrid Nussbaum gains are developed in the proposed technique to deal with uncertain high-power dynamics for consensus of multiagent systems [12]. The aforementioned schemes are relatively stringent, as they require all control directions to be aligned.

To overcome these limitations and achieve the desired stability analysis, this work introduces novel Nussbaum-type functions, modifying their properties to adapt to different system designs and enhance their applicability. First, the intent focuses on steady-state behavior to ensure that the system achieves accuracy and stability over time, and attaining asymptotic tracking stability is a crucial step to ensure steady-state performance. Second, it examines the dynamic performance, which deals with how quickly and efficiently the system converges to its desired state, utilizing the concept of fixed-time convergence. Third, the book addresses system complexity, exploring the challenges of applying control strategies to complex systems, such as multiagent systems and nonholonomic systems. These three parts form the core of the monograph, with each chapter dedicated to exploring strategies and solutions for these specific control challenges.

1.2 Asymptotic Tracking

In the first part of this monograph, we focus on steady-state behavior, with a particular emphasis on asymptotic stability as a central theme. Asymptotic stability refers to the property where the system's output converges to a desired reference signal as time approaches infinity, with the tracking error gradually diminishing to zero [19]. This is a more stringent requirement than merely ensuring that the system output remains within bounded limits, known as uniform ultimate boundedness (UUB), which confines the error within a specific range without guaranteeing a precise convergence to the desired trajectory [17]. Asymptotic stability emphasizes the precision of tracking over time, ensuring that the error diminishes completely as time progresses, rather than merely staying bounded [20]. The motivation for studying asymptotic stability comes from the growing need for high-precision control in modern systems [21, 22]. In applications such as robotics [23] and aircraft [24], it is increasingly insufficient only to prevent instability or keep outputs bounded within acceptable limits. Therefore, asymptotic tracking becomes a critical design goal, as it ensures that the system output aligns with the reference signal in the long term, with no steady-state error [25]. Recent studies have shown the importance of asymptotic stability in handling real-world systems where achieving zero steady-state error is necessary [26]. To achieve asymptotic stability, the design of control laws must ensure system stability and facilitate the convergence of the tracking error to zero [27]. This requires enhanced control strategies that adapt to the system's current state and reference signal.

Moreover, another significant challenge lies in the state constraints on system states [28]. Many systems have inherent limitations, such as physical bounds on velocities, positions, or energy consumption, that must be respected while ensuring the system's output converges to the desired trajectory. These constraints complicate the design of control laws and require careful consideration to avoid violating system limits while achieving asymptotic stability.

1.3 Fixed-Time Stability

In the second part of this work, we focus on the dynamic performance. The Nussbaum-type design results only guarantee stability in infinite time, which might not meet the requirement in some real-world applications. To improve the infinite settling time for a conventional stabilizing design, Bhat and Bernstein presented the concept of finite-time stabilization [29]. To achieve attitude tracking within a finite time interval, Gao et al. developed a novel fault-tolerant attitude controller for spacecraft attitude tracking to deal with actuator saturation and the fault with the Nussbaum technique [30]. Chen et al. established switching rules based on Lyapunov-based logic for the lower-triangular nonlinear systems under the condition that the control direction is unknown, where sufficient conditions for global finite-time stability are investigated [31]. Wu et al. developed an adaptive switching control based on a power integrator design for the nonlinearly parametrized systems with unknown control directions [32]. Dong et al. utilized the Nussbaum function to address the unknown control directions problem to ensure the finite-time convergence [33]. Song et al. developed a modified fractional-order design to achieve finite-time tracking performance while guaranteeing the full-state constraints [34]. Liu et al. investigated multiple Nussbaum gains to cope with multiple unknown control directions in the stochastic system [35]. Nonetheless, the finite-time performance depends on the beginning state, which may limit its viability. This might hinder the practical application of finite-time design because the initial state of engineering systems is difficult to obtain in advance.

By extending the global finite-time stability with guaranteed settling time, which is independent of the system's initial conditions, Polyakov first proposed the fixed-time stability in [36]. Since then, fixed-time stability has been extensively investigated to improve the control performance for complex nonlinear systems. Tao et al. dealt with the singularity issues in the conventional fixed-time design by developing a nonsingular sliding mode surface [37]. Mei et al. investigated the fixed-time stabilization problem with consideration of the output constraint [38]. Hou et al. distributed formation control protocol is proposed for heterogeneous nonlinear multiagent systems with fixed time stability and global constraints [39]. Cheng et al. introduce a finite-time performance function to guarantee the predetermined formation error convergence within a fixed time [40]. Kong et al. establish new fixed-time stability lemmas to analyze a class of discontinuous fuzzy inertial neural networks and the stability of time-delayed discontinuous systems in [41, 42], respectively.

While fixed-time stability offers significant improvements over conventional stability methods by guaranteeing convergence within a predetermined timeframe, other factors may still affect the performance of these systems. For instance, random disturbances and external noise can significantly impact the accuracy and reliability of fixed-time convergence. Additionally, modeling inaccuracies and unknown dynamic interactions pose further challenges in ensuring robust performance under fixed-time control. Therefore, designing controllers that are resilient to such uncertainties and capable of maintaining desired performance in the presence of stochastic influences remains a critical area for ongoing research.

1.4 Advanced Topics

In the third part of this work, we focus on the complexity of systems, particularly in terms of the interaction between multiple systems and the constraints of the control strategies. These two challenges are prevalent in many practical engineering applications, and managing them effectively is crucial for system performance. The typical systems that illustrate these issues are multiagent systems (MASs) and nonholonomic systems.

MASs have been extensively studied recently due to their wide applications in the field of engineering systems, such as microgrid systems [43], chemical reactor [44], and vehicle networks [45], to name a few. The nonlinearity commonly exists in the agent dynamics in real-world applications, such as circuit systems [46], robotic manipulators [47]. In the case of MASs, the complexity arises from the need to coordinate multiple agents that may have different dynamics and states. These systems are inherently distributed, with each agent making local decisions based on its observations and the information it receives from other agents [48, 49]. The agent dynamics in MASs under consideration have evolved from first- and second-order dynamical systems to general high-order linear systems [50–52]. Within multiagent coordination, there are several strategies to achieve synchronization or consensus among agents [45]. The two most widely studied approaches are consensus and bipartite consensus. Consensus refers to the process by which all agents reach a common value or state, such as agreeing on a trajectory or state variable. This approach is particularly useful when agents need to align their behavior toward a common goal, such as information control or resource allocation [50]. On the other hand, bipartite consensus involves a situation where the agents are divided into two groups, each with a distinct objective. However, the overall system must ensure that agents within each group reach consensus independently while maintaining certain relations across the groups [17]. Though conceptually similar, these approaches differ in their structural requirements and are applicable to different types of multiagent systems. The subsequent chapters will discuss both consensus and bipartite consensus in more detail.

Multiagent systems (MASs) have been extensively studied in recent years due to their wide applications across various engineering domains, such as microgrid systems [43], chemical reactors [44], and vehicular networks [45]. In real-world scenarios, the agent dynamics in MASs are often nonlinear, as observed in circuit systems [46], robotic manipulators [47], and many other practical applications. The inherent complexity of MASs stems not only from the nonlinearities in agent models but also from the distributed nature of decision-making, where each agent operates based on local information and exchanges data with neighboring agents [48, 49]. Over the past decades, research on MASs has progressed significantly, evolving from studies on first- and second-order dynamical systems to more general high-order linear and nonlinear systems [50–52]. A central challenge in MASs is achieving coordinated behavior among agents despite their heterogeneous dynamics and limited communication capabilities. To this end, several coordination strategies have been proposed, with consensus and bipartite consensus being the two most widely studied frameworks [45]. Consensus refers to the process by which all agents asymptotically reach an agreement on a common value or trajectory. This approach is fundamental in applications where agents are required to synchronize their states to accomplish tasks such as trajectory tracking, information fusion, or resource distribution [50]. In a typical consensus protocol, each agent adjusts its state based on the relative information received from its neighbors, leading to a collective agreement over time. Bipartite consensus, on the other hand, addresses coordination problems in networks characterized by competitive or antagonistic interactions. In this framework, agents are divided into two distinct groups, with each group reaching an internal consensus while maintaining a predefined relational structure—typically opposite in sign—with respect to the other group [17]. Such coordination patterns are particularly relevant in applications involving adversarial behaviors or competitive clustering. Although consensus and bipartite consensus share the common objective of achieving coordinated behavior, they differ significantly in their structural assumptions and convergence mechanisms. This part systematically investigates both strategies under the context of nonlinear MASs with multiple unknown control directions, aiming to develop robust control schemes capable of handling the complexities inherent in distributed nonlinear systems.

Moving from MASs to nonholonomic systems, we encounter another class of systems characterized by intrinsic motion constraints. Nonholonomic systems are defined by non-integrable, velocity-dependent constraints that restrict the system's ability to move freely in certain directions [53]. Such constraints are commonly encountered in practical robotic applications, including autonomous vehicles [54] and wheeled mobile robots [55], where lateral movements are inherently limited. The presence of these nonholonomic constraints necessitates careful control design to ensure that system trajectories remain feasible while achieving desired objectives. Although nonholonomic systems are controllable, it is well established that continuous, pure-state feedback is difficult to stabilize the closed-loop system [56]. This limitation stems from the fact that nonholonomic systems violate Brockett's necessary condition for smooth feedback stabilization [57], prompting significant research into alternative control methods. Various approaches have been explored, including time-varying feedback [58, 59], discontinuous time-invariant feedback [60, 61], and hybrid feedback control [62]. Beyond the inherent control challenges, nonholonomic systems are particularly vulnerable to adversarial attacks due to their underactuated and strongly coupled nature. External attacks, such as false data injection, can change the system inputs and manifest as uncertainties in control directions. In the presence of such attacks, conventional control strategies often fail to guarantee stability. To address this issue, this part investigates the application of Nussbaum-type function techniques to nonholonomic systems under attack. By treating the unknown control direction as an uncertainty induced by disturbances, the proposed methods utilize Nussbaum functions to achieve robust stabilization, thus enhancing the system's resilience against malicious attacks.

In multiagent systems, the interactions among agents introduce multiple unknown control directions, which are often mutually influential. Traditional control methods face significant challenges in ensuring system stability under such conditions. In nonholonomic systems, structural motion constraints inherently limit control authority. When subjected to external attacks, the control directions may become uncertain, further exacerbating the difficulty of maintaining system stability. Moreover, accurate system modeling is often unattainable in real-world applications due to inherent complexity and pervasive uncertainties. These challenges underscore the necessity for control design strategies that are resilient to model inaccuracies and unknown control directions. This monograph addresses these issues by developing a adaptive Nussbaum control capable of ensuring stability and robustness in complex nonlinear systems.

1.5 Preliminary Knowledge

1.5.1 Nussbaum Design

We begin by studying the Nussbaum-type functions, which are crucial in dealing with control direction uncertainties in nonlinear systems. In this subsection, we first present the fundamental definition and related lemmas to establish the foundational properties of Nussbaum-type functions.

Definition 1.1 ([15]) The Nussbaum-type function is defined as a smooth function $N(\xi) : \mathbb{R} \to \mathbb{R}$ satisfying

$$\lim_{s\to\infty} \sup \frac{1}{s} \int_0^s N(\xi) \mathrm{d}\xi = \infty,$$
$$\lim_{s\to\infty} \inf \frac{1}{s} \int_0^s N(\xi) \mathrm{d}\xi = -\infty.$$

Lemma 1.1 ([15]) *Suppose $\zeta(t)$ and $V(t)$ are smooth functions on $[0, t_f)$ with $V(t) \geq 0, \forall t \in [0, t_f)$. If the following inequality holds*

$$V(t) \leq c_0 + \int_0^t (g(\tau) N(\zeta(\tau)) + 1) \dot{\zeta}(\tau) d\tau, \tag{1.1}$$

with unknown continuous function $g(\cdot)$ and positive constants c_0. Then, $V(t)$, $\zeta(t)$, and $\int_0^t g(\tau) N(\zeta(\tau)) \dot{\zeta}(\tau) \mathrm{d}\tau$ are bounded on $[0, t_f)$.

Lemma 1.2 ([63]) *Suppose that there exist smooth functions $\zeta(t)$ and $V_\zeta(t)$ satisfy $V_\zeta(t) \geq 0, \forall t \in [0, t_f)$ and $N(\zeta(t))$ is a smooth Nussbaum-type function. If the following inequality holds*

$$V_\zeta(t) \leq c_0 + \mathrm{e}^{-c_1 t} \int_0^t (g(\tau) N(\zeta(\tau)) + 1) \dot{\zeta}(\tau) \mathrm{e}^{c_1 \tau} \mathrm{d}\tau,$$

with unknown continuous function $g(\cdot)$ and positive constants c_0 and c_1. Then, $V_\zeta(t)$, $\zeta(t)$, and $\int_0^t g(\tau) N(\zeta(\tau)) \dot{\zeta}(\tau) \mathrm{d}\tau$ are bounded on $[0, t_f)$.

To investigate the application of Nussbaum-type functions in control systems, we present the foundational definitions and essential lemmas that establish their key properties. The commonly used Nussbaum-type function selections include $N(\xi) = \xi^2 \sin(\xi)$, $\xi^2 \cos(\xi)$, and $e^{\xi^2} \cos\left(\frac{\pi}{2}\xi\right)$.

Nonetheless, as highlighted by Lemmas 1.1 and 1.2, these conventional Nussbaum functions encounter significant challenges when addressing scenarios with multiple unknown control directions. Consequently, the traditional Nussbaum designs presented in [15] are not directly applicable to systems characterized by multiple unknown control directions. To overcome this limitation, the subsequent sections of this work introduce novel Nussbaum functions tailored for complex systems, along with the corresponding theoretical proofs that validate its effectiveness.

1.5.2 Fixed-Time Stability

To comprehensively understand fixed-time convergence in control systems, the foundational definitions for deterministic systems are established. These definitions lay the groundwork for analyzing the stability and convergence properties within a finite or fixed time frame.

Definition 1.2 ([36]) For the nonlinear dynamical system $\dot{x} = F(x)$ with Lipschitz function $F(x)$ and initial conditions $x_0 = x(t_0)$. The equilibrium of this system is said to be finite-time stable if the following conditions can be satisfied.

1. The equilibrium of this system is asymptotically stable;
2. There exists a mapping $T : \mathbb{R}^n \to \mathbb{R}^+ \cup \{0\}$ such that $x(t) = 0$ for all $t \geq T(x_0)$.

In addition, $T(x_0)$ is called settling time.

Definition 1.3 ([36]) For the nonlinear dynamical system $\dot{x} = F(x)$ with Lipschitz function $F(x)$ and initial conditions $x_0 = x(t_0)$. The equilibrium of this system is said to be fixed-time stable if the following conditions can be satisfied.

1. The equilibrium of this system is finite-time stable;
2. The settling time $T(x_0)$ is upper bounded in the sense that $T(x_0) \leq T_{max}$ for $\forall x_0$.

Definition 1.4 ([64]) For the nonlinear dynamical system $\dot{x} = F(x)$ with Lipschitz function $F(x)$ and initial conditions $x_0 = x(t_0)$. The equilibrium of this system is said to be practically fixed-time stable if the following conditions can be satisfied.

1. For arbitrary positive constant ε, there exists a settling time $T(\varepsilon, x_0)$ such that the state satisfies $\|x(t)\| \leq \varepsilon$ for all $t \geq T(\varepsilon, x_0)$;
2. The settling time is upper bounded for all x_0.

However, real-world systems often exhibit stochastic behaviors due to inherent uncertainties and external disturbances. We introduce the necessary definitions for stochastic systems to extend the applicability of fixed-time stability concepts to such scenarios. The following definitions pertain to the fixed-time convergence properties of stochastic dynamical systems, ensuring robustness in the presence of randomness.

Consider the stochastic system evolving in continuous time as

$$dx = f(x, t)\,dt + g(x, t)\,d\omega, \tag{1.2}$$

where$x(t_0) = x_0 \in \mathbb{R}^n$is the initial condition, $x \in \mathbb{R}^n$ is the system state, $f : \mathbb{R}^n \times \mathbb{R} \to \mathbb{R}^n$ is the drift term, and $g : \mathbb{R}^n \times \mathbb{R} \to \mathbb{R}^{n \times r}$ is the diffusion term. In addition, $f(\cdot)$ and $g(\cdot)$ are locally Lipschitz functions satisfying $f(0, t) = 0$ and $g(0, t) = 0$, ω is an r-dimensional standard Brownian motion defined on the complete probability space $\left(\Omega, F, \{F_t\}_{t \geq 0}, P\right)$ with a sample space Ω, a σ-field F, a filtration $\{F_t\}_{t \geq 0}$, and a probability measure P.

The following definitions about system (1.2) are required for subsequent discussions for the stochastic systems.

Definition 1.5 Consider the $\mathbb{C}^{2,1}$ function $V(t)$, for system (1.2), the operator $\mathcal{L}$ of V can be donated as

$$\mathcal{L}V = \frac{1}{2}\mathrm{Tr}\left\{g^{\mathrm{T}}\frac{\partial^2 V}{\partial x^2}g\right\} + \frac{\partial V}{\partial x}f + \frac{\partial V}{\partial t}. \tag{1.3}$$

Definition 1.6 The stochastic system (1.2) is globally finite-time stable in probability if the following statements hold for any initial state $x_0 \neq 0$.

1. $P\{T(x_0) < \infty\} = 1$ with $T(x_0) = \inf\{t; x(t) = 0\}$;
2. $P\{\|x(t)\| \leq \beta\} = 1$ with a positive constant β for $t \geq T(x_0)$.

Definition 1.7 The stochastic system (1.2) is globally fixed-time stable in probability if the following statements hold for any initial state x_0.

1. The trivial solution $x(t)$ is globally finite-time stable in probability;
2. The settling time $T(x_0)$ satisfies $E[T(x_0)] \leq T_{\max}$ with the positive bound $T_{\max}$.

Definition 1.8 The stochastic system (1.2) is practically fixed-time stable in probability, if there exists a positive constant ς such that

1. For any $t > t_0 + T(\varsigma, x_0)$, one has $P\{\|x(t)\| \leq \varsigma\} = 1$, where $T(\varsigma, x_0)$ is the settling time defined as $T(\varsigma, x_0) = \inf\{t; \|x(t)\| = \varsigma\}$;
2. The settling time $T(\varsigma, x_0)$ is upper-bounded in the sense that $E[T(\varsigma, x_0)] \leq T_{\max}$, where $T_{\max} > 0$, $x_0 \in \Theta$ with Θ being a compact set.

1.5.3 Graph Theory

In this section, we explore the foundational concepts of graph theory as applied to MASs. We begin by discussing consensus in directed graphs and subsequently delve into bipartite consensus using signed graphs. These frameworks provide the necessary mathematical underpinnings for analyzing leader-follower dynamics and achieving desired coordination among agents.

1.5.3.1 Unsigned Directed Graphs for Consensus

Consider the multiagent systems consisting of m followers and a leader. The leader is labeled as 0 and the followers set are denoted as $\mathcal{F} = \{1, 2, ..., m\}$. The communication network among the followers is modeled by a directed graph $\mathcal{G}_F = \{\mathcal{F}, \mathcal{E}, \mathcal{A}\}$, where $\mathcal{F}$ is the follower set, $\mathcal{E} = \{(i, k) \in \mathcal{F} \times \mathcal{F}\}$ is the edge set, and $\mathcal{A} = [a_{ik}] \in \mathbb{R}^{m \times m}$ is the weighted adjacency matrix of the graph, with $a_{ik} = 1$ if there is an edge from $k\text{-}th$ follower to $i\text{-}th$ follower, and $a_{ik} = 0$ otherwise. We assume that the graph $\mathcal{G}_F$ is simple, i.e., $a_{ii} = 0, \forall i \in \mathcal{F}$. Denote the neighbor set of $i\text{-}th$ follower as $\mathcal{N}_i = \{k \,|a_{ik} = 1, k \in \mathcal{F}\}$. Define the in-degree matrix as $\mathcal{D} = \mathrm{diag}[d_i, ..., d_m]$, where $d_i = \sum_{k \in \mathcal{N}_i} a_{ik}$. The Laplacian matrix is defined as $\mathcal{L} = \mathcal{D} - \mathcal{A}$. Denote $\bar{\mathcal{G}}$ as the augmented graph that includes both the leader $\{0\}$ and the follower set $\mathcal{F}$ and all associated edges. The pinning gain matrix $\mathcal{B} = \mathrm{diag}[b_i, ..., b_m]$, where b_i is the pinning gain connecting the leader and the $i\text{-}th$ follower with $b_i = 1$ if the $i\text{-}th$ follower is connected to the leader and $b_i = 0$ otherwise. Denote $\Delta = \mathcal{L} + \mathcal{B}$ with $(i, k)\text{-}th$ entry $\delta_{i,k}$, where $\delta_{i,k} = -a_{i,k}$ if $i \neq k$ and $\delta_{i,k} = b_i$ if $i = k$.

1.5.3.2 Signed Undirected Graphs for Bipartite Consensus

In scenarios where agents may exhibit antagonistic interactions, signed graphs become a suitable model to capture both cooperative and adversarial relationships. A signed graph allows edges to have positive or negative weights, representing friendly or hostile interactions, respectively.

In this study, we explore the communication of leader-follower MASs utilizing graph theory. The graph $\mathcal{G}$ comprises the node set $\mathcal{V} = \{0, \mathcal{F}\}$ and the edge set $\mathcal{E}$. Each node in the nodes set $\mathcal{V}$ stands for each agent, where 0 denotes the leader and $\mathcal{F} = \{1, \ldots, N\}$ is the set of followers. Each edge denotes the interconnection between two agents, meaning there exists a link between the follower κ and follower

ℓ if $(\kappa, \ell) \in \mathcal{E}$. In a signed graph, the signed weighed matrix $\mathcal{A} \in \mathbb{R}^{N \times N}$ is defined with element $a_{i,j}$, where $a_{i,j} = 1$ or $a_{i,j} = -1$ if $(j, i) \in \mathcal{E}$, and $a_{i,j} = 0$ otherwise. Furthermore, the leader adjacency matrix between the leader and followers is represented by the adjacency matrix $\mathcal{B}$ with $\mathcal{B} = \text{diag}\,(b_1, b_2, \ldots, b_N)$ with $b_i > 0$ if agent i can receive information from the leader and $b_i = 0$ otherwise. The neighbor set of follower κ is expressed as $\mathcal{N}_\kappa \triangleq \left\{ \ell \in \mathcal{F} \,|\, a_{\kappa,\ell} = \pm 1 \right\}$. Define the in-degree matrix $\mathcal{D}$ as $\mathcal{D} = \text{diag}\,(d_1, d_2, \ldots, d_N)$ with $d_i = \sum\limits_{k \in \mathcal{N}_i} \left| a_{i,k} \right|$. Then, the Laplacian matrix of the graphs $\mathcal{G}$ is defined as $\mathcal{L} = \mathcal{D} - \mathcal{A}$.

The following lemma and definitions are required for the subsequent discussions.

Definition 1.9 ([65]) A graph $\mathcal{G}$ is said to be connected if there exists at least one root node such that all the other nodes can obtain information from the root node.

To facilitate bipartite consensus, where agents agree on opposite values based on their respective partitions, we introduce the concept of structural balance in signed graphs.

Definition 1.10 ([66]) The signed graph $\mathcal{G}$ is considered structurally balanced if there exist two sets Υ_1 and Υ_2 satisfying $\Upsilon_1 \cup \Upsilon_2 = \mathcal{G}$, $\Upsilon_1 \cap \Upsilon_2 = \emptyset$ such that $a_{i,j} > 0$ if i and j belong to the same set and $a_{i,j} < 0$ if they belong to different sets.

Lemma 1.3 ([66]) *The signed graph $\mathcal{G}$ is considered structurally balanced if and only if there exists a matrix Q from the set $Q = \{diag\,[\rho_1, \ldots, \rho_N] \,|\, \rho_i \in \{1, -1\}\}$ such that the product $Q\mathcal{A}Q$ has all nonnegative entries.*

1.5.4 Radial Basis Function Neural Networks

To deal with the unknown functions, function approximation techniques can be employed, such as the radial basis function neural networks (RBFNNs). For example, to approximate the unknown function $F(\cdot) : \mathbb{R}^n \to \mathbb{R}$, the RBFNN is considered as

$$\hat{F}(x) = W^{\mathrm{T}} \varphi(x), \tag{1.4}$$

with the weight vector $W \in \mathbb{R}^N$, and the RBF basis function $\varphi : \mathbb{R}^n \to \mathbb{R}^N$ defined as

$$\varphi(x) = \begin{bmatrix} \varphi_1(x) \\ \vdots \\ \varphi_N(x) \end{bmatrix} \in \mathbb{R}^N,$$

$$\varphi_j(x) = \exp\left[-\frac{\left(x - \iota_j\right)^{\mathrm{T}} \left(x - \iota_j\right)}{2o_j^2} \right] \in \mathbb{R},$$

$$j = 1, 2, \ldots, N.$$

where $\iota_i \in \Theta \subseteq \mathbb{R}^n$ is the center and $o_i \in \mathbb{R}$ is the width of the Gaussian function. The function approximation property of the RBFNN is investigated in the following lemma.

Lemma 1.4 ([63]) *For any continuous function $F : \Theta \subseteq \mathbb{R}^n \mapsto \mathbb{R}$ defined on the compact set Θ, there exists an RBFNN in the form of* (1.4) *such that with arbitrary accuracy $\varepsilon_m > 0$ and sufficient many hidden nodes number N, the function approximation residual $\varepsilon^* (x)$ is bounded on Θ in the sense that $|\varepsilon^* (x)| \leq \varepsilon_m$, where*

$$\begin{aligned}\varepsilon^* (x) &= F (x) - (W^*)^{\mathrm{T}}\varphi (x), \\ W^* &= \arg\min_{W} \left\{\sup_{x} \left|F (x) - W^{\mathrm{T}}\varphi (x)\right|\right\}.\end{aligned}$$

1.5.5 Fuzzy Logic Systems

A fuzzy set is a mathematical representation of a vague concept where the elements are assigned degrees of the membership function, indicating the extent to which an element belongs to the set. A linguistic variable is where the values are expressed using linguistic labels or words. A membership function assigns degrees of membership to linguistic terms within a fuzzy set. The fuzzy logic system includes M IF-THEN rules with the mth rule R^m being

IF x_1 is F_1^m and x_2 is F_2^m and $\cdots$ and x_n is F_n^m,
THEN y is G^m.

$\{x_i\}_{i=1}^n$ and y are the input and output of FLSs, respectively, and $\left\{F_1^m\right\}_{m=1}^M$ and $\{G^m\}_{m=1}^M$ are the fuzzy sets for linguistic variables $\{x_i\}_{i=1}^n$. By incorporating the fuzzifier, fuzzy inference engine, fuzzy rules base, and defuzzifier, the output can be reformulated as

$$y(x) = \frac{\sum_{m=1}^M \max\limits_{y\in R} \mu_{G^m}(y) \prod_{i=1}^n \mu_{F_i^m} (x_i)}{\sum_{m=1}^M \left[\prod_{i=1}^n \mu_{F_i^m} (x_i)\right]},$$

where $x = \{x_1 (t), x_2 (t), \ldots, x_n (t)\}^{\mathrm{T}}$ and $\mu_{F_i^m} (x_i)$ and $\mu_{G^m} (y)$ are the membership functions of fuzzy sets F_i^m and G^m, respectively. Equivalently, the FLS can also be written in a compact form as an expansion of fuzzy basis functions (FBFs), i.e., $y(x) = W^{\mathrm{T}}\varphi(x)$ with

$$\begin{aligned}W &= \left[\max_{y\in R} \mu_{G^1}(y), \max_{y\in R} \mu_{G^2}(y), \ldots, \max_{y\in R} \mu_{G^M}(y)\right]^{\mathrm{T}}, \\ \varphi (x) &=[\varphi_1 (x), \varphi_2 (x), \ldots, \varphi_M (x)]^{\mathrm{T}},\end{aligned}$$

where $\varphi_m(x) = \frac{\prod_{i=1}^{n} \mu_{F_i^m}(x_i)}{\sum_{m=1}^{M}\left[\prod_{i=1}^{n} \mu_{F_i^m}(x_i)\right]}$ is the FBF.

Consider the set of all the FBF expansions Y with the sup-metric $d_\infty(f_1, f_2) = \sup_{x\in\Omega} \|f_1(x) - f_2(x)\|$. Then, (Y, d_∞) is a metric space with the following fact.

Lemma 1.5 ([67]) *For any real continuous function $f(\cdot)$ defined on a compact set Ω. Given an arbitrary $\varepsilon > 0$, there exists a function $y(\cdot) \in Y$ such that $\sup_{x\in\Omega} |f(x) - y(x)| \le \varepsilon$.*

1.5.6 Other Inequalities

Lemma 1.6 ([63]) *There exist constants $m > 0$, $p > 0$ and $q > 0$ such that*

$$|x|^p|y|^q \le \frac{p}{p+q} m|x|^{p+q} + \frac{q}{p+q} m^{-\frac{p}{q}}|y|^{p+q}, \forall x, y \in \mathbb{R}.$$

Lemma 1.7 ([68]) *For any positive real numbers $\{\varsigma, \xi, p, q, k\}$, one can obtain*

$$\left(\sum_{i=1}^{n} \Upsilon\right)^{\frac{3}{4}} \le \sum_{i=1}^{n} \Upsilon^{\frac{3}{4}}, \left(\sum_{i=1}^{n} \Upsilon\right)^{2} \le n \sum_{i=1}^{n} \Upsilon^2.$$

Lemma 1.8 ([22]) *Denote $\kappa = 0.2785$, one has*

$$0 \le |x| - x \tanh\left(\frac{|x|}{\epsilon}\right) \le \kappa\epsilon$$

for all $\epsilon > 0$ and $x \in \mathbb{R}$.

Lemma 1.9 ([69]) *For any $y > 0$ and $x \in \mathbb{R}$, the following inequality holds:*

$$|x| - y \le \frac{x^2}{\sqrt{x^2 + y^2}}.$$

Lemma 1.10 ([20]) *For uniformly continuous scalar function f defined on $[0, +\infty)$, $\lim_{t\to\infty} f(t) = 0$ if $\lim_{t\to\infty} \int_0^t |f(\tau)|d\tau$ exists and is finite.*

Lemma 1.11 ([70]) *For $|\eta| < 1$, the following inequality holds*

$$\log \frac{1}{1-\eta^2} < \frac{\eta^2}{1-\eta^2},$$

References

1. Wen, C., Zhou, J., Liu, Z., Su, H.: Robust adaptive control of uncertain nonlinear systems in the presence of input saturation and external disturbance. IEEE Trans. Autom. Control **56**(7), 1672–1678 (2011)
2. Nussbaum, R.D.: Some remarks on a conjecture in parameter adaptive control. Syst. Control Lett. **3**(5), 243–246 (1983)
3. Ge, S.S., Hong, F., Lee, T.H.: Adaptive neural control of nonlinear time-delay systems with unknown virtual control coefficients. IEEE Trans. Syst. Man Cybern. Part B (Cybern.) **34**(1), 499–516 (2004)
4. Chen, C., Liu, Z., Xie, K., Liu, Y., Zhang, Y., Chen, C.L.P.: Adaptive fuzzy asymptotic control of MIMO systems with unknown input coefficients via a robust Nussbaum gain-based approach. IEEE Trans. Fuzzy Syst. **25**(5), 1252–1263 (2017)
5. Hua, C., Li, H., Li, K., Ding, W.: Low-computation tracking control of nonlinear systems with asymmetric full-state constraints and unknown control directions. IEEE Trans. Syst. Man Cybern.: Syst. **53**(2), 1051–1059 (2023)
6. Li, J., Du, J., Chen, C.L.P.: Command-filtered robust adaptive NN control with the prescribed performance for the 3-D trajectory tracking of underactuated AUVs. IEEE Trans. Neural Netw. Learn. Syst. **33**(11), 6545–6557 (2022)
7. Zhou, Q., Wang, W., Ma, H., Li, H.: Event-triggered fuzzy adaptive containment control for nonlinear multiagent systems with unknown Bouc-Wen hysteresis input. IEEE Trans. Fuzzy Syst. **29**(4), 731–741 (2021)
8. Dong, G., Cao, L., Yao, D., Li, H., Lu, R.: Adaptive attitude control for multi-MUAV systems with output dead-zone and actuator fault. IEEE/CAA J. Autom. Sinica. **8**(9), 1567–1575 (2021)
9. Li, D., Han, H.G., Qiao, J.F.: Adaptive NN controller of nonlinear state-dependent constrained systems with unknown control direction. IEEE Trans. Neural Netw. Learn. Syst. **35**(1), 913–922 (2024)
10. Ryan, E.P.: A nonlinear universal servomechanism. IEEE Trans. Autom. Control **39**(4), 753–761 (1994)
11. Tao, G., Tang, X., Chen, S., Fei, J., Joshi, S.M.: Adaptive Failure Compensation of Two-State Aircraft Morphing Actuators. IEEE Trans. Control Syst. Technol. **14**(1), 157–164 (2006)
12. Lv, M., Yu, W., Cao, J., Baldi, S.: Consensus in High-Power Multiagent Systems With Mixed Unknown Control Directions via Hybrid Nussbaum-based control. IEEE Trans. Cybern. **52**(6), 5184–5196 (2022)
13. Yang, Y., Liu, G., Li, Q., Ahn, C.K.: Multiple adaptive fuzzy Nussbaum-type functions design for stochastic nonlinear systems with fixed-time performance. Fuzzy Sets Syst. **476**, 108767 (2024)
14. Na, J.: Adaptive prescribed performance control of nonlinear systems with unknown dead zone. Int. J. Adapt. Control Signal Process. **27**(5), 426–446 (2013)
15. Ye, X., Jiang, J.: Adaptive nonlinear design without a priori knowledge of control directions. IEEE Trans. Autom. Control **43**(11), 1617–1621 (1998)
16. Zhao, K., Song, Y., Chen, C.L.P., Chen, L.: Adaptive asymptotic tracking with global performance for nonlinear systems with unknown control directions. IEEE Trans. Autom. Control **67**(3), 1566–1573 (2022)
17. Liu, G., Yang, Y., Zhao, X., Ahn, C.K.: Adaptive fuzzy practical bipartite synchronization for multiagent systems with intermittent feedback under multiple unknown control directions. IEEE Trans. Fuzzy Syst. **32**(5), 3211–3225 (2024)
18. Lv, M., De Schutter, B., Wang, Y., Shen, D.: Fuzzy adaptive zero-error-constrained tracking control for HFVs in the presence of multiple unknown control directions. IEEE Trans. Cybern. **53**(5), 2779–2790 (2023)
19. Praly, L.: Asymptotic stabilization via output feedback for lower triangular systems with output dependent incremental rate. IEEE Trans. Autom. Control **48**(6), 1103–1108 (2003)
20. Liang, Y., Li, Y.X., Che, W.W., Hou, Z.: Adaptive fuzzy asymptotic tracking for nonlinear systems with nonstrict-feedback structure. IEEE Trans. Cybern. **51**(2), 853–861 (2021)

21. Zhang, X., Lin, Y.: A new approach to global asymptotic tracking for a class of low-triangular nonlinear systems via output feedback. IEEE Trans. Autom. Control **57**(12), 3192–3196 (2012)
22. Li, Y.X., Hu, X., Che, W., Hou, Z.: Event-based adaptive fuzzy asymptotic tracking control of uncertain nonlinear systems. IEEE Trans. Fuzzy Syst. **29**(10), 3003–3013 (2021)
23. Chen, C., Liu, Z., Zhang, Y., Chen, C.L.P., Xie, S.: Saturated Nussbaum function based approach for robotic systems with unknown actuator dynamics. IEEE Trans. Cybern. **46**(10), 2311–2322 (2016)
24. Tang, X., Tao, G., Joshi, S.M.: Adaptive actuator failure compensation for parametric strict feedback systems and an aircraft application. Automatica **39**(11), 1975–1982 (2003)
25. Li, Y.X.: Command filter adaptive asymptotic tracking of uncertain nonlinear systems with time-varying parameters and disturbances. IEEE Trans. Autom. Control **67**(6), 2973–2980 (2022)
26. Wu, L.B., Park, J.H., Xie, X.P., Li, Y.F.: Adaptive asymptotic tracking control of uncertain nonlinear systems based on Taylor decoupling and event-trigger. IEEE Trans. Syst. Man Cybern.: Syst. **52**(4), 2053–2060 (2022)
27. Cheng, T.T., Niu, B., Zhang, J.M., Wang, D., Wang, Z.H.: Time-/event-triggered adaptive neural asymptotic tracking control of nonlinear interconnected systems with unmodeled dynamics and prescribed performance. IEEE Trans. Neural Netw. Learn. Syst. **34**(9), 6557–6567 (2023)
28. Yang, Y., Ding, D.W., Xiong, H., Yin, Y., Wunsch, D.C.: Online Barrier-actor-critic learning for H_∞ control with full-state constraints and input saturation. J. Franklin Inst. **357**(6), 3316–3344 (2020)
29. Bhat, S.P., Bernstein, D.S.: Continuous finite-time stabilization of the translational and rotational double integrators. IEEE Trans. Autom. Control **43**(5), 678–682 (1998)
30. Gao, S., Liu, X., Jing, Y., Dimirovski, G.M.: Finite-time prescribed performance control for spacecraft attitude tracking. IEEE/ASME Trans. Mechatron. **27**(5), 3087–3098 (2022)
31. Chen, W., Wen, C., Wu, J.: Global exponential/finite-time stability of nonlinear adaptive switching systems with applications in controlling systems with unknown control direction. IEEE Trans. Autom. Control **63**(8), 2738–2744 (2018)
32. Wu, J., Li, J., Zong, G., Chen, W.: Global finite-time adaptive stabilization of nonlinearly parametrized systems with multiple unknown control directions. IEEE Trans. Syst. Man Cybern.: Syst. **47**(7), 1405–1414 (2017)
33. Dong, G., Li, H., Ma, H., Lu, R.: Finite-time consensus tracking neural network FTC of multi-agent systems. IEEE Trans. Neural Netw. Learn. Syst. **32**(2), 653–662 (2021)
34. Song, S., Park, J.H., Zhang, B., Song, X.: Composite adaptive fuzzy finite-time quantized control for full state-constrained nonlinear systems and its application. IEEE Trans. Syst. Man Cybern.: Syst. **52**(4), 2479–2490 (2022)
35. Liu, Y., Zhang, H., Wang, Y., Ren, H., Li, Q.: Adaptive fuzzy prescribed finite-time tracking control for nonlinear system with unknown control directions. IEEE Trans. Fuzzy Syst. **30**(6), 1993–2003 (2022)
36. Polyakov, A.: Nonlinear feedback design for fixed-time stabilization of linear control systems. IEEE Trans. Autom. Control **57**(8), 2106–2110 (2012)
37. Tao, M., Chen, Q., He, X., Xie, S.: Fixed-time filtered adaptive parameter estimation and attitude control for quadrotor UAVs. IEEE Trans. Aerosp. Electron. Syst. **58**(5), 4135–4146 (2022)
38. Mei, K., Ding, S., Chen, C.C.: Fixed-time stabilization for a class of output-constrained nonlinear systems. IEEE Trans. Syst. Man Cybern.: Syst. **52**(10), 6498–6510 (2022)
39. Hou, H.Q., Liu, Y.J., Lan, J., Liu, L.: Adaptive fuzzy fixed-time time-varying formation control for heterogeneous multi-agent systems with full state constraints. IEEE Trans. Fuzzy Syst. **31**(4), 1152–1162 (2023)
40. Cheng, W., Zhang, K., Jiang, B.: Fixed-time fault-tolerant formation control for a cooperative heterogeneous multiagent system with prescribed performance. IEEE Trans. Syst. Man Cybern.: Syst. **53**(1), 462–474 (2023)
41. Kong, F., Zhu, Q.: Fixed-time stabilization of discontinuous neutral neural networks with proportional delays via new fixed-time stability lemmas. IEEE Trans. Neural Netw. Learn. Syst. **34**(2), 775–785 (2023)

42. Kong, F., Zhu, Q., Huang, T.: New fixed-time stability lemmas and applications to the discontinuous fuzzy inertial neural networks. IEEE Trans. Fuzzy Syst. **29**(12), 3711–3722 (2021)
43. Abhinav, S., Modares, H., Lewis, F.L., Davoudi, A.: Resilient cooperative control of DC microgrids. IEEE Trans. Smart Grid. **10**(1), 1083–1085 (2019)
44. Li, K., Hua, C.C., You, X., Guan, X.P.: Output feedback-based consensus control for nonlinear time delay multiagent systems. Automatica **111**, 108669 (2020)
45. Yang, Y., Modares, H., Wunsch, D.C., Yin, Y.: Optimal containment control of unknown heterogeneous systems with active leaders. IEEE Trans. Control Syst. Technol. **27**(3), 1228–1236 (2019)
46. Yu, W., Chen, G., Cao, M., Kurths, J.: Second-order consensus for multiagent systems with directed topologies and nonlinear dynamics. IEEE Trans. Syst. Man Cybern. Part B (Cybern.) **40**(3), 881–891 (2010)
47. Hou, Z., Cheng, L., Tan, M.: Decentralized robust adaptive control for the multiagent system consensus problem using neural networks. IEEE Trans. Syst. Man Cybern. Part B (Cybern.) **39**(3), 636–647 (2009)
48. Yang, Y., Xu, C.Z.: Adaptive fuzzy leader-follower synchronization of constrained heterogeneous multiagent systems. IEEE Trans. Fuzzy Syst. **30**(1), 205–219 (2022)
49. Yang, Y., Gao, W., Modares, H., Xu, C.Z.: Robust actor-critic learning for continuous-time nonlinear systems with unmodeled dynamics. IEEE Trans. Fuzzy Syst. **30**(6), 2101–2112 (2022)
50. Lewis, F., Zhang, H., Hengster-Movric, K., Das, A.: Cooperative Control of Multi-Agent Systems: Optimal and Adaptive Design Approaches. Springer, London (2013)
51. Li, Z., Duan, Z.: Cooperative Control Of Multi-agent Systems: A Consensus Region Approach. CRC Press, Boca Raton (2017)
52. Yang, Y., Modares, H., Wunsch, D.C., Yin, Y.: Leader-follower output synchronization of linear heterogeneous systems with active leader using reinforcement learning. IEEE Trans. Neural Netw. Learn. Syst. **29**(6), 2139–2153 (2018)
53. Liu, G., Yang, Y., Gao, W., Wunsch, D.C.: Adaptive Nussbaum design for nonholonomic systems with asymptotic stabilization against false data injection. IEEE Trans. Cybern. **55**(11), 5189–5201 (2025)
54. Wang, Y., Miao, Z., Zhong, H., Pan, Q.: Simultaneous stabilization and tracking of nonholonomic mobile robots: a Lyapunov-based approach. IEEE Trans. Control Syst. Technol. **23**(4), 1440–1450 (2015)
55. Sun, W., Su, S.F., Xia, J., Wu, Y.: Adaptive tracking control of wheeled inverted pendulums with periodic disturbances. IEEE Trans. Cybern. **50**(5), 1867–1876 (2020)
56. Ge, S.S., Wang, Z., Lee, T.H.: Adaptive stabilization of uncertain nonholonomic systems by state and output feedback. Automatica **39**(8), 1451–1460 (2003)
57. Brockett, R.W.: Asymptotic Stability and Feedback Stabilization. Birkhauser, Boston (1983)
58. Samson, C.: Control of chained systems application to path following and time-varying point-stabilization of mobile robots. IEEE Trans. Autom. Control **40**(1), 64–77 (1995)
59. Tian, Y.P., Li, S.: Exponential stabilization of nonholonomic dynamic systems by smooth time-varying control. Automatica **38**(7), 1139–1146 (2002)
60. M'Closkey, R.T., Murray, R.M.: Exponential stabilization of driftless nonlinear control systems using homogeneous feedback. IEEE Trans. Autom. Control **42**(5), 614–628 (1997)
61. Astolfi, A.: Discontinuous control of nonholonomic systems. Syst. & Control Lett. **27**(1), 37–45 (1996)
62. Kolmanovsky, I., McClamroch, N.H.: Hybrid feedback laws for a class of cascade nonlinear control systems. IEEE Trans. Autom. Control **41**(9), 1271–1282 (1996)
63. Tang, L., Yang, Y., Chen, X., Li, Q.: Nussbaum-based robust adaptive control for nonlinear switching systems with switching unknown control coefficients. Int. J. Robust Nonlinear Control **33**(10), 5566–5582 (2023)
64. Wang, F., Lai, G.: Fixed-time control design for nonlinear uncertain systems via adaptive method. Syst. & Control Lett. **140**, 104704 (2020)

65. Wu, Y., Ma, H., Chen, M., Li, H.: Observer-based fixed-time adaptive fuzzy bipartite containment control for multiagent systems with unknown hysteresis. IEEE Trans. Fuzzy Syst. **30**(5), 1302–1312 (2022)
66. Altafini, C.: Consensus problems on networks with antagonistic interactions. IEEE Trans. Autom. Control **58**(4), 935–946 (2013)
67. Zhou, Q., Wang, W., Liang, H., Basin, M.V., Wang, B.: Observer-based event-triggered fuzzy adaptive bipartite containment control of multiagent systems with input quantization. IEEE Trans. Fuzzy Syst. **29**(2), 372–384 (2021)
68. Ibrahim, A., Dragomir, S.S., Darus, M.: Power series inequalities via Young's inequality with applications. J. Inequal. Appl. **2013**, 314 (2013)
69. Li, Y.-X.: Barrier Lyapunov function-based adaptive asymptotic tracking of nonlinear systems with unknown virtual control coefficients. Automatica **121**, 109181 (2020)
70. Tee, K.P., Ren, B., Ge, S.S.: Control of nonlinear systems with time-varying output constraints. Automatica **47**(11), 2511–2516 (2011)

Part I
Adaptive Nussbaum Design with Asymptotic Stability

Chapter 2
Adaptive Nussbaum Design for Tracking Control with Asymptotic Stability

2.1 Introduction

Strict feedback systems are a class of control systems characterized by their cascaded structure [1, 2]. The cascaded structure of strict feedback systems implies sequential dependence on successive subsystems, enabling a recursive framework for system analysis [3] and control strategy design [4]. A commonly employed method for designing such recursive frameworks is the backstepping approach [5, 6], which leverages recursive virtual control design for each subsystem to achieve an overall control objective [7]. In backstepping control, precise tracking is crucial, as tracking errors within subsystems can degrade overall system performance or even lead to instability [8]. Therefore, considerable attention must be devoted to designing and adjusting system dynamics and control strategies to ensure fast and accurate reference tracking. However, in practical scenarios, the control directions may not always be explicitly available for control design [9]. In such cases, designing asymptotic tracking control systems becomes significantly more challenging, as the system cannot precisely determine or measure the influence of control signals on its behavior. Alternative methods or techniques must therefore be adopted to address these limitations.

Although Nussbaum-type functions have shown effectiveness in dealing with single unknown control direction [10, 11], their applicability becomes severely limited in strict feedback systems with multiple such uncertainties [12]. In these scenarios, the opposing signs of control directions can lead to mutual cancellation of the associated Nussbaum gains, rendering traditional designs incapable of reliably distinguishing or compensating for the control influence [13]. As a result, existing Nussbaum-based techniques are not directly applicable to multi-directional unknown settings [14, 15]. Some attempts have been made to circumvent this difficulty by presuming uniformity in the sign of all unknown control directions, assuming either all positive or all negative [16]. Nonetheless, such assumptions rarely hold in practical systems, where the directions are often sign-indefinite or mixed [17]. To address these chal-

Y. Yang et al., *Adaptive Nussbaum Design for Nonlinear Dynamical Systems*,
https://doi.org/10.1007/978-981-95-4997-9_2

lenges, this chapter introduces a novel class of Nussbaum-type functions designed with enhanced structural adaptability.

Compared to existing single Nussbaum-type methods [18–20], the summation of multiple Nussbaum-type gains in traditional designs may result in cancellation. To address this issue, this chapter proposes a novel Nussbaum-type function that eliminates the need for uniform positivity or negativity of all control directions. This improvement for conventional assumptions significantly enhances the adaptability and effectiveness of the Nussbaum design in handling systems with multiple unknown control directions.

The remainder of this chapter is structured as follows. Section 2.2 presents the problem formation along with necessary assumptions. Section 2.3 delves into the design of the control schemes within the recursive backstepping framework. Section 2.4 analyzes the stability of asymptotic tracking. In Sect. 2.5, a simulation example is conducted to validate the effectiveness of the proposed adaptive robust design for uncertain nonlinear systems with unknown control directions. Finally, Sect. 2.6 offers concluding remarks.

2.2 Problem Formulation

A class of continuous-time strict feedback nonlinear systems considered is described by the state space form as

$$
\begin{aligned}
\dot{x}_i &= g_i\left(\bar{x}_i\right) x_{i+1} + f_i\left(\bar{x}_i\right), \; 1 \leq i \leq n-1 \\
\dot{x}_n &= g_n\left(\bar{x}_n\right) u\left(t\right) + f_n\left(\bar{x}_n\right) \\
y &= x_1
\end{aligned}
\tag{2.1}
$$

with states $\bar{x}_i = [x_1, x_2, \ldots, x_i]^{\mathrm{T}} \in \mathbb{R}^i$, $i = 1, 2, \ldots, n$, control input $u \in \mathbb{R}$, output $y = x_1 \in \mathbb{R}$, and unknown smooth nonlinear functions $f_i\left(\cdot\right) : \mathbb{R}^i \to \mathbb{R}$ and $g_i\left(\cdot\right) : \mathbb{R}^i \to \mathbb{R}$.

This chapter aims to design an adaptive tracking controller for the system (2.1) such that the asymptotic convergence of system output tracking error is ensured and all signals in closed-loop systems are bounded. Based on the above purpose, the necessary assumptions are presented as follows.

Assumption 2.1 The control directions $g_i\left(\cdot\right)$ are bounded in the sense that $0 < g_{i,q} \leq |g_i| \leq g_{i,p}$ with positive constants $g_{i,p}$ and $g_{i,q}$.

Assumption 2.2 The reference signal $x_0(t)$ and its derivatives are continuous and bounded.

Lemma 2.1 *Under Assumption 2.1, consider a non-negative function $V(t)$ defined on $[0, t_f)$ with the Nussbaum-type functions $\{N_j(\cdot)\}_{j=1}^{M}$ as*

$$N_j(\xi) = 2\xi e^{\xi^2} \cos(2^{j-1}\xi) - 2^{j-1}(e^{\xi^2} - 1)\sin(2^{j-1}\xi), \tag{2.2}$$

for all $\xi \in \mathbb{R}$. *Then,* $V(t)$ *and* $\xi(t)$ *remain bounded on* $[0, t_f)$ *provided that*

$$V(t) \leq \sum_{j=1}^{M} \int_0^t \left(g_j(\sigma)N_j(\xi_j(\sigma)) + 1\right)\dot{\xi}_j(\sigma)\mathrm{d}\sigma + c_0, \tag{2.3}$$

with constant $c_0 > 0$ *and unknown continuous functions* $\{g_j(\cdot)\}_{j=1}^M$ *bounded as* $g_{j,\min} \leq \|g_j(\cdot)\| \leq g_{j,\max}$ *for* $g_{j,\min} > 0$ *and* $g_{j,\max} > 0$.

Proof Without loss of generality, let $\xi_1, \ldots, \xi_\Upsilon$ and $\xi_{\Upsilon+1}, \ldots, \xi_M$ denote the unbounded signals and bounded signals, respectively. The fact in (2.3) implies

$$V(t) \leq \sum_{j=1}^{\Upsilon} \int_0^t \left(g_j(\sigma)N_j(\xi_j(\sigma)) + 1\right)\dot{\xi}_j(\sigma)\mathrm{d}\sigma + c_1, \tag{2.4}$$

with

$$c_1 = c_0 + \sum_{j=\Upsilon+1}^{M} \int_0^t \left(g_j(\sigma)\, N_j\left(\xi_j(\sigma)\right) + 1\right)\dot{\xi}_j(\sigma)\,\mathrm{d}\sigma.$$

Based on the assumption that $\xi_{\Upsilon+1}, \ldots, \xi_M$ are bounded, one can know that c_1 is bounded. Next, we will prove the boundedness of $V(t)$ and $\left\{\xi_j(t)\right\}_{j=1}^M$ by means of contradiction.

First, suppose that $\xi_1, \ldots, \xi_\Upsilon$ is not upper-bound on $\left[0, t_f\right)$. Define monotone-increasing sequences $\left\{\Psi_{j,r}\right\}_{r=1}^\infty$ and $\left\{\psi_{j,r}\right\}_{r=1}^\infty$ as

$$\begin{aligned}\Psi_{j,r} &= \frac{(4r+1)\,\pi}{2^j} + v_j\pi \\ \psi_{j,r} &= \frac{(4r-1)\,\pi}{2^j} + v_j\pi\end{aligned}$$

with

$$v_j = \sum_{k=1}^{j} \left(2^{-k} + 2^{-k}\mathrm{sign}\,(g_k)\right) \geq 0.$$

Consider the increasing sequences $\left\{t_{j,r}\right\}_{r=1}^\infty$ satisfying $\xi_j\left(t_{j,r}\right) = \Psi_{j,r} - \frac{\zeta_j}{2^{j-1}}$ with $0 \leq \zeta_j \leq \frac{\pi}{2}$. Then, one has

$$\lim_{r\to\infty} t_{j,r} = t_f$$
$$\lim_{r\to\infty} \left(\Psi_{j,r} - \frac{\zeta_j}{2^{j-1}}\right) = \infty. \tag{2.5}$$

Integrating the term $\operatorname{sign}\left(g_j\right) N_j\left(\xi_j\right)$ over $\xi_j \in \left[0, \psi_{j,r}\right]$, one can obtain

$$\begin{aligned}&\int_0^{\psi_{j,r}} \operatorname{sign}\left(g_j\right) N_j\left(\xi_j\right) \mathrm{d}\xi_j \\ &= \operatorname{sign}\left(g_j\right)\left(\mathrm{e}^{\psi_{j,r}^2} - 1\right)\cos\left(2r\pi - \frac{\pi}{2} + 2^{j-1}v_j\pi\right).\end{aligned} \tag{2.6}$$

Note that $2^{j-1}v_j$ is a positive integer. Then, one has

$$\begin{aligned}\int_0^{\psi_{j,r}} \operatorname{sign}\left(g_j\right) N_j\left(\xi_j\right) \mathrm{d}\xi_j &= \int_0^{\Psi_{j,r}} \operatorname{sign}\left(g_j\right) N_j\left(\xi_j\right) \mathrm{d}\xi_j \\ &= 0.\end{aligned} \tag{2.7}$$

Therefore, the summation of $\int_0^{\psi_{j,r}} \operatorname{sign}\left(g_j\right) N_j\left(\xi_j\right) \mathrm{d}\xi_j$ satisfies

$$\sum_{j=1}^{\Upsilon} \int_0^{\psi_{j,r}} \left|g_j\right| \operatorname{sign}\left(g_j\right) N_j\left(\xi_j\right) \mathrm{d}\xi_j \le 0. \tag{2.8}$$

Furthermore, for the term $2^{j-1}v_j$, one has

$$2^{j-1}v_j = 2^{j-1}\sum_{k=1}^{j-1}\left[2^{-k} + 2^{-k}\operatorname{sign}\left(g_k\right)\right] + \frac{1}{2}\left[1 + \operatorname{sign}\left(g_j\right)\right]$$

with $2^{j-1}\sum\limits_{k=1}^{j-1}\left(2^{-k} + 2^{-k}\operatorname{sign}\left(g_k\right)\right)$ being an even integer. Noted that

$$\frac{1}{2}\left[1 + \operatorname{sign}\left(g_j\right)\right]\begin{cases}\text{is odd,} & \text{if } \operatorname{sign}\left(g_j\right) = 1\\ \text{is even,} & \text{if } \operatorname{sign}\left(g_j\right) = -1.\end{cases}$$

Then, $2^{j-1}v_j$ is odd if $\operatorname{sign}\left(g_j\right) = 1$ and is even if $\operatorname{sign}\left(g_j\right) = -1$. On the interval $\left[0, \frac{r\pi}{2^{j-2}} + v_j\pi\right]$, one has

$$\begin{aligned}&\int_0^{\frac{r\pi}{2^{j-2}} + v_j\pi} \operatorname{sign}\left(g_j\right) N_j\left(\xi_j\right) \mathrm{d}\xi_j \\ &= \operatorname{sign}\left(g_j\right)\left(\mathrm{e}^{\left(\frac{r\pi}{2^{j-2}} + v_j\pi\right)^2} - 1\right)\cos\left(2^{j-1}v_j\pi\right) \\ &< 0,\end{aligned} \tag{2.9}$$

which implies that $\int_0^{\xi_j} \operatorname{sign}(g_j) N_j(\sigma) \mathrm{d}\sigma \le 0$ on the interval $[\psi_{j,r}, \Psi_{j,r}]$. Therefore, for $\xi_j \in \left[\psi_{j,r} + \frac{\zeta_j}{2^{j-1}}, \Psi_{j,r} - \frac{\zeta_j}{2^{j-1}}\right]$, one has

$$
\begin{aligned}
&\sum_{j=1}^{\Upsilon} \int_{\psi_{j,r}+\frac{\zeta_j}{2^{j-1}}}^{\Psi_{j,r}-\frac{\zeta_j}{2^{j-1}}} |g_j| \operatorname{sign}(g_j) N_j(\xi_j) \mathrm{d}\xi_j \\
&\le \sum_{j=1}^{\Upsilon} G_{\min} \int_{\psi_{j,r}+\frac{\zeta_j}{2^{j-1}}}^{\Psi_{j,r}-\frac{\zeta_j}{2^{j-1}}} \operatorname{sign}(g_j) N_j(\xi_j) \mathrm{d}\xi_j \\
&= -\sum_{j=1}^{\Upsilon} G_{\min} \sin(\zeta_j) \mathrm{e}^{\left(\Psi_{j,r}-\frac{\zeta_j}{2^{j-1}}\right)^2} \\
&\quad -\sum_{j=1}^{\Upsilon} G_{\min} \sin(\zeta_j) \mathrm{e}^{\left(\psi_{j,r}+\frac{\zeta_j}{2^{j-1}}\right)^2},
\end{aligned} \tag{2.10}
$$

with $G_{\min} = \min\limits_j g_{j,\min}$.

Moreover, for $\xi_j \in \left[\psi_{j,r}, \psi_{j,r} + \frac{\zeta_j}{2^{j-1}}\right]$, one can obtain

$$
\begin{aligned}
&\sum_{j=1}^{\Upsilon} \int_{\psi_{j,r}}^{\psi_{j,r}+\frac{\zeta_j}{2^{j-1}}} |g_j| \operatorname{sign}(g_j) N_j(\xi_j) \mathrm{d}\xi_j \\
&\le G_{\min} \sum_{j=1}^{\Upsilon} \int_{\psi_{j,r}}^{\psi_{j,r}+\frac{\zeta_j}{2^{j-1}}} \operatorname{sign}(g_j) N_j(\xi_j) \mathrm{d}\xi_j \\
&\le 0.
\end{aligned} \tag{2.11}
$$

Considering the term $\sum\limits_{j=1}^{\Upsilon} \int_0^{\psi_{j,r}} |g_j| \operatorname{sign}(g_j) N_j(\xi_j) \mathrm{d}\xi_j$ in (2.3) for $\xi_j \in [0, \psi_{j,r}]$, one has

$$
\begin{aligned}
&\sum_{j=1}^{\Upsilon} \int_0^{\psi_{j,r}} |g_j| \operatorname{sign}(g_j) N_j(\xi_j) \mathrm{d}\xi_j \\
\le &\sum_{j=1}^{\Upsilon} \sum_{w=0}^{2r+2^{j-1}v_j-3} g_{j,\max} \int_{\frac{(2w+3)\pi}{2^j}}^{\frac{(2w+5)\pi}{2^j}} \operatorname{sign}(g_j) N_j(\xi_j) \mathrm{d}\xi_j \\
&+\sum_{j=1}^{\Upsilon} \sum_{w=0}^{2r+2^{j-1}v_j-3} g_{j,\min} \int_{\frac{(2w+1)\pi}{2^j}}^{\frac{(2w+3)\pi}{2^j}} \operatorname{sign}(g_j) N_j(\xi_j) \mathrm{d}\xi_j \\
&+\sum_{j=1}^{\Upsilon} g_{j,\max} \int_0^{\frac{\pi}{2^j}} \operatorname{sign}(g_j) N_j(\xi_j) \mathrm{d}\xi_j .
\end{aligned}
$$

One can divide $\sum\limits_{j=1}^{\Upsilon} \int_0^{t_{j,r}} |g_j| \operatorname{sign}(g_j) N_j(\xi_j) \dot{\xi}_j \mathrm{d}\sigma$ into

$$
\begin{aligned}
&\sum_{j=1}^{\Upsilon} \int_0^{t_{j,r}} |g_j| \operatorname{sign}(g_j) N_j(\xi_j) \dot{\xi}_j \mathrm{d}\sigma \\
&= \sum_{j=1}^{\Upsilon} \int_{\psi_{j,r}+\frac{\zeta_j}{2^{j-1}}}^{\Psi_{j,r}-\frac{\zeta_j}{2^{j-1}}} |g_j| \operatorname{sign}(g_j) N_j(\xi_j) \mathrm{d}\xi_j \\
&\quad + \sum_{j=1}^{\Upsilon} \int_{\psi_{j,r}}^{\psi_{j,r}+\frac{\zeta_j}{2^{j-1}}} |g_j| \operatorname{sign}(g_j) N_j(\xi_j) \mathrm{d}\xi_j \\
&\quad + \sum_{j=1}^{\Upsilon} \int_0^{\psi_{j,r}} |g_j| \operatorname{sign}(g_j) N_j(\xi_j) \mathrm{d}\xi_j .
\end{aligned} \tag{2.12}
$$

Substituting (2.8), (2.10), (2.11) into (2.12), one has

$$
\begin{aligned}
&\sum_{j=1}^{\Upsilon} \int_0^{t_{j,r}} |g_j| \operatorname{sign}(g_j) N_j(\xi_j) \dot{\xi}_j \mathrm{d}\sigma \\
&\leq -\sum_{j=1}^{\Upsilon} G_{\min} \sin(\zeta_j) \left[\mathrm{e}^{\left(\Psi_{j,r}-\frac{\zeta_j}{2^{j-1}}\right)^2} + \mathrm{e}^{\left(\psi_{j,r}+\frac{\zeta_j}{2^{j-1}}\right)^2} \right].
\end{aligned} \tag{2.13}
$$

Based on the fact

$$
\sum_{j=1}^{\Upsilon} \int_0^{t_{j,r}} \dot{\xi}_j \mathrm{d}\sigma \leq \sum_{j=1}^{\Upsilon} \left(\Psi_{j,r} - \frac{\zeta_j}{2^{j-1}} \right) - \sum_{j=1}^{\Upsilon} \xi_j(0), \tag{2.14}
$$

the Lyapunov candidate satisfies

$$
V(t) \leq -\sum_{j=1}^{\Upsilon} G_{\min} \sin(\zeta_j) \Gamma_{j,r} \mathrm{e}^{\left(\Psi_{j,r}-\frac{\zeta_j}{2^{j-1}}\right)^2} + c_2,
$$

with $c_2 = -\sum\limits_{i=1}^{\Upsilon} \xi_i(0) + c_1$ and

$$
\Gamma_{j,r} = 1 + \frac{\mathrm{e}^{\left(\Psi_{j,r}-\frac{\zeta_j}{2^{j-1}}\right)^2}}{\mathrm{e}^{\left(\psi_{j,r}+\frac{\zeta_j}{2^{j-1}}\right)^2}} - \frac{\Psi_{j,r} - \frac{\zeta_j}{2^{j-1}}}{G_{\min} \sin(\zeta_j) \mathrm{e}^{\left(\Psi_{j,r}-\frac{\zeta_j}{2^{j-1}}\right)^2}}.
$$

One can observe that $\lim_{r\to\infty} G_{\min}\sin\left(\zeta_j\right)\Gamma_{j,r} > 0$. Therefore, $\lim_{t\to t_f} V(t) = -\infty$, which contradicts the condition that $V(t) \geq 0$. Therefore, the terms $\xi_1, \ldots, \xi_\Upsilon$ are upper-bounded on $\left[0, t_f\right)$.

Second, suppose that $\xi_1, \ldots, \xi_\Upsilon$ is not lower bounded on $\left[0, t_f\right)$, i.e., $-\xi_1, \ldots, -\xi_\Upsilon$ are not upper bounded on $\left[0, t_f\right)$. Note that the integral term $\int_0^{\xi_j} \operatorname{sign}\left(g_j\right) N_j(\tau)\,\mathrm{d}\tau$ is an even function. Thus, the Lyapunov candidate satisfies

$$V(t) \leq \sum_{j=1}^{\Upsilon} \int_0^t \left(g_j(\sigma) N_j\left(\xi_j(\sigma)\right) - 1\right)\dot{\xi}_j(\sigma)\,\mathrm{d}\sigma + l_1, \tag{2.15}$$

where $l_1 = l_0 + \sum_{j=\Upsilon+1}^{M} \int_0^t \left(g_j(\sigma) N_j\left(\xi_j(\sigma)\right) + 1\right)\dot{\xi}_j(\sigma)\,\mathrm{d}\sigma$ is bounded. Following the discussions in the first case, one has

$$\begin{aligned} V(t) \leq & -\sum_{j=1}^{\Upsilon} G_{\min}\sin\left(\zeta_j\right)\mathrm{e}^{\left(\Psi_{j,r}-\frac{\zeta_j}{2^{j-1}}\right)^2} \\ & -\sum_{j=1}^{\Upsilon} G_{\min}\sin\left(\zeta_j\right)\mathrm{e}^{\left(\psi_{j,r}+\frac{\zeta_j}{2^{j-1}}\right)^2} \\ & +\sum_{j=1}^{\Upsilon}\xi_j(0) - \sum_{j=1}^{\Upsilon}\left(\Psi_{j,r} - \frac{\zeta_j}{2^{j-1}}\right) + l_1. \end{aligned} \tag{2.16}$$

Therefore, one can obtain

$$V(t) \leq -\sum_{j=1}^{\Upsilon} G_{\min}\sin\left(\zeta_j\right)\bar{\Gamma}_{j,r}\mathrm{e}^{\left(\Psi_{j,r}-\frac{\zeta_j}{2^{j-1}}\right)^2} + l_2,$$

with $l_2 = \sum_{j=1}^{\Upsilon}\xi_j(0) + l_1$ and

$$\bar{\Gamma}_{j,r} = 1 + \frac{\mathrm{e}^{\left(\Psi_{j,r}-\frac{\zeta_j}{2^{j-1}}\right)^2}}{\mathrm{e}^{\left(\psi_{j,r}+\frac{\zeta_j}{2^{j-1}}\right)^2}} + \frac{\Psi_{j,r} - \frac{\zeta_j}{2^{j-1}}}{G_{\min}\sin\left(\zeta_j\right)\mathrm{e}^{\left(\Psi_{j,r}-\frac{\zeta_j}{2^{j-1}}\right)^2}}.$$

Thus, on $\left[0, t_f\right)$, the terms $\xi_1, \ldots, \xi_\Upsilon$ are lower-bounded. It can be observed that $\lim_{r\to\infty} G_{\min}\sin\left(\zeta_j\right)\bar{\Gamma}_{j,r} > 0$, which implies $\lim_{t\to t_f} V(t) = -\infty$, which contradicts the fact that $V(t) \geq 0$. Therefore, the terms $\xi_1, \ldots, \xi_\Upsilon$ are lower-bounded on $\left[0, t_f\right)$. Finally, we can conclude the boundedness of $V(t)$ and $\left\{\xi_j(t)\right\}_{j=1}^{M}$. This completes the proof. □

2.3 Adaptive Nussbaum Design

2.3.1 Step 1

Following the backstepping approach, define the tracking error for the first subsystem as $z_1 = x_1 - x_0$ and for the second subsystem as $z_2 = x_2 - \alpha_1$, where α_1 represents the virtual control of the second subsystem. Differentiating the tracking error z_1 yields

$$\begin{aligned} \dot{z}_1 &= \dot{x}_1 - \dot{x}_0 \\ &= g_1(x_1) z_2 + g_1(x_1) \alpha_1 + f_1(x_1) - \dot{x}_0. \end{aligned} \tag{2.17}$$

Since $f_1(x_1)$ is unknown, define $F_1(X_1) = f_1(x_1) - \dot{x}_0$ where $X_1 = [x_1, \dot{x}_0]^{\mathrm{T}}$. Using the RBFNN, $F_1(X_1)$ can be approximated as $F_1(X_1) = W_1^{\mathrm{T}} \Phi_1(X_1) + \varepsilon_1(t)$, with the approximation error $\varepsilon_1(t) \le \bar{\varepsilon}_1$. Define $\theta_1 = \|W_1\|^2$ and $\vartheta_1 = \bar{\varepsilon}_1^2$, both of which are unknown. Adaptive estimates $\hat{\theta}_1$ and $\hat{\vartheta}_1$ are introduced, with $\tilde{\theta}_1 = \theta_1 - \hat{\theta}_1$ and $\tilde{\vartheta}_1 = \vartheta_1 - \hat{\vartheta}_1$ are the parameter estimation errors.

Consider the following Lyapunov candidate

$$V_1 = \frac{1}{2} z_1^2 + \frac{1}{2\gamma_1} \tilde{\theta}_1^2 + \frac{1}{2\lambda_1} \tilde{\vartheta}_1^2, \tag{2.18}$$

where $\gamma_1 > 0$ and $\lambda_1 > 0$ are known constants to be chosen. The derivative of V_1 can be deduced as

$$\begin{aligned} \dot{V}_1 &= z_1 \dot{z}_1 + \frac{1}{\gamma_1} \tilde{\theta}_1 \dot{\tilde{\theta}}_1 + \frac{1}{\lambda_1} \tilde{\vartheta}_1 \dot{\tilde{\vartheta}}_1 \\ &= g_1(x_1) z_1 z_2 + g_1(x_1) z_1 \alpha_1 + z_1 W_1^{\mathrm{T}} \Phi_1 \\ &\quad + z_1 \varepsilon_1 - \frac{1}{\lambda_1} \tilde{\vartheta}_1 \dot{\hat{\vartheta}}_1 - \frac{1}{\gamma_1} \tilde{\theta}_1 \dot{\hat{\theta}}_1. \end{aligned} \tag{2.19}$$

Applying Young's inequality to the terms in (2.19), one has

$$\begin{aligned} z_1 W_1^{\mathrm{T}} \Phi_1 &\le \frac{1}{q_1} |z_1| \theta_1 \Phi_1^{\mathrm{T}} \Phi_1 + q_1 |z_1|, \\ z_1 \varepsilon_1(t) &\le \frac{1}{p_1} |z_1| \vartheta_1 + p_1 |z_1|, \end{aligned} \tag{2.20}$$

where $q_1, p_1 > 0$. From Lemma 1.8, one can obtain

$$\begin{aligned} |z_1| \Phi_1^T \Phi_1 &\le \kappa \sigma_1 + z_1 \Phi_1^{\mathrm{T}} \Phi_1 \tanh\left(\frac{z_1 \Phi_1^{\mathrm{T}} \Phi_1}{\sigma_1}\right), \\ |z_1| &\le \kappa \sigma_1 + z_1 \tanh\left(\frac{z_1}{\sigma_1}\right), \end{aligned} \tag{2.21}$$

where $\sigma_1 = d_1 \mathrm{e}^{-r_1 t}$, $d_1 > 0$, $r_1 > 0$ is considered to satisfy $\sigma_1(t) > 0$ and

$$\lim_{t \to \infty} \int_{t_0}^{t} \sigma_1(\iota)\,\mathrm{d}\iota \le \bar{\sigma}_1 < \infty, \quad \bar{\sigma}_1 > 0. \tag{2.22}$$

Design the function $\eta_{1,1}$ and $\eta_{1,2}$ as

$$\begin{aligned} \eta_{1,1} &= \Phi_1^{\mathrm{T}} \Phi_1 \tanh\left(\frac{z_1 \Phi_1^{\mathrm{T}} \Phi_1}{\sigma_1}\right), \\ \eta_{1,2} &= \tanh\left(\frac{z_1}{\sigma_1}\right). \end{aligned} \tag{2.23}$$

Then, one has

$$\begin{aligned} |z_1|\, \Phi_1^{\mathrm{T}} \Phi_1 &\le \kappa \sigma_1 + z_1 \eta_{1,1}, \\ |z_1| &\le \kappa \sigma_1 + z_1 \eta_{1,2}. \end{aligned} \tag{2.24}$$

Based on Lemma 1.8, the following facts hold

$$\begin{aligned} z_1 W_1^{\mathrm{T}} \Phi_1 &\le \kappa \sigma_1 \left(\frac{1}{q_1}\theta_1 + q_1\right) + \frac{1}{q_1} z_1 \theta_1 \eta_{1,1} + q_1 z_1 \eta_{1,2}, \\ z_1 \varepsilon_1(t) &\le \kappa \sigma_1 \left(\frac{1}{p_1}\vartheta_1 + p_1\right) + \frac{1}{p_1} z_1 \vartheta_1 \eta_{1,2} + p_1 z_1 \eta_{1,2}. \end{aligned} \tag{2.25}$$

Inserting (2.25) into (2.19) yields

$$\begin{aligned} \dot{V}_1 \le{} & \kappa \sigma_1 \left(\frac{1}{q_1}\theta_1 + q_1 + \frac{1}{p_1}\vartheta_1 + p_1\right) \\ & + \frac{1}{q_1} z_1 \theta_1 \eta_{1,1} + q_1 z_1 \eta_{1,2} + \frac{1}{p_1} z_1 \vartheta_1 \eta_{1,2} \\ & + p_1 z_1 \eta_{1,2} + g_1(x_1) z_1 z_2 \\ & + g_1(x_1) z_1 \alpha_1 - \frac{1}{\gamma_1} \tilde{\theta}_1 \dot{\hat{\theta}}_1 - \frac{1}{\lambda_1} \tilde{\vartheta}_1 \dot{\hat{\vartheta}}_1. \end{aligned} \tag{2.26}$$

In addition, adding and subtracting $k_1 z_1^2$ with $k_1 > 0$, one has

$$\begin{aligned} \dot{V}_1 \le{} & \kappa \sigma_1 \left(\frac{1}{q_1}\theta_1 + q_1 + \frac{1}{p_1}\vartheta_1 + p_1\right) \\ & + z_1 \left[k_1 z_1 + \frac{1}{q_1} \hat{\theta}_1 \eta_{1,1} + \left(\frac{\hat{\vartheta}_1}{p_1} + q_1 + p_1\right) \eta_{1,2}\right] \\ & + \frac{\tilde{\vartheta}_1}{\lambda_1} \left(\frac{\lambda_1}{p_1} z_1 \eta_{1,2} - \dot{\hat{\vartheta}}_1\right) + \frac{\tilde{\theta}_1}{\gamma_1} \left(\frac{\gamma_1}{q_1} z_1 \eta_{1,1} - \dot{\hat{\theta}}_1\right) \\ & + g_1(x_1) z_1 z_2 + g_1(x_1) z_1 \alpha_1 - k_1 z_1^2. \end{aligned} \tag{2.27}$$

Consider the adaptive Nussbaum-type gain ξ_1 and virtual control design α_1 as

$$\begin{aligned}\dot{\xi}_1 &= z_1\left[k_1 z_1 + \frac{1}{q_1}\hat{\theta}_1\eta_{1,1} + \left(\frac{\hat{\vartheta}_1}{p_1} + q_1 + p_1\right)\eta_{1,2}\right],\\ \alpha_1 &= N_1(\xi_1)\left[k_1 z_1 + \frac{1}{q_1}\hat{\theta}_1\eta_{1,1} + \left(\frac{\hat{\vartheta}_1}{p_1} + q_1 + p_1\right)\eta_{1,2}\right].\end{aligned} \tag{2.28}$$

Define $s_1 = k_1 z_1 + \frac{\hat{\theta}_1}{q_1}\eta_{1,1} + \left(\frac{\hat{\vartheta}_1}{p_1} + p_1 + q_1\right)\eta_{1,2}$, one can get

$$\begin{aligned}\dot{\xi}_1 &= z_1 s_1,\\ \alpha_1 &= N_1(\xi_1)\, s_1.\end{aligned} \tag{2.29}$$

Inserting (2.29) into (2.27) yields

$$\begin{aligned}\dot{V}_1 \le\ & \kappa\sigma_1\left(\frac{1}{q_1}\theta_1 + q_1 + \frac{1}{p_1}\vartheta_1 + p_1\right)\\ &+\frac{\tilde{\theta}_1}{\gamma_1}\left(\frac{\gamma_1}{q_1}z_1\eta_{1,1} - \dot{\hat{\theta}}_1\right) + \frac{\tilde{\vartheta}_1}{\lambda_1}\left(\frac{\lambda_1}{p_1}z_1\eta_{1,2} - \dot{\hat{\vartheta}}_1\right)\\ &+\left[g_1(x_1)N_1(\xi_1)+1\right]\dot{\xi}_1 - k_1 z_1^2 + g_1(x_1)z_1 z_2.\end{aligned} \tag{2.30}$$

To compensate for the unknown dynamics, we consider

$$\begin{aligned}\dot{\hat{\theta}}_1 &= \gamma_1\left(\frac{\eta_{1,1}z_1}{q_1} - \sigma_1\hat{\theta}_1\right),\\ \dot{\hat{\vartheta}}_1 &= \lambda_1\left(\frac{\eta_{1,2}z_1}{p_1} - \sigma_1\hat{\vartheta}_1\right).\end{aligned} \tag{2.31}$$

With (2.31), $\dot{V}_1$ further satisfies

$$\begin{aligned}\dot{V}_1 \le\ & \kappa\sigma_1\left(\frac{1}{q_1}\theta_1 + q_1 + \frac{1}{p_1}\vartheta_1 + p_1\right) + \left[g_1(x_1)N_1(\xi_1)+1\right]\dot{\xi}_1\\ &-k_1 z_1^2 + g_1(x_1)z_1 z_2 + \sigma_1\left(\tilde{\theta}_1\hat{\theta}_1 + \tilde{\vartheta}_1\hat{\vartheta}_1\right).\end{aligned} \tag{2.32}$$

2.3.2 *Step i* $(2 \le i \le n-1)$

Define the tracking error for the ith subsystem as $z_i = x_i - \alpha_{i-1}$, where α_{i-1} is the virtual control of the ith subsystem. Differentiating z_i yields

$$\dot{z}_i = \dot{x}_i - \dot{\alpha}_{i-1} = g_i(\bar{x}_i) z_{i+1} + g_i(\bar{x}_i)\alpha_i + f_i(\bar{x}_i) - \dot{\alpha}_{i-1}. \tag{2.33}$$

Define $F_i(X_i) = f_i(\bar{x}_i) + g_{i-1}(\bar{x}_{i-1}) z_{i-1} - \dot{\alpha}_{i-1}$ with $X_i = [x_1, x_2, \ldots, x_i, \dot{\alpha}_{i-1}]^{\mathrm{T}}$. From Lemma 1.4, $F_i(X_i)$ can be approximated as $F_i(X_i) = W_i^{\mathrm{T}}\Phi_i(X_i) + \varepsilon_i(t)$, with the approximation error $\varepsilon_i(t) \le \bar{\varepsilon}_i$. Define $\theta_i = \|W_i\|^2$ and $\vartheta_i = \bar{\varepsilon}_i^2$, both of which are unknown. Adaptive estimates $\hat{\theta}_i$ and $\hat{\vartheta}_i$ are introduced, with $\tilde{\theta}_i = \theta_i - \hat{\theta}_i$ and $\tilde{\vartheta}_i = \vartheta_i - \hat{\vartheta}_i$ are the parameter estimation errors.

Consider the following Lyapunov function

$$V_i = \frac{1}{2}z_i^2 + \frac{1}{2\gamma_i}\tilde{\theta}_i^2 + \frac{1}{2\lambda_i}\tilde{\vartheta}_i^2 + V_{i-1}, \tag{2.34}$$

where $\gamma_i > 0$ and $\lambda_i > 0$ are known constants to be chosen. Then, one has

$$\begin{aligned}\dot{V}_i &= z_i\dot{z}_i + \frac{1}{\gamma_i}\tilde{\theta}_i\dot{\tilde{\theta}}_i + \frac{1}{\lambda_i}\tilde{\vartheta}_i\dot{\tilde{\vartheta}}_i + \dot{V}_{i-1}\\ &= \sum_{j=1}^{i}\left[g_j(\bar{x}_j) z_j\alpha_j + z_j W_j^{\mathrm{T}}\Phi_j + z_j\varepsilon_j(t) - \frac{1}{\gamma_j}\tilde{\theta}_j\dot{\hat{\theta}}_j - \frac{1}{\lambda_j}\tilde{\vartheta}_j\dot{\hat{\vartheta}}_j\right]\\ &\quad + g_i(\bar{x}_i) z_i z_{i+1}.\end{aligned} \tag{2.35}$$

Similar to Step 1, denote

$$\begin{aligned}\eta_{j,1} &= \Phi_j^{\mathrm{T}}\Phi_j \tanh\left(\frac{z_j\Phi_j^{\mathrm{T}}\Phi_j}{\sigma_j}\right),\\ \eta_{j,2} &= \tanh\left(\frac{z_j}{\sigma_j}\right)\end{aligned} \tag{2.36}$$

with $\sigma_j(t) = d_j e^{-r_j t}$, $d_j > 0$, $r_j > 0$ satisfying $\lim_{t\to\infty}\int_{t_0}^{t}\sigma_j(\iota)\,d\iota \le \bar{\sigma}_j < \infty$ and positive constant $\bar{\sigma}_j$. Then, one has

$$\begin{aligned}z_j W_j^{\mathrm{T}}\Phi_j &\le \kappa\sigma_j\left(\frac{1}{q_j}\theta_j + q_j\right) + \frac{1}{q_j}z_j\theta_j\eta_{j,1} + q_j z_j\eta_{j,2},\\ z_j\varepsilon_j(t) &\le \kappa\sigma_j\left(\frac{1}{p_j}\vartheta_j + p_j\right) + \frac{1}{p_j}z_j\vartheta_j\eta_{j,2} + p_j z_j\eta_{j,2},\end{aligned} \tag{2.37}$$

where $q_j, p_j > 0$. Substituting (2.37) into (2.35), one can obtain

$$\begin{aligned}\dot{V}_i \leq & \sum_{j=1}^{i}\left[\kappa\sigma_j\left(\frac{1}{q_j}\theta_j + q_j\frac{1}{p_j}\vartheta_j + p_j\right) + \frac{\tilde{\theta}_j}{\gamma_j}\left(\frac{\gamma_j}{q_j}z_j\eta_{j,1} - \dot{\hat{\theta}}_j\right)\right] \\ & + \sum_{j=1}^{i}\left[\frac{\tilde{\vartheta}_j}{\lambda_j}\left(\frac{\lambda_j}{p_j}z_j\eta_{j,2} - \dot{\hat{\vartheta}}_j\right)\right] \\ & + \sum_{j=1}^{i}\left[z_j\left(k_jz_j + \frac{1}{q_j}\hat{\theta}_j\eta_{j,1} + \left(\frac{\hat{\vartheta}_j}{p_j} + q_j + p_j\right)\eta_{j,2}\right)\right] \\ & + \sum_{j=1}^{i}\left[g_j\left(\bar{x}_j\right)z_j\alpha_j - k_jz_j^2\right] + g_i\left(\bar{x}_i\right)z_iz_{i+1}. \end{aligned} \tag{2.38}$$

The Nussbaum-type gain ξ_j and the virtual control α_j are designed as

$$\begin{aligned}\dot{\xi}_j &= z_js_j, \\ \alpha_j &= N_j\left(\xi_j\right)s_j\end{aligned} \tag{2.39}$$

with $s_j = k_jz_j + \frac{1}{q_j}\hat{\theta}_j\eta_{j,1} + \left(\frac{1}{p_j}\hat{\vartheta}_j + p_j + q_j\right)\eta_{j,2}$. Then, one has

$$\begin{aligned}\dot{V}_i \leq & \sum_{j=1}^{i}\left[g_j\left(\bar{x}_j\right)N_j\left(\xi_j\right) + 1\right]\dot{\xi}_j - \sum_{j=1}^{i}k_jz_j^2 \\ & + \sum_{j=1}^{i}\left[\kappa\sigma_j\left(\frac{1}{q_j}\theta_j + q_j\frac{1}{p_j}\vartheta_j + p_j\right)\right] \\ & + \sum_{j=1}^{i}\left[\frac{\tilde{\theta}_j}{\gamma_j}\left(\frac{\gamma_j}{q_j}z_j\eta_{j,1} - \dot{\hat{\theta}}_j\right) + \frac{\tilde{\vartheta}_j}{\lambda_j}\left(\frac{\lambda_j}{p_j}z_j\eta_{j,2} - \dot{\hat{\vartheta}}_j\right)\right] \\ & + g_i\left(\bar{x}_i\right)z_iz_{i+1}.\end{aligned} \tag{2.40}$$

To compensate for the unknown dynamics, we consider

$$\begin{aligned}\dot{\hat{\theta}}_j &= \gamma_j\left(\frac{\eta_{j,1}z_j}{q_j} - \sigma_j\hat{\theta}_j\right), \\ \dot{\hat{\vartheta}}_j &= \lambda_j\left(\frac{\eta_{j,2}z_j}{p_j} - \sigma_j\hat{\vartheta}_j\right).\end{aligned} \tag{2.41}$$

From (2.41), $\dot{V}_i$ further satisfies

$$\begin{aligned}\dot{V}_i \le& \sum_{j=1}^{i}\left[g_j\left(\bar{x}_j\right)N_j\left(\xi_j\right)+1\right]\dot{\xi}_j-\sum_{j=1}^{i}k_jz_j^2\\&+\sum_{j=1}^{i}\left[\kappa\sigma_j\left(\frac{1}{q_j}\theta_j+q_j+\frac{1}{p_j}\theta_j+p_j\right)\right]\\&+\sum_{j=1}^{i}\sigma_j\left(\tilde{\theta}_j\hat{\theta}_j+\tilde{\vartheta}_j\hat{\vartheta}_j\right)+g_i\left(\bar{x}_i\right)z_iz_{i+1}.\end{aligned}\tag{2.42}$$

2.3.3 Step n

Denote the tracking error for the n-th subsystem as $z_n = x_n - \alpha_{n-1}$. Differentiating z_n yields

$$\begin{aligned}\dot{z}_n &= \dot{x}_n - \dot{\alpha}_{n-1}\\&= g_n\left(\bar{x}_n\right)u\left(t\right)+f_n\left(\bar{x}_n\right)-\dot{\alpha}_{n-1}.\end{aligned}\tag{2.43}$$

Define

$$F_n\left(X_n\right)=f_n\left(\bar{x}_n\right)+g_{n-1}\left(\bar{x}_{n-1}\right)z_{n-1}-\dot{\alpha}_{n-1}\tag{2.44}$$

with $X_n = \left[x_1, x_2, \ldots, x_n, \dot{\alpha}_{n-1}\right]^{\mathrm{T}}$. From Lemma 1.4, the unknown function can be expressed as $F_n\left(X_n\right) = W_n^{\mathrm{T}}\Phi_n\left(X_n\right)+\varepsilon_n\left(t\right)$ with $\varepsilon_n\left(t\right)\le\bar{\varepsilon}_n$. Define $\theta_n = \|W_n\|^2$ and $\vartheta_n = \bar{\varepsilon}_n^2$, both of which are unknown. Then, we consider introduce adaptive designs $\hat{\theta}_n$ and $\hat{\vartheta}_n$ with $\tilde{\theta}_n = \theta_n - \hat{\theta}_n$ and $\tilde{\vartheta}_n = \vartheta_n - \hat{\vartheta}_n$ are the parameter estimation errors.

Design

$$\begin{aligned}\eta_{n,1} &= \Phi_n^{\mathrm{T}}\Phi_n\tanh\left(\frac{z_n\Phi_n^{\mathrm{T}}\Phi_n}{\sigma_n}\right),\\\eta_{n,2} &= \tanh\left(\frac{z_n}{\sigma_n}\right)\end{aligned}\tag{2.45}$$

with $\sigma_n(t) = d_n\mathrm{e}^{-r_nt}$, $d_n > 0$, $r_n > 0$ satisfying $\lim_{t\to\infty}\int_{t_0}^{t}\sigma_n\left(\iota\right)\mathrm{d}\iota \le \bar{\sigma}_n < \infty$, $\bar{\sigma}_n > 0$. The adaptive Nussbaum-based design u is considered as

$$\begin{aligned}u\left(t\right) &= N_n\left(\xi_n\right)s_n,\\\dot{\xi}_n &= z_ns_n\end{aligned}\tag{2.46}$$

with

$$s_n = k_n z_n + \frac{1}{q_n}\hat{\theta}_n \eta_{n,1} + \left(\frac{1}{p_n}\hat{\vartheta}_n + p_n + q_n\right)\eta_{n,2} \tag{2.47}$$

where $k_n, q_n, p_n > 0$ are constants. To deal with unknown dynamics, consider the adaptive compensation law as

$$\begin{aligned}\dot{\hat{\theta}}_n &= \gamma_n\left(\frac{\eta_{n,1}z_n}{q_n} - \sigma_n\hat{\theta}_n\right),\\ \dot{\hat{\vartheta}}_n &= \lambda_n\left(\frac{\eta_{n,2}z_n}{p_n} - \sigma_n\hat{\vartheta}_n\right),\end{aligned} \tag{2.48}$$

where $\lambda_n, \gamma_n > 0$ are constants.

2.4 Asymptotic Stability Analysis

In this subsection, the adaptive Nussbaum design is investigated in terms of stability analysis as follows.

Theorem 2.1 *Under Assumptions 2.1–2.2, consider the system in* (2.1) *with the adaptive Nussbaum design in Sect. 2.3. One can ensure that the system* (2.1) *can obtain asymptotic tracking performance.*

Proof Define $V_n = \frac{1}{2}z_n^2 + \frac{1}{2\gamma_n}\tilde{\theta}_n^2 + \frac{1}{2\lambda_n}\tilde{\vartheta}_n^2 + V_{n-1}$. Differentiating V_n yields

$$\dot{V}_n = z_n\dot{z}_n + \frac{1}{\gamma_n}\tilde{\theta}_n\dot{\hat{\theta}}_n + \frac{1}{\lambda_n}\tilde{\vartheta}_n\dot{\hat{\vartheta}}_n + \dot{V}_{n-1}. \tag{2.49}$$

From (2.42), one has

$$\begin{aligned}\dot{V}_{n-1} \le& \sum_{i=1}^{n-1}\left[g_i(\bar{x}_i)N_i(\xi_i) + 1\right]\dot{\xi}_i - \sum_{i=1}^{n-1}k_i z_i^2\\ &+\sum_{i=1}^{n-1}\left[\kappa\sigma_i\left(\frac{1}{q_i}\theta_i + q_i + \frac{1}{p_i}\theta_i + p_i\right)\right]\\ &+\sum_{i=1}^{n-1}\sigma_i\left(\tilde{\theta}_i\hat{\theta}_i + \tilde{\vartheta}_i\hat{\vartheta}_i\right) + g_{n-1}(\bar{x}_{n-1})z_{n-1}z_n.\end{aligned} \tag{2.50}$$

Based on this and (2.43), $\dot{V}_n$ satisfies

$$\begin{aligned}\dot{V}_n \leq\ & z_n g_n(\bar{x}_n)\, u + z_n W_n^{\mathrm{T}} \Phi_n(X_n) + z_n \varepsilon_n(t) \\ & + \sum_{i=1}^{n-1} [g_i(\bar{x}_i) N_i(\xi_i) + 1]\, \dot{\xi}_i - \sum_{i=1}^{n-1} k_i z_i^2 \\ & + \sum_{i=1}^{n-1} \left[\kappa\sigma_i \left(\frac{1}{q_i}\theta_i + q_i + \frac{1}{p_i}\theta_i + p_i\right)\right] \\ & + \sum_{i=1}^{n-1} \sigma_i \left(\tilde{\theta}_i \hat{\theta}_i + \tilde{\vartheta}_i \hat{\vartheta}_i\right) - \frac{1}{\gamma_n} \tilde{\theta}_n \dot{\hat{\theta}}_n - \frac{1}{\lambda_n} \tilde{\vartheta}_n \dot{\hat{\vartheta}}_n. \end{aligned} \tag{2.51}$$

Considering the inequality

$$\begin{aligned} z_n W_n^{\mathrm{T}} \Phi_n &\leq \kappa\sigma_n \left(\frac{1}{q_n}\theta_n + q_n\right) + \frac{1}{q_n} z_n \theta_n \eta_{n,1} + q_n z_n \eta_{n,2}, \\ z_n \varepsilon_n(t) &\leq \kappa\sigma_n \left(\frac{1}{p_n}\vartheta_n + p_n\right) + \frac{1}{p_n} z_n \vartheta_n \eta_{n,2} + p_n z_n \eta_{n,2}, \end{aligned} \tag{2.52}$$

with the fact in (2.46) and (2.48), one has

$$\begin{aligned}\dot{V}_n \leq\ & \sum_{i=1}^{n} [g_i(\bar{x}_i) N_i(\xi_i) + 1]\, \dot{\xi}_i - \sum_{i=1}^{n} k_i z_i^2 \\ & + \sum_{i=1}^{n} \left[\kappa\sigma_i \left(\frac{1}{q_i}\theta_i + q_i + \frac{1}{p_i}\theta_i + p_i\right)\right] \\ & + \sum_{i=1}^{n} \sigma_i \left(\tilde{\theta}_i \hat{\theta}_i + \tilde{\vartheta}_i \hat{\vartheta}_i\right). \end{aligned} \tag{2.53}$$

Based on the completion of squares, one can obtain

$$\begin{aligned} \tilde{\theta}_j \hat{\theta}_j &= \tilde{\theta}_j(\theta_j - \tilde{\theta}_j) = -\tilde{\theta}_j^2 + \tilde{\theta}_j \theta_j \\ &= -\left(\tilde{\theta}_j - \frac{1}{2}\theta_j\right)^2 + \frac{\theta_j^2}{4} \\ &\leq \frac{\theta_j^2}{4}, \\ \tilde{\vartheta}_j \hat{\vartheta}_j &= \tilde{\vartheta}_j(\vartheta_j - \tilde{\vartheta}_j) = -\tilde{\vartheta}_j^2 + \tilde{\vartheta}_j \vartheta_j \\ &= -\left(\tilde{\vartheta}_j - \frac{1}{2}\vartheta_j\right)^2 + \frac{\vartheta_j^2}{4} \\ &\leq \frac{\vartheta_j^2}{4}. \end{aligned} \tag{2.54}$$

Substituting the above fact into (2.53), one has

$$
\begin{aligned}
\dot{V}_n \leq & \sum_{i=1}^{n} \left[g_i\left(\bar{x}_i\right) N_i\left(\xi_i\right) + 1 \right] \dot{\xi}_i - \sum_{i=1}^{n} k_i z_i^2 \\
& + \sum_{i=1}^{n} \sigma_i \left[\kappa \left(\frac{1}{q_i}\theta_i + q_i + \frac{1}{p_i}\theta_i + p_i \right) + \frac{1}{4}\theta_i^2 + \frac{1}{4}\vartheta_i^2 \right] \\
\leq & \sum_{i=1}^{n} \left[g_i\left(\bar{x}_i\right) N_i\left(\xi_i\right) + 1 \right] \dot{\xi}_i - \sum_{i=1}^{n} k_i z_i^2 + \sum_{i=1}^{n} \sigma_i \chi_i ,
\end{aligned} \tag{2.55}
$$

with

$$
\chi_i = \kappa \left(\frac{1}{q_i}\theta_i + q_i + \frac{1}{p_i}\theta_i + p_i \right) + \frac{1}{4}\theta_i^2 + \frac{1}{4}\vartheta_i^2 .
$$

Integrating (2.55) from 0 to t, one can obtain

$$
\begin{aligned}
V_n(t) \leq & \sum_{i=1}^{n} \int_0^t \left[g_i\left(\bar{x}_i\right) N_i\left(\xi_i(\tau)\right) + 1 \right] \dot{\xi}_i(\tau)\, \mathrm{d}\tau \\
& + \sum_{i=1}^{n} \int_0^t \sigma_i(\tau) \chi_i \mathrm{d}\tau + V_n(0) - \sum_{i=1}^{n} \int_0^t k_i z_i^2(\tau)\, \mathrm{d}\tau \\
\leq & \sum_{i=1}^{n} \int_0^t \left[g_i\left(\bar{x}_i\right) N_i\left(\xi_i(\tau)\right) + 1 \right] \dot{\xi}_i(\tau)\, \mathrm{d}\tau \\
& + \sum_{i=1}^{n} \bar{\sigma}_i \chi_i + V_n(0) .
\end{aligned} \tag{2.56}
$$

From Lemma 2.1, one can obtain that $V_n(t)$ and $\xi_i(t)$ remain bounded on $[0, t)$, which implies that z_i, $\tilde{\theta}_i$, $\tilde{\vartheta}_i$ are uniformly bounded for $i = 1, \ldots, n$. From $x_1 = z_1 + x_0$ and x_0 is bounded, we can obtain x_1 is bounded. Note that the virtual control signal α_1 is the function of bounded signals $\xi_1, x_1, \hat{\theta}_1$ and $\hat{\vartheta}_1$,it is straightforward to show that α_1 is uniformly bounded, which in turn implies that $x_2 = z_2 + \alpha_1$ is bounded. By using a similar analysis as the system state x_2, one can conclude that the system states x_i are bounded with the help of boundedness of virtual control signals $\alpha_i, i = 1, 2, \ldots, n-1$. Moreover, considering (2.46), it is easy to verify that the actual control signal u is also uniformly bounded. Thus, all the signals in the closed-loop system are bounded at any time. Furthermore, (2.56) also implies

$$
\begin{aligned}
\sum_{i=1}^{n} \int_0^t k_i z_i^2(\tau)\, \mathrm{d}\tau \leq & \sum_{i=1}^{n} \int_0^t \left[g_i\left(\bar{x}_i\right) N_i\left(\xi_i(\tau)\right) + 1 \right] \dot{\xi}_i(\tau)\, \mathrm{d}\tau \\
& + \sum_{i=1}^{n} \bar{\sigma}_j \chi_i + V_n(0) .
\end{aligned} \tag{2.57}
$$

Therefore, z_i^2 is uniformly continuous and $\dot{z}_i$ is bounded on $[0, \infty)$. Based on Lemma 1.10, z_i converges asymptotically to the origin for $i = 1, \ldots, n$. This completes the proof. □

2.5 Simulation Example

In this section, the simulation study on a marine surface vehicle is conducted to verify the presented adaptive Nussbaum design. The marine surface vehicle system adopted from [12] is described by a nonlinear Norrbin dynamical model as

$$T\ddot{\psi} + \dot{\psi} + K\dot{\psi}^3 = \phi\chi,$$

where ψ represents the course angle, χ is the command rudder angle, T denotes the positive time constant, K stands for the Norrbin coefficient, and ϕ is the gain constant. Furthermore, define $x_1 = \psi$, $x_2 = \dot{\psi}$ and $u = \chi$. Then, one can transform the Norrbin dynamics into the following form as

$$\begin{aligned} \dot{x}_1 &= x_2 \\ \dot{x}_2 &= -\frac{1}{T}x_2 - \frac{K}{T}x_2^3 + \frac{\phi}{T}u \end{aligned}$$

with $T = 7\,\mathrm{s}$, $K = 0.6\,\mathrm{s}^2$, and $\phi = 0.32\,\mathrm{s}^{-1}$. The objective is for x_1 to track the desired trajectory $x_0 = 0.6 + 0.8\sin t$.

The initial conditions are $x_1(0) = 1$ and $x_2(0) = 0$. The design parameters are selected as

$$k_1 = k_2 = 9, \lambda_1 = \lambda_2 = 66, \gamma_1 = \gamma_2 = 88, \sigma_1 = 6\mathrm{e}^{-0.03\mathrm{t}}, \sigma_2 = 8\mathrm{e}^{-0.03\mathrm{t}}.$$

The results with the Nussbaum-based asymptotic tracking design are shown in Figs. 2.1 and 2.2. The Evolutions of the system states x_1 and x_2, the reference signal x_d, the virtual control α_1, and the tracking errors z_i are shown in Fig. 2.1. One can observe that the system output can track the reference signal precisely, indicating asymptotic tracking performance with the Nussbaum-based control design. The evolution of the adaptive Nussbaum design, including $\hat{\theta}_i$, $\hat{\vartheta}_i$ and ξ_i are shown in Fig. 2.2, which are bounded during the online process. The Nussbaum-type gains ξ_i also remain bounded, which verifies the discussions in Lemma 2.1.

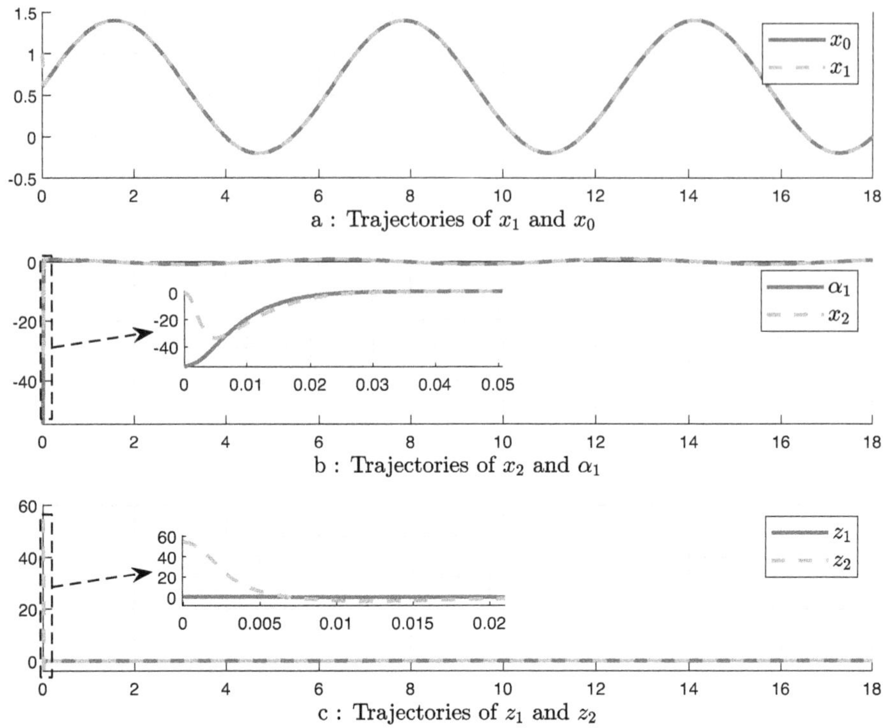

Fig. 2.1 System states, virtual control signal, and tracking errors

2.6 Conclusion

This chapter proposed an asymptotic tracking design for nonlinear strict feedback systems that encounter the challenge of multiple unknown control directions. To address these complexities, a novel Nussbaum-type function was developed, providing an adaptive mechanism to handle the uncertainties associated with control directionality. The proposed approach offers a robust and systematic solution, ensuring reliable asymptotic tracking performance even in the presence of significant unknowns. The effectiveness of the proposed design was rigorously validated through a simulation study on a marine surface vehicle. The results demonstrated the controller's ability to achieve accurate tracking under challenging conditions, confirming the practical applicability and robustness of the proposed scheme.

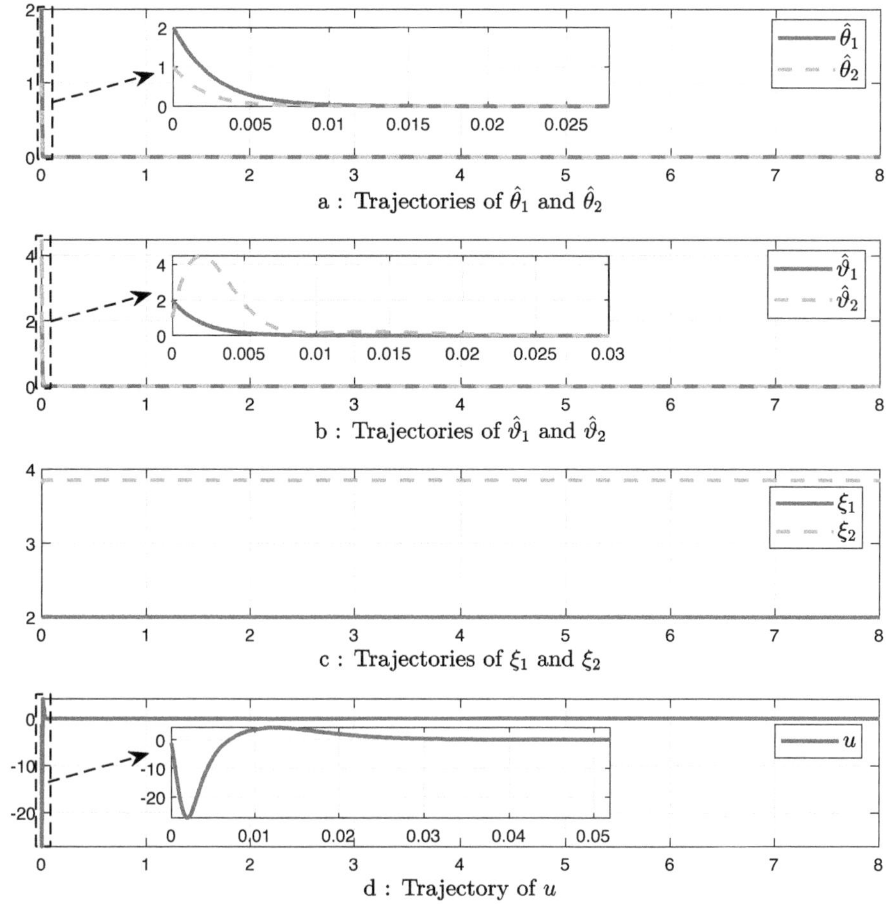

Fig. 2.2 Parameter learning laws, adaptive Nussbaum gains, and the control signal

References

1. Yang, Y., Tang, L., Zou, W., Ahn, C.K.: Novel command-filtered Nussbaum design for continuous-time nonlinear dynamical systems with multiple unknown high-frequency gains. Nonlinear Dyn. **111**, 4313–4323 (2023)
2. Li, Y., Liu, Y., Tong, S.: Observer-based neuro-adaptive optimized control of strict-feedback nonlinear systems with state constraints. IEEE Trans. Neural Netw. Learn. Syst. **33**(7), 3131–3145 (2022)
3. Vaidyanathan, S., Azar, A.T.: Backstepping Control of Nonlinear Dynamical Systems. Academic (2020)
4. Chen, B., Liu, X., Liu, K., Lin, C.: Direct adaptive fuzzy control of nonlinear strict-feedback systems. Automatica **45**(6), 1530–1535 (2009)
5. Liu, H., Pan, Y., Li, S., Chen, Y.: Adaptive fuzzy backstepping control of fractional-order nonlinear systems. IEEE Trans. Syst. Man Cybern.: Syst. **47**(8), 2209–2217 (2017)

6. Du, Z., Liang, H., Ahn, C.K.: Adaptive fuzzy control for multi-agent systems with unknown measurement sensitivity via a simplified backstepping approach. IEEE Trans. Circuits Syst. II Express Briefs **69**(6), 2862–2866 (2022)
7. Yang, Y., Liu, G., Li, Q., Ahn, C.K.: Multiple adaptive fuzzy nussbaum-type functions design for stochastic nonlinear systems with fixed-time performance. Fuzzy Sets Syst. **476**, 108767 (2024)
8. Xia, Y., Fu, M., Shi, P., Wu, Z., Zhang, J.: Adaptive backstepping controller design for stochastic jump systems. IEEE Trans. Autom. Control **54**(12), 2853–2859 (2009)
9. Liu, G., Yang, Y., Gao, W., Wunsch, D.C.: Adaptive Nussbaum design for nonholonomic systems with asymptotic stabilization against false data injection. IEEE Trans. Cybern. **55**(11), 5189–5201 (2025)
10. Cheng, S., Xin, B., Wang, Q., Chen, J., Deng, F.: Command filtered neuroadaptive fault-tolerant control for nonlinear systems with input saturation and unknown control direction. IEEE Trans. Neural Netw. Learn. Syst.. **35**(6), 7963–7973 (2024)
11. Wang, H., Karimi, H.R., Liu, P.X., Yang, H.: Adaptive neural control of nonlinear systems with unknown control directions and input dead-zone. IEEE Trans. Syst. Man Cybern.: Syst. **48**(11), 1897–1907 (2018)
12. Liu, G., Yang, Y., Zhao, X., Ahn, C.K.: Adaptive fuzzy practical bipartite synchronization for multiagent systems with intermittent feedback under multiple unknown control directions. IEEE Trans. Fuzzy Syst. **32**(5), 3211–3225 (2024)
13. Liu, Y., Ma, H.: Adaptive fuzzy tracking control of nonlinear switched stochastic systems with prescribed performance and unknown control directions. IEEE Trans. Syst. Man Cybern.: Syst. **50**(2), 590–599 (2020)
14. Lv, M., Yu, W., Cao, J., Baldi, S.: Consensus in high-power multiagent systems with mixed unknown control directions via hybrid Nussbaum-based control. IEEE Trans. Cybern. **52**(6), 5184–5196 (2022)
15. Rezaee, H., Abdollahi, F.: Adaptive leaderless consensus control of strict-feedback nonlinear multiagent systems with unknown control directions. IEEE Trans. Syst. Man Cybern.: Syst. **51**(10), 6435–6444 (2021)
16. Li, Y.X.: Barrier Lyapunov function-based adaptive asymptotic tracking of nonlinear systems with unknown virtual control coefficients. Automatica **121**, 109181 (2020)
17. Ye, X.: Adaptive nonlinear output-feedback control with unknown high-frequency gain sign. IEEE Trans. Autom. Control **46**(1), 112–115 (2001)
18. Yu, J., Shi, P., Lin, C., Yu, H.: Adaptive neural command filtering control for nonlinear MIMO systems with saturation input and unknown control direction. IEEE Trans. Cybern. **50**(6), 2536–2545 (2020)
19. Wang, C., Guo, L., Wen, C., Hu, Q., Qiao, J.: Adaptive neural network control for a class of nonlinear systems with unknown control direction. IEEE Trans. Syst. Man Cybern.: Syst. **50**(11), 4708–4718 (2020)
20. Lei, Y., Wang, Y.W., Guan, Z.H., Shen, Y.J.: Event-triggered adaptive output regulation for a class of nonlinear systems with unknown control direction. IEEE Trans. Syst. Man Cybern.: Syst. **50**(9), 3181–3188 (2020)

Chapter 3
Adaptive Nussbaum Design for Asymptotic Tracking Control with Full-State Constraints

3.1 Introduction

State constraint is a critical challenge in control system design. Due to physical and performance limitations, many nonlinear systems require that certain states be confined within specified amplitude ranges [1]. For instance, in scenarios with large initial condition offsets or under heavy external disturbances, abrupt amplitude changes in state trajectories must be avoided to ensure safety and system reliability [2]. Violating these constraints in practice may lead to safety incidents, performance degradation, instability, or even permanent damage. However, some existing literature fails to consider state constraints, limiting the real-world applicability of their proposed control strategies [3]. To address this challenge, various control strategies have been developed, such as set-invariance-based methods [4] and reference governor techniques [5]. Model predictive control provides another solution by formulating constraints into an optimization framework and solving the control law online [6]. More recently, the barrier function method has been introduced for the control of nonlinear systems with state and output constraints [7, 8]. This approach exploits the key property that a barrier function approaches infinity as its argument nears predefined artificial bounds [9, 10]. By designing the control law such that the BLF remains bounded, these bounds are never violated, and state constraints can be rigorously guaranteed. Benefiting from its normative design structure, the BLF-based technique has been widely applied to many system classes [1]. To solve state/output-constrained control problems, the Barrier Lyapunov Function (BLF)-based control method proposed in [11] has gradually become a powerful technique [12]. However, most of these methods are applicable only to feedback-linearized nonlinear systems. Although some results [13–15] have addressed stabilization, finite-time stabilization, and tracking control of output-constrained high-order nonlinear systems, others have discussed full-state-constrained stabilization and tracking [16, 17]. However, most of these results assume known system powers or do not ensure asymptotic convergence of the system states. Based on the above analysis, designing controllers for

Y. Yang et al., *Adaptive Nussbaum Design for Nonlinear Dynamical Systems*,
https://doi.org/10.1007/978-981-95-4997-9_3

nonlinear systems under both modeling uncertainties and state constraints remains a novel and valuable research topic. It is even more attractive to further achieve asymptotic tracking, which brings inspiration to this chapter.

Compared to existing single Nussbaum-type methods without constraints [19, 20], this chapter focuses on the tracking control problem for uncertain nonlinear systems with multiple unknown control directions under state constraints. To handle these constraints while maintaining tracking performance, a BLF-based adaptive control scheme is developed. The proposed design not only ensures that the system states remain within prescribed bounds but also achieves asymptotic convergence of the tracking error.

The chapter proceeds as follows. Section 3.2 outlines the problem formation with the state constraint. Section 3.3 introduces the control scheme constructed via a BLF-based recursive backstepping methodology. Stability properties related to asymptotic tracking are examined in Sect. 3.4. Section 3.5 provides a simulation study to illustrate the performance of the proposed adaptive Nussbaum design under uncertain nonlinear dynamics with unknown control directions. The chapter concludes with a summary in Sect. 3.6.

3.2 Problem Formulation

Consider the strict feedback nonlinear systems as

$$\begin{aligned} \dot{x}_i &= g_i\left(\bar{x}_i\right) x_{i+1} + f_i\left(\bar{x}_i\right), \ 1 \le i \le n-1, \\ \dot{x}_n &= g_n\left(\bar{x}_n\right) u\left(t\right) + f_n\left(\bar{x}_n\right), \\ y &= x_1, \end{aligned} \tag{3.1}$$

where $\bar{x}_i = [x_1, x_2, \ldots, x_i]^{\mathrm{T}} \in \mathbb{R}^i, i = 1, 2, \ldots, n$ are the systems states, $u \in \mathbb{R}$ denotes the control input, $y \in \mathbb{R}$ is the output of the system, $f_i\left(\cdot\right) : \mathbb{R}^i \to \mathbb{R}$ and $g_i\left(\cdot\right) : \mathbb{R}^i \to \mathbb{R}$ are unknown smooth nonlinear functions. In addition, the function $g_i\left(\cdot\right)$ is referred to as the time-varying control direction. The output $y\left(t\right)$ and the system states are required to remain in the set $|x_i| \le k_{c_i}, \forall t \ge 0$ with k_{c_i} being known positive constants.

Assumption 3.1 The control direction $g_i\left(\cdot\right)$ is bounded, i.e., $0 < g_{i,q} \le |g_i| \le g_{i,p}$ with constants $g_{i,p}$ and $g_{i,q}$.

Assumption 3.2 The desired trajectory $x_d(t)$ and its first derivative $\dot{x}_d(t)$ are continuous and bounded, satisfying $-k_{c1} < -A_0 \le \underline{x}_d \le x_d(t) \le \bar{x}_d \le A_0 < k_{c1}$ and $|\dot{x}_d(t)| \le B_1$.

To avoid the mutual cancellation problem of Nussbaum-type gains [18], the Nussbaum-type design with state constraints in this chapter employs the following lemma.

Lemma 3.1 *If there exists a positive definite, radially unbounded function $V(t)$ for all $t \in [0, t_f)$ with the Nussbaum-type functions $\{N_j(\cdot)\}_{j=1}^{M}$ defining as*

$$N_j(\xi) = 2\xi e^{\xi^2} \cos(2^{j-1}\xi) - 2^{j-1}(e^{\xi^2} - 1)\sin(2^{j-1}\xi), \tag{3.2}$$

for all $\xi \in \mathbb{R}$. Then, $\sum\limits_{j=1}^{M} \int\limits_{0}^{t} \left(g_j(\sigma) N_j(\xi_j(\sigma)) + 1\right) \dot{\xi}_j(\sigma) d\sigma$, $V(t)$, and ξ_j are bounded if

$$V(t) \leq \sum_{j=1}^{M} \int_{0}^{t} \left(g_j(\sigma) N_j(\xi_j(\sigma)) + 1\right) \dot{\xi}_j(\sigma) d\sigma + c_0, \tag{3.3}$$

where $a > 0$, $c_0 > 0$, and $\{g_j(\cdot)\}_{j=1}^{M}$ are continuous functions satisfying $|g_j(\cdot)| \in [g_{j,\min}, g_{j,\max}]$ with constants $g_{j,\min} > 0$ and $g_{j,\max} > 0$.

Proof The proof of this lemma can be found in Lemma 2.1. □

3.3 Adaptive Nussbaum Design

In this section, a novel asymmetric positive definite Lyapunov function is constructed to deal with the full state constraints. First, the tracking errors z_i are defined for each subsystem as

$$z_i = x_i - \alpha_{i-1},\ i = 1, 2, \ldots, n, \tag{3.4}$$

where α_i is the virtual control signal for $i = 1, ..., n-1$ and $\alpha_0 = x_d$ is the desired signal. From (3.1) and (3.4), differentiating the tracking errors z_i yields

$$\begin{aligned} \dot{z}_i &= \dot{x}_i - \dot{\alpha}_{i-1} = g_i(\bar{x}_i) x_{i+1} + f_i(\bar{x}_i) - \dot{\alpha}_{i-1}, i = 1, 2, \ldots, n-1, \\ \dot{z}_n &= \dot{x}_n - \dot{\alpha}_{n-1} = g_n(\bar{x}_n) u(t) + f_n(\bar{x}_n) - \dot{\alpha}_{n-1}. \end{aligned} \tag{3.5}$$

Define $F_i(X_i) = f_i(\bar{x}_i) - \dot{\alpha}_{i-1}$ where $X_i = [x_1, x_2, \ldots, x_i, \dot{\alpha}_{i-1}]^{\mathrm{T}}, i = 1, 2, \ldots, n$. Using the RBFNN, $F_i(X_i)$ can be approximated as $F_i(X_i) = W_i^{\mathrm{T}} \Phi_i(X_i) + \varepsilon_i(t)$ with the approximation error $\varepsilon_i(t) \leq \vartheta_i$. Define $\theta_i = \|W_i\|$, both θ_i and ϑ_i are unknown positive constants. Then, one can use $\hat{\theta}_i$ and $\hat{\vartheta}_i$ to learn the unknown parameter with $\tilde{\theta}_i = \theta_i - \hat{\theta}_i$ and $\tilde{\vartheta}_i = \vartheta_i - \hat{\vartheta}_i$ being the parameter estimation errors. Based on this, (3.5) is equivalent to

$$\begin{aligned} \dot{z}_i &= g_i(\bar{x}_i) z_{i+1} + g_i(\bar{x}_i) \alpha_i + W_i^T \Phi_i + \varepsilon_i(t), i = 1, 2, \ldots, n-1 \\ \dot{z}_n &= g_n(\bar{x}_n) u(t) + W_n^T \Phi_n + \varepsilon_n(t). \end{aligned} \tag{3.6}$$

3.3.1 Step 1

From (3.6), we can obtain

$$\dot{z}_1 = g_1(x_1) z_2 + g_1(x_1) \alpha_1 + W_1^T \Phi_1 + \varepsilon_1(t). \tag{3.7}$$

Consider the Lyapunov function as

$$V_1 = \frac{1}{2} \log \left(\frac{\delta_1^2}{\delta_1^2 - z_1^2} \right) + \frac{1}{2\gamma_1} \tilde{\theta}_1^2 + \frac{1}{2\lambda_1} \tilde{\vartheta}_1^2, \tag{3.8}$$

where $\gamma_1 > 0$ and $\lambda_1 > 0$ are known constants to be chosen. Then, one can get

$$\dot{V}_1 = g_1(x_1) \kappa_1 z_2 + g_1(x_1) \kappa_1 \alpha_1 + \kappa_1 W_1^{\mathrm{T}} \Phi_1 + \kappa_1 \varepsilon_1 - \frac{1}{\gamma_1} \tilde{\theta}_1 \dot{\hat{\theta}}_1 - \frac{1}{\lambda_1} \tilde{\vartheta}_1 \dot{\hat{\vartheta}}_1, \tag{3.9}$$

with $\kappa_1 = \frac{z_1}{\delta_1^2 - z_1^2}$. Applying Young's inequality to the term in (3.9) with Assumption 3.1, one has

$$\begin{aligned} g_1(x_1) \kappa_1 z_2 &= \frac{g_1(x_1) z_1 z_2}{\delta_1^2 - z_1^2} \\ &\leq \frac{1}{2} \frac{z_1^2}{\delta_1^2 - z_1^2} g_{1,\max}^2 + \frac{1}{2} \frac{1}{\delta_1^2 - z_1^2} z_2^2. \end{aligned} \tag{3.10}$$

Based on Lemma 1.9, one can obtain the following fact as

$$\begin{aligned} \kappa_1 W_1^T \Phi_1 &\leq |\kappa_1| \left\| W_1^T \Phi_1 \right\| \leq \theta_1 \left(\frac{\kappa_1^2 \Phi_1^T \Phi_1}{\sqrt{\kappa_1^2 \Phi_1^T \Phi_1 + \sigma_1^2}} + \sigma_1 \right), \\ \kappa_1 \varepsilon_1 &\leq |\kappa_1| \vartheta_1 \leq \vartheta_1 \left(\frac{\kappa_1^2}{\sqrt{\kappa_1^2 + \sigma_1^2}} + \sigma_1 \right). \end{aligned} \tag{3.11}$$

Define

$$\begin{aligned} \varsigma_1 &= \frac{\kappa_1 \Phi_1^{\mathrm{T}} \Phi_1}{\sqrt{\kappa_1^2 \Phi_1^{\mathrm{T}} \Phi_1 + \sigma_1^2}}, \\ \zeta_1 &= \frac{\kappa_1}{\sqrt{\kappa_1^2 + \sigma_1^2}}, \end{aligned}$$

with any positive integrable function $\sigma_1(t)$, which satisfies $\lim_{t \to \infty} \int_{t_0}^{t} \sigma_1(\iota)\, \mathrm{d}\iota \leq \bar{\sigma}_1 < \infty$ with positive constant $\bar{\sigma}_1$. In this chapter, σ_1 is chosen as $\sigma_1 = d_1 \mathrm{e}^{-r_1 t}$ with $r_1 > 0$.

Then, one has

$$
\begin{aligned}
\kappa_1 W_1^{\mathrm{T}} \Phi_1 &\leq \kappa_1 \theta_1 \varsigma_1 + \sigma_1 \theta_1, \\
\kappa_1 \varepsilon_1 (t) &\leq \kappa_1 \vartheta_1 \zeta_1 + \sigma_1 \vartheta_1.
\end{aligned} \tag{3.12}
$$

Substituting (3.10) and (3.12) into (3.9), one can obtain

$$
\begin{aligned}
\dot{V}_1 \leq & \frac{1}{2} \frac{z_1^2}{\delta_1^2 - z_1^2} g_{1,\max}^2 + \frac{1}{2} \frac{1}{\delta_1^2 - z_1^2} z_2^2 + g_1 (x_1) \kappa_1 \alpha_1 \\
& + \frac{1}{\gamma_1} \tilde{\theta}_1 \left(\gamma_1 \kappa_1 \varsigma_1 - \dot{\hat{\theta}}_1 \right) + \frac{1}{\lambda_1} \tilde{\vartheta}_1 \left(\lambda_1 \kappa_1 \zeta_1 - \dot{\hat{\vartheta}}_1 \right) \\
& + \kappa_1 \left(\hat{\theta}_1 \varsigma_1 + \hat{\vartheta}_1 \zeta_1 \right) + \sigma_1 (\theta_1 + \vartheta_1).
\end{aligned} \tag{3.13}
$$

In addition, adding and subtracting $\kappa_1 c_1 z_1$ with $c_1 > 0$, one has

$$
\begin{aligned}
\dot{V}_1 \leq & \frac{1}{2} \frac{1}{\delta_1^2 - z_1^2} z_2^2 + g_1 (x_1) \kappa_1 \alpha_1 \\
& + \frac{1}{\gamma_1} \tilde{\theta}_1 \left(\kappa_1 \gamma_1 \varsigma_1 - \dot{\hat{\theta}}_1 \right) + \frac{1}{\lambda_1} \tilde{\vartheta}_1 \left(\kappa_1 \lambda_1 \zeta_1 - \dot{\hat{\vartheta}}_1 \right) \\
& + \kappa_1 \left(c_1 z_1 + \hat{\theta}_1 \varsigma_1 + \hat{\vartheta}_1 \zeta_1 \right) + \sigma_1 (\theta_1 + \vartheta_1) \\
& + \frac{1}{2} \frac{z_1^2 g_{1,\max}^2}{\delta_1^2 - z_1^2} - \frac{c_1 z_1^2}{\delta_1^2 - z_1^2}.
\end{aligned} \tag{3.14}
$$

Design the Nussbaum-type gain ξ_1 and the virtual control α_1 as

$$
\begin{aligned}
\dot{\xi}_1 &= \kappa_1 \left(c_1 z_1 + \hat{\theta}_1 \varsigma_1 + \hat{\vartheta}_1 \zeta_1 \right), \\
\alpha_1 &= N_1 (\xi_1) \left(c_1 z_1 + \hat{\theta}_1 \varsigma_1 + \hat{\vartheta}_1 \zeta_1 \right).
\end{aligned} \tag{3.15}
$$

By defining $s_1 = c_1 z_1 + \hat{\theta}_1 \varsigma_1 + \hat{\vartheta}_1 \zeta_1$, one can obtain

$$
\begin{aligned}
\dot{\xi}_1 &= \kappa_1 s_1, \\
\alpha_1 &= N_1 (\xi_1) s_1.
\end{aligned} \tag{3.16}
$$

To compensate for the unknown dynamics, the adaptive laws for $\hat{\theta}_1$ and $\hat{\vartheta}_1$ are designed as

$$
\begin{aligned}
\dot{\hat{\theta}}_1 &= \gamma_1 \left(\varsigma_1 \kappa_1 - \sigma_1 \hat{\theta}_1 \right), \\
\dot{\hat{\vartheta}}_1 &= \lambda_1 \left(\zeta_1 \kappa_1 - \sigma_1 \hat{\vartheta}_1 \right).
\end{aligned} \tag{3.17}
$$

Consider (3.16) and (3.17), $\dot{V}_1$ further satisfies

$$\begin{aligned}\dot{V}_1 \leq & \frac{1}{2} \frac{1}{\delta_1^2 - z_1^2} z_2^2 + [g_1(x_1) N_1(\xi_1) + 1] \dot{\xi}_1 \\ & + \sigma_1 \left(\tilde{\theta}_1 \hat{\theta}_1 + \tilde{\vartheta}_1 \hat{\vartheta}_1 + \theta_1 + \vartheta_1\right) \\ & + \frac{1}{2} \frac{z_1^2}{\delta_1^2 - z_1^2} g_{1,\max}^2 - c_1 \frac{z_1^2}{\delta_1^2 - z_1^2}. \end{aligned} \tag{3.18}$$

3.3.2 Step i $(2 \leq i \leq n-1)$

From (3.5), one can obtain

$$\dot{z}_i = g_i(\bar{x}_i) z_{i+1} + g_i(\bar{x}_i) \alpha_i + f_i(\bar{x}_i) - \dot{\alpha}_{i-1}. \tag{3.19}$$

Consider the Lyapunov function as

$$V_i = \frac{1}{2} \log\left(\frac{\delta_i^2}{\delta_i^2 - z_i^2}\right) + \frac{1}{2\gamma_i} \tilde{\theta}_i^2 + \frac{1}{2\lambda_i} \tilde{\vartheta}_i^2 + V_{i-1}, \tag{3.20}$$

where $\gamma_i > 0$ and $\lambda_i > 0$ are known constants to be chosen. Then, the derivative of V_i can be expressed as

$$\begin{aligned}\dot{V}_i = & \frac{\eta_i \dot{\eta}_i}{1 - \eta_i^2} - \frac{1}{\gamma_i} \tilde{\theta}_i \dot{\hat{\theta}}_i - \frac{1}{\lambda_i} \tilde{\vartheta}_i \dot{\hat{\vartheta}}_i + \dot{V}_{i-1} \\ = & \sum_{j=1}^{i} \left[g_j(\bar{x}_j) \kappa_j z_{j+1} + g_j(\bar{x}_j) \kappa_j \alpha_j + \kappa_j W_j^T \Phi_j\right] \\ & + \sum_{j=1}^{i} \left[\kappa_j \varepsilon_j(t) - \frac{1}{\gamma_j} \tilde{\theta}_j \dot{\hat{\theta}}_j - \frac{1}{\lambda_j} \tilde{\vartheta}_j \dot{\hat{\vartheta}}_j\right] \end{aligned} \tag{3.21}$$

with $\kappa_j = \frac{z_j}{\delta_j^2 - z_j^2}$. Similarly to Step 1, applying Young's inequality to the term in (3.21) with Assumption 3.1, one can obtain

$$\begin{aligned} g_j(x_j) \kappa_j z_{j+1} = & \frac{g_j(x_j) z_j z_{j+1}}{\delta_j^2 - z_j^2} \\ \leq & \frac{1}{2} \frac{z_j^2}{\delta_j^2 - z_j^2} g_{j,\max}^2 + \frac{1}{2} \frac{1}{\delta_j^2 - z_j^2} z_{j+1}^2. \end{aligned} \tag{3.22}$$

Define $\varsigma_j = \frac{\kappa_j \Phi_j^{\mathrm{T}} \Phi_j}{\sqrt{\kappa_j^2 \Phi_j^{\mathrm{T}} \Phi_j + \sigma_j^2}}$ and $\zeta_j = \frac{\kappa_j}{\sqrt{\kappa_j^2 + \sigma_j^2}}$ with any positive integrable time-varying functions $\sigma_j(t) = d_j \mathrm{e}^{-r_j t}$ with $r_j > 0$, which satisfies $\lim_{t \to \infty} \int_{t_0}^{t} \sigma_j(\iota)\, \mathrm{d}\iota \le \bar{\sigma}_j < \infty$ with positive constant $\bar{\sigma}_j$. Similarly, one has

$$
\begin{aligned}
\kappa_j W_j^{\mathrm{T}} \Phi_j &\le \kappa_j \theta_j \varsigma_j + \sigma_j \theta_j, \\
\kappa_j \varepsilon_j(t) &\le \kappa_j \vartheta_j \zeta_j + \sigma_j \vartheta_j.
\end{aligned} \tag{3.23}
$$

Then, one has

$$
\begin{aligned}
\dot{V}_i \le & \sum_{j=1}^{i} \left[\frac{1}{2} \frac{z_j^2}{\delta_j^2 - z_j^2} g_{j,\max}^2 + \frac{1}{2} \frac{1}{\delta_j^2 - z_j^2} z_{j+1}^2 + g_j(\bar{x}_j) \kappa_j \alpha_j \right] \\
& + \sum_{j=1}^{i} \left[\frac{1}{\gamma_j} \tilde{\theta}_j \left(\kappa_j \gamma_j \varsigma_j - \dot{\hat{\theta}}_j \right) + \frac{1}{\lambda_j} \tilde{\vartheta}_j \left(\kappa_j \lambda_j \zeta_j - \dot{\hat{\vartheta}}_j \right) \right] \\
& + \sum_{j=1}^{i} \left[\kappa_j \left(\hat{\theta}_j \varsigma_j + \hat{\vartheta}_j \zeta_j \right) + \sigma_j \left(\theta_j + \vartheta_j \right) \right].
\end{aligned} \tag{3.24}
$$

In addition, adding and subtracting $\kappa_j c_j z_j$ with $c_j > 0$, one can obtain

$$
\begin{aligned}
\dot{V}_i \le & [g_1(x_1) N_1(\xi_1) + 1] \dot{\xi}_1 + \sigma_1 \left(\tilde{\theta}_1 \hat{\theta}_1 + \tilde{\vartheta}_1 \hat{\vartheta}_1 + \theta_1 + \vartheta_1 \right) \\
& + \sum_{j=2}^{i} \left[g_j(\bar{x}_j) \kappa_j \alpha_j + \frac{1}{\gamma_j} \tilde{\theta}_j \left(\kappa_j \gamma_j \varsigma_j - \dot{\hat{\theta}}_j \right) + \frac{1}{\lambda_j} \tilde{\vartheta}_j \left(\kappa_j \lambda_j \zeta_j - \dot{\hat{\vartheta}}_j \right) \right] \\
& + \sum_{j=1}^{i} \left[\kappa_j \left(c_j z_j + \hat{\theta}_j \varsigma_j + \hat{\vartheta}_j \zeta_j + \frac{1}{2} \frac{\delta_j^2 - z_j^2}{\delta_{j-1}^2 - z_{j-1}^2} z_j \right) + \sigma_j \left(\theta_j + \vartheta_j \right) \right] \\
& + \sum_{j=2}^{i} \left[\frac{1}{2} \frac{z_j^2}{\delta_j^2 - z_j^2} g_{j,\max}^2 - c_j \frac{z_j^2}{\delta_j^2 - z_j^2} \right] + \frac{1}{2} \frac{1}{\delta_i^2 - z_i^2} z_{i+1}^2.
\end{aligned} \tag{3.25}
$$

Design the Nussbaum-type gain ξ_j and the virtual control α_j as

$$
\begin{aligned}
\dot{\xi}_j &= \kappa_j s_j, \\
\alpha_j &= N_j(\xi_j) s_j
\end{aligned} \tag{3.26}
$$

with $s_j = c_j z_j + \hat{\theta}_j \varsigma_j + \hat{\vartheta}_j \zeta_j + \frac{1}{2} \frac{\delta_j^2 - z_j^2}{\delta_{j-1}^2 - z_{j-1}^2} z_j$. To compensate for the unknown dynamics, the adaptive law for θ_j is designed as

$$\dot{\hat{\theta}}_j = \gamma_j \left(\kappa_j \varsigma_j - \sigma_j \hat{\theta}_j \right),$$
$$\dot{\hat{\vartheta}}_j = \lambda_j \left(\kappa_j \zeta_j - \sigma_j \hat{\vartheta}_j \right). \tag{3.27}$$

Then, $\dot{V}_i$ further satisfies

$$\begin{aligned}\dot{V}_i \leq & \sum_{j=1}^{i} \left[\left[g_j(x_j) N_j(\xi_j) + 1 \right] \dot{\xi}_j + \sigma_j \left(\tilde{\theta}_j \hat{\theta}_j + \tilde{\vartheta}_j \hat{\vartheta}_j + \theta_j + \vartheta_j \right) \right] \\ & + \sum_{j=2}^{i} \left[\frac{1}{2} \frac{z_j^2}{\delta_j^2 - z_j^2} g_{j,\max}^2 - c_j \frac{z_j^2}{\delta_j^2 - z_j^2} \right] + \frac{1}{2} \frac{1}{\delta_i^2 - z_i^2} z_{i+1}^2. \end{aligned} \tag{3.28}$$

3.3.3 Step n

Consider the Lyapunov function as

$$V_n = \frac{1}{2} \log \left(\frac{\delta_n^2}{\delta_n^2 - z_n^2} \right) + \frac{1}{2\gamma_n} \tilde{\theta}_n^2 + \frac{1}{2\lambda_n} \tilde{\vartheta}_n^2 + V_{n-1}, \tag{3.29}$$

where $\gamma_n > 0$ and $\lambda_n > 0$ are known constants to be chosen. Differentiating V_n yields

$$\dot{V}_n = g_n(\bar{x}_n) \kappa_n u(t) + \kappa_n W_n^T \Phi_n + \kappa_n \varepsilon_n - \frac{1}{\gamma_n} \tilde{\theta}_n \dot{\hat{\theta}}_n - \frac{1}{\lambda_n} \tilde{\vartheta}_n \dot{\hat{\vartheta}}_n + \dot{V}_{n-1} \tag{3.30}$$

with $\kappa_n = \frac{z_n}{\delta_n^2 - z_n^2}$. From (3.28), one has

$$\begin{aligned}\dot{V}_{n-1} \leq & \sum_{i=1}^{n-1} \left[\left[g_i(x_i) N_i(\xi_i) + 1 \right] \dot{\xi}_i + \sigma_i \left(\tilde{\theta}_i \hat{\theta}_i + \tilde{\vartheta}_i \hat{\vartheta}_i + \theta_i + \vartheta_i \right) \right] \\ & + \sum_{i=2}^{n-1} \left[\frac{1}{2} \frac{z_i^2}{\delta_i^2 - z_i^2} g_{i,\max}^2 - c_i \frac{z_i^2}{\delta_i^2 - z_i^2} \right] + \frac{1}{2} \frac{1}{\delta_{n-1}^2 - z_{n-1}^2} z_n^2. \end{aligned} \tag{3.31}$$

Define $\sigma_n(t) = d_n \mathrm{e}^{-r_n t}$ satisfying $\lim_{t \to \infty} \int_{t_0}^{t} \sigma_n(\iota) \mathrm{d}\iota \leq \bar{\sigma}_n < \infty$ with positive constant $\bar{\sigma}_n$ and $r_n > 0$. Similar to Step 1, one can utilize Lemmas 1.6 and 1.9 to obtain

$$\kappa_n W_n^{\mathrm{T}} \Phi_n \leq \kappa_n \theta_n \varsigma_n + \sigma_n \theta_n$$
$$\kappa_n \varepsilon_n(t) \leq \kappa_n \vartheta_n \zeta_n + \sigma_n \vartheta_n \tag{3.32}$$

with $\varsigma_n = \frac{\kappa_n \Phi_n^{\mathrm{T}} \Phi_n}{\sqrt{\kappa_n^2 \Phi_n^{\mathrm{T}} \Phi_n + \sigma_n^2}}$ and $\zeta_n = \frac{\kappa_n}{\sqrt{\kappa_n^2 + \sigma_n^2}}$. Based on (3.32), one has

$$\begin{aligned}\dot{V}_n \leq\ & g_n(\bar{x}_n)\kappa_n u(t) + \sigma_n(\theta_n + \vartheta_n) \\ & + \frac{1}{\gamma_n}\tilde{\theta}_n\left(\kappa_n\gamma_n\varsigma_n - \dot{\hat{\theta}}_n\right) + \frac{1}{\lambda_n}\tilde{\vartheta}_n\left(\kappa_n\lambda_n\zeta_n - \dot{\hat{\vartheta}}_n\right) \\ & + \sum_{i=1}^{n-1}\left[[g_i(x_i)N_i(\xi_i) + 1]\dot{\xi}_i + \sigma_i\left(\tilde{\theta}_i\hat{\theta}_i + \tilde{\vartheta}_i\hat{\vartheta}_i + \theta_i + \vartheta_i\right)\right] \\ & + \sum_{i=2}^{n-1}\left[\frac{1}{2}\frac{z_i^2}{\delta_i^2 - z_i^2}g_{i,\max}^2 - c_i\frac{z_i^2}{\delta_i^2 - z_i^2}\right] \\ & + \kappa_n\left(\hat{\theta}_n\varsigma_n + \hat{\vartheta}_n\zeta_n\right) + \frac{1}{2}\frac{1}{\delta_{n-1}^2 - z_{n-1}^2}z_n^2. \end{aligned} \tag{3.33}$$

In addition, adding and subtracting $\kappa_n c_n z_n$ with $c_n > 0$, one can obtain

$$\begin{aligned}\dot{V}_n \leq\ & g_n(\bar{x}_n)\kappa_n u(t) + \frac{1}{\gamma_n}\tilde{\theta}_n\left(\kappa_n\gamma_n\varsigma_n - \dot{\hat{\theta}}_n\right) + \frac{1}{\lambda_n}\tilde{\vartheta}_n\left(\kappa_n\lambda_n\zeta_n - \dot{\hat{\vartheta}}_n\right) \\ & + \sum_{i=1}^{n-1}\left[[g_i(x_i)N_i(\xi_i) + 1]\dot{\xi}_i + \sigma_i\left(\tilde{\theta}_i\hat{\theta}_i + \tilde{\vartheta}_i\hat{\vartheta}_i + \theta_i + \vartheta_i\right)\right] \\ & + \sum_{i=2}^{n-1}\left[\frac{1}{2}\frac{z_i^2}{\delta_i^2 - z_i^2}g_{i,\max}^2 - c_i\frac{z_i^2}{\delta_i^2 - z_i^2}\right] + \sigma_n(\theta_n + \vartheta_n) \\ & + \kappa_n\left(c_n z_n + \hat{\theta}_n\varsigma_n + \hat{\vartheta}_n\zeta_n + \frac{1}{2}\frac{\delta_n^2 - z_n^2}{\delta_{n-1}^2 - z_{n-1}^2}z_n\right) - c_n\frac{z_n^2}{\delta_n^2 - z_n^2}. \end{aligned} \tag{3.34}$$

Design the Nussbaum-type gain ξ_n and the control input u as

$$\begin{aligned}\dot{\xi}_n &= \kappa_n s_n, \\ u(t) &= N_n(\xi_n)s_n \end{aligned} \tag{3.35}$$

with $s_n = c_n z_n + \hat{\theta}_n\varsigma_n + \hat{\vartheta}_n\zeta_n + \frac{1}{2}\frac{\delta_n^2 - z_n^2}{\delta_{n-1}^2 - z_{n-1}^2}z_n$. The adaptive laws for θ_n and ϑ_n are designed as

$$\begin{aligned}\dot{\hat{\theta}}_n &= \gamma_n\left(\kappa_n\varsigma_n - \sigma_n\hat{\theta}_n\right), \\ \dot{\hat{\vartheta}}_n &= \lambda_n\left(\kappa_n\zeta_n - \sigma_n\hat{\vartheta}_n\right). \end{aligned} \tag{3.36}$$

3.4 Asymptotic Stability Analysis

In this section, we analyze the asymptotic tracking stability of the closed-loop system (3.1) under state constraint.

Theorem 3.1 *Under Assumptions 3.1–3.2, consider the system in* (3.1)*, with the virtual control* (3.15) *and* (3.26)*, the adaptive laws* (3.17)*,* (3.27) *and* (3.36)*, and the controller design* (3.35)*, one can ensure that the output tracking errors converge to zero asymptotically and the boundedness of the closed-loop system states is guaranteed.*

Proof Substituting (3.36) into $\dot{V}_n$ yields $\dot{V}_n$ further satisfies

$$\begin{aligned}\dot{V}_n \leq & \sum_{i=1}^{n} \left[g_i\left(\bar{x}_i\right) N_i\left(\xi_i\right)+1\right] \dot{\xi}_i+\sum_{i=1}^{n-1} \frac{1}{2} \frac{z_i^2}{\delta_i^2-z_i^2} g_{i,\max}^2 \\ & +\sum_{i=1}^{n}\left[\sigma_i\left(\tilde{\theta}_i \hat{\theta}_i+\tilde{\vartheta}_i \hat{\vartheta}_i+\theta_i+\vartheta_i\right)-c_i \frac{z_i^2}{\delta_i^2-z_i^2}\right].\end{aligned} \tag{3.37}$$

For the terms in (3.37), based on the completion of squares, one has

$$\begin{aligned}\tilde{\theta}_j \hat{\theta}_j & =\tilde{\theta}_j(\theta_j-\tilde{\theta}_j)=-\tilde{\theta}_j^2+\tilde{\theta}_j \theta_j \\ & =-\left(\tilde{\theta}_j-\frac{1}{2} \theta_j\right)^2+\frac{\theta_j^2}{4} \\ & \leq \frac{\theta_j^2}{4}, \\ \tilde{\vartheta}_j \hat{\vartheta}_j & =\tilde{\vartheta}_j(\vartheta_j-\tilde{\vartheta}_j)=-\tilde{\vartheta}_j^2+\tilde{\vartheta}_j \vartheta_j \\ & =-\left(\tilde{\vartheta}_j-\frac{1}{2} \vartheta_j\right)^2+\frac{\vartheta_j^2}{4} \\ & \leq \frac{\vartheta_j^2}{4}.\end{aligned} \tag{3.38}$$

Substituting (3.38) into (3.37), one can obtain

$$\begin{aligned}\dot{V}_n \leq & \sum_{i=1}^{n}\left[\left[g_i\left(\bar{x}_i\right) N_i\left(\xi_i\right)+1\right] \dot{\xi}_i+\sigma_i\left(\frac{\theta_j^2}{4}+\frac{\vartheta_j^2}{4}+\theta_i+\vartheta_i\right)-c_i \frac{z_i^2}{\delta_i^2-z_i^2}\right] \\ & +\sum_{i=1}^{n} \frac{1}{2} \frac{z_i^2}{\delta_i^2-z_i^2} g_{i,\max}^2 \\ = & \sum_{i=1}^{n}\left[\left[g_i\left(\bar{x}_i\right) N_i\left(\xi_i\right)+1\right] \dot{\xi}_i+\sigma_i \chi_i-\left(c_i-\frac{1}{2} g_{i,\max}^2\right) \frac{z_i^2}{\delta_i^2-z_i^2}\right],\end{aligned} \tag{3.39}$$

with $\chi_i = \frac{\theta_i^2}{4} + \frac{\vartheta_i^2}{4} + \theta_i + \vartheta_i$ for $j = 1, 2, \ldots, n$. From Lemma 1.11, V_n satisfies

$$\begin{aligned}
V_n(t) \leq & \sum_{i=1}^{n} \int_0^t [g_i(\bar{x}_i) N_i(\xi_i(\tau)) + 1] \dot{\xi}_i(\tau) d\tau + \sum_{i=1}^{n} \int_0^t \sigma_i(\tau) \chi_i d\tau \\
& - \sum_{i=1}^{n} \int_0^t \left(c_i - \frac{1}{2} g_{i,\max}^2\right) \frac{z_i^2(\tau)}{\delta_i^2 - z_i^2(\tau)} d\tau \\
\leq & \sum_{i=1}^{n} \int_0^t [g_i(\bar{x}_i) N_i(\xi_i(\tau)) + 1] \dot{\xi}_i(\tau) d\tau + \sum_{i=1}^{n} \int_0^t \sigma_i(\tau) \chi_i d\tau \\
& - \sum_{i=1}^{n} \int_0^t \left(c_i - \frac{1}{2} g_{i,\max}^2\right) \log \frac{\delta_i^2}{\delta_i^2 - z_i^2(\tau)} d\tau \\
\leq & \sum_{i=1}^{n} \int_0^t [g_i(\bar{x}_i) N_i(\xi_i(\tau)) + 1] \dot{\xi}_i(\tau) d\tau + \sum_{i=1}^{n} \bar{\sigma}_i \chi_i,
\end{aligned} \tag{3.40}$$

with $c_i \geq \frac{1}{2} g_{i,\max}^2$, $i = 1, 2, \ldots, n$. From (3.40), we can obtain that z_i, ξ_i, $\tilde{\theta}_i$, and $\tilde{\vartheta}_i$ are bounded. From Eq. (3.40), it can be concluded that z_i, ξ_i, $\tilde{\theta}_i$, and $\tilde{\vartheta}_i$ are all bounded. To verify whether the system state constraints are violated, we analyze the tracking error $z_1 = x_1 - x_d$. Given that the desired trajectory satisfies $|x_d(t)| \leq A_0$ in Assumption 3.2, one has $|x_1| = |z_1 + x_d| \leq |z_1| + |x_d| < \delta_1 + A_0 = k_{c1}$ where δ_1 is the upper bound of $|z_1|$. The virtual control input α_1 is the function of the bounded signals x_1, x_d, $\dot{x}_d$, and parameter estimates. Hence, it is bounded as $|\alpha_1| \leq A_1$. Furthermore, for each state variable x_i, $i = 2, 3, \ldots, n$, the boundedness of the corresponding virtual control input α_i, i.e., $|\alpha_i| \leq A_i$, which ensures that $|x_i| < k_{ci}$. Therefore, all signals in the closed-loop system remain bounded, and the system state constraints are not violated throughout the entire operation.

From the definition of V, together with the boundedness of all the closed-loop signals, we can obtain

$$\begin{aligned}
& \sum_{i=1}^{n} \int_0^t \left(c_i - \frac{1}{2} g_{i,\max}^2\right) \frac{z_i^2(\tau)}{\delta_i^2 - z_i^2(\tau)} d\tau \\
\leq & \sum_{i=1}^{n} \int_0^t [g_i(\bar{x}_i) N_i(\xi_i(\tau)) + 1] \dot{\xi}_i(\tau) d\tau \\
& + \sum_{i=1}^{n} \bar{\sigma}_i \chi_i
\end{aligned} \tag{3.41}$$

which implies that $\frac{z_i^2}{\delta_i - z_i^2}$ and its time derivative are uniformly bounded. This implies that $\frac{z_i^2}{\delta_i - z_i^2}$ is uniformly continuous. Consequently, by applying Barbalat's Lemma, we can assert that $\lim_{t\to\infty} \frac{z_j^2}{\delta_i - z_i^2} = 0$ for $j = 1, 2, \ldots, n$. Then, it follows that $\lim_{t\to\infty} z_i = 0$ for $i = 1, 2, \ldots, n$, which implies that asymptotic tracking is achieved. This completes the proof. □

3.5 Simulation Example

In this section, the simulation results are presented to illustrate the effectiveness of the proposed Nussbaum-based adaptive control method. Consider the following second-order nonlinear system

$$
\begin{aligned}
\dot{x}_1 &= g_1 x_2 + f_1(x_1), \\
\dot{x}_2 &= g_2 u + f_2(x_1, x_2), \\
y &= x_1,
\end{aligned}
\tag{3.42}
$$

where $g_1 = 2$, $g_2 = -1.5$, $f_1 = -x_1^2$ and $f_2 = -7.3x_1 - 4x_2 - x_1^2 - x_2^2$. The objective is for x_1 to track the desired trajectory $x_{\mathrm{d}} = sin(t)$ while ensuring that all closed signals are bounded and that the full-state constraints are not violated. The initial conditions are $x_1(0) = 0.15$ and $x_2(0) = 0.1$. The design parameters are selected as

$$
\begin{aligned}
k_1 &= 10, k_2 = 9, \\
\lambda_1 &= \lambda_2 = \gamma_1 = \gamma_2 = 1, \\
\sigma_1 &= 5\mathrm{e}^{-0.1\mathrm{t}}, \sigma_2 = 10\mathrm{e}^{-0.1\mathrm{t}}, \\
\delta_1 &= 0.5, \delta_2 = 1.5.
\end{aligned}
$$

The results obtained from the Nussbaum-based asymptotic tracking design subject to full-state constraints are shown in Figs. 3.1, 3.2, and 3.3. Figure 3.1a illustrates the trajectory of the system output x_1 and the desired signal x_d. Figure 3.1b shows the trajectories of the system state x_2 and the virtual control α_1. One can observe that the system output tracks the reference signal precisely. More specifically, the tracking errors z_i, which are shown in Fig. 3.2a and b, asymptotically converge to zero. Moreover, the constraints δ_i of z_i are not violated during the whole process. It can also be observed from Fig. 3.2c that as the boundary is approached, the control input will increase to ensure that the boundaries will not be crossed. The trajectory of the adaptive Nussbaum design, including $\hat{\theta}_i$, $\hat{\vartheta}_i$ and ξ_i are shown in Fig. 3.3, which are bounded during the online process. The Nussbaum-type gains ξ_i also remain bounded, which verifies the discussions in Lemma 3.1.

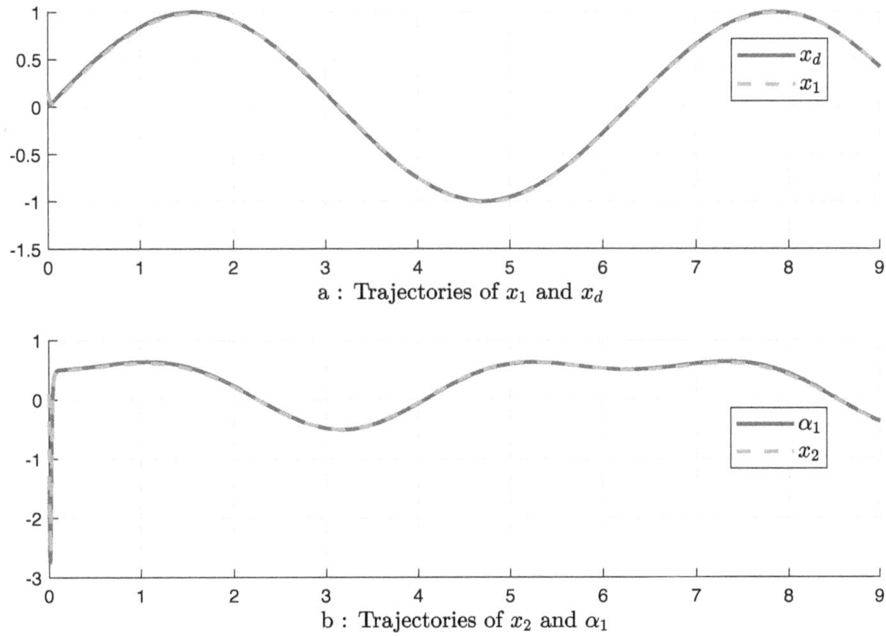

Fig. 3.1 System states and virtual control signal

3.6 Conclusion

This chapter has addressed the challenging problem of enforcing transient state constraints in nonlinear systems with unknown control directions. While existing BLF-based methods have shown effectiveness in constraint handling, most are restricted to systems with known control directions and do not account for the complexity introduced by direction uncertainty. By integrating BLF with Nussbaum-type function techniques, a novel control strategy has been proposed to ensure both constraint satisfaction and system stability under unknown control directions. The proposed framework not only guarantees that transient constraints are rigorously respected during the entire dynamic evolution, but also facilitates the asymptotic convergence of system states.

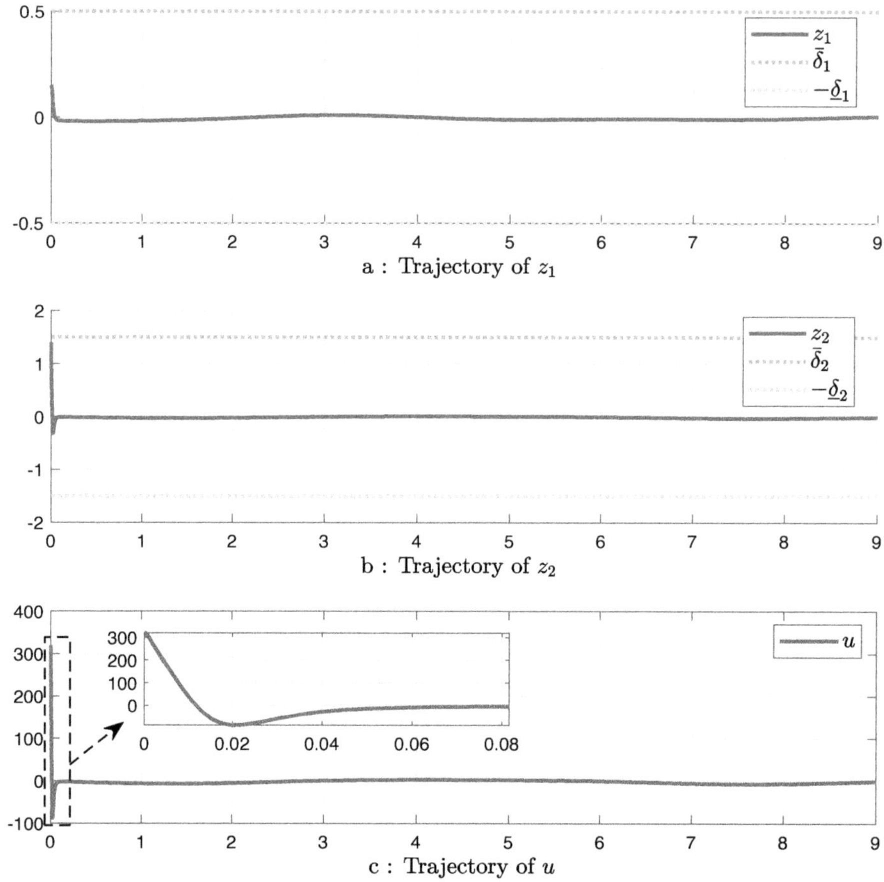

Fig. 3.2 Tracking errors and the control signal

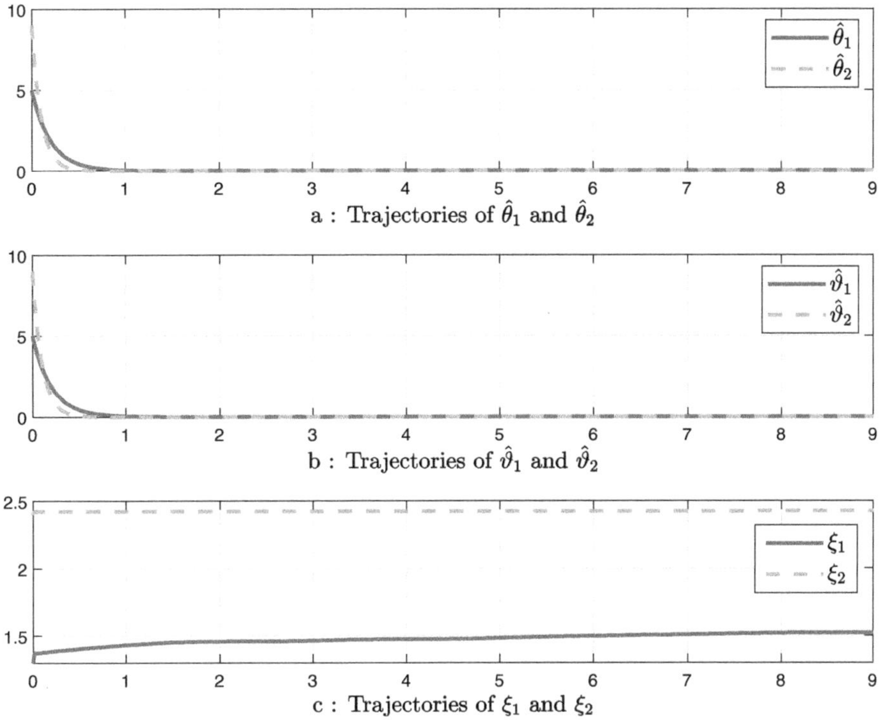

Fig. 3.3 Adaptive Nussbaum gains and parameter learning laws

References

1. Ren, B., Ge, S.S., Tee, K.P., Lee, T.H.: Adaptive neural control for output feedback nonlinear systems using a barrier Lyapunov function. IEEE Trans. Neural Netw. **21**(8), 1339–1345 (2010)
2. Yang, Y., Ding, D.W., Xiong, H., Yin, Y., Wunsch, D.C.: Online barrier-actor-critic learning for H_∞ control with full-state constraints and input saturation. J. Franklin Inst. **357**(6), 3316–3344 (2020)
3. Liu, G., Yang, Y., Gao, W., Wunsch, D.C.: Adaptive Nussbaum design for nonholonomic systems with asymptotic stabilization against false data injection. IEEE Trans. Cybern. **55**(11), 5189–5201 (2025)
4. Gilbert, E., Kolmanovsky, I.: Nonlinear tracking control in the presence of state and control constraints: a generalized reference governor. Automatica **38**(12), 2063–2073 (2002)
5. Blanchini, F.: Set invariance in control. Automatica. **35**(11), 1747–1767 (1999)
6. Mayne, D.Q., Rawlings, J.B., Rao, C.V., Scokaert, P.O.M.: Constrained model predictive control: stability and optimality. Automatica **36**(6), 789–814 (2000)
7. Tee, K.P., Ren, B., Ge, S.S.: Control of nonlinear systems with time-varying output constraints. Automatica **47**(11), 2511–2516 (2011)
8. Sun, W., Su, S.-F., Wu, Y., Xia, J., Nguyen, V.-T.: Adaptive fuzzy control with high-order barrier Lyapunov functions for high-order uncertain nonlinear systems with full-state constraints. IEEE Trans. Cybern. **50**(8), 3424–3432 (2020)

9. Yuan, F., Liu, Y.-J., Liu, L., Lan, J., Li, D., Tong, S., Chen, C.L.P.: Adaptive neural consensus tracking control for nonlinear multiagent systems using integral barrier Lyapunov functionals. IEEE Trans. Neural Netw. Learn. Syst. **34**(8), 4544–4554 (2023)
10. Tang, F., Niu, B., Wang, H., Zhang, L., Zhao, X.: Adaptive fuzzy tracking control of switched MIMO nonlinear systems with full state constraints and unknown control directions. IEEE Trans. Circuits Syst. II Express Briefs **69**(6), 2912–2916 (2022)
11. Tee, K.P., Ge, S.S., Tay, E.H.: Barrier Lyapunov functions for the control of output-constrained nonlinear systems. Automatica **45**(4), 918–927 (2009)
12. Li, D.P., Li, D.J., Liu, Y.J., Tong, S., Chen, C.L.P.: Approximation-based adaptive neural tracking control of nonlinear MIMO unknown time-varying delay systems with full state constraints. IEEE Trans. Cybern. **47**(10), 3100–3109 (2017)
13. Sun, H., Kong, X., Hou, L., Yang, D., Wei, Y.: Adaptive dynamic event-triggered asymptotic tracking control for strict-feedback nonlinear systems with a self-adjusting performance function. IEEE Trans. Autom. Sci. Eng. **22**, 8200–8214 (2025)
14. Zhang, J., Niu, B., Wang, D., Wang, H., Zhao, P., Zong, G.: Time-/event-triggered adaptive neural asymptotic tracking control for nonlinear systems with full-state constraints and application to a single-link robot. IEEE Trans. Neural Netw. Learn. Syst. **33**(11), 6690–6700 (2022)
15. Xie, R., Guo, C., Xie, X.J.: Asymptotic tracking control of state-constrained nonlinear systems with time-varying powers. IEEE Trans. Cybern. **52**(5), 4073–4078 (2022)
16. Wu, Y., Xie, X.J.: Adaptive fuzzy control for high-order nonlinear time-delay systems with full-state constraints and input saturation. IEEE Trans. Fuzzy Syst. **28**(8), 1652–1663 (2020)
17. Shojaei, F., Arefi, M.M., Khayatian, A., Karimi, H.R.: Observer-based fuzzy adaptive dynamic surface control of uncertain nonstrict feedback systems with unknown control direction and unknown dead-zone. IEEE Trans. Syst. Man Cybern.: Syst. **49**(11), 2340–2351 (2019)
18. Liu, G., Yang, Y., Zhao, X., Ahn, C.K.: Adaptive fuzzy practical bipartite synchronization for multi-agent systems with intermittent feedback under multiple unknown control directions. IEEE Trans. Fuzzy Syst. **32**(5), 3211–3225 (2024)
19. Wang, C., Guo, L., Wen, C., Hu, Q., Qiao, J.: Adaptive neural network control for a class of nonlinear systems with unknown control direction. IEEE Trans. Syst. Man Cybern.: Syst. **50**(11), 4708–4718 (2020)
20. Lei, Y., Wang, Y.W., Guan, Z.H., Shen, Y.J.: Event-triggered adaptive output regulation for a class of nonlinear systems with unknown control direction. IEEE Trans. Syst. Man Cybern.: Syst. **50**(9), 3181–3188 (2020)

Part II
Adaptive Nussbaum Design with Fixed-Time Performance

Chapter 4
Adaptive Fixed-Time Nussbaum Design for Deterministic Nonlinear Systems

4.1 Introduction

Due to the challenge of complex nonlinear dynamical systems, neural networks and fuzzy logic systems based intelligent control techniques have been extensively studied for stability analysis and control synthesis for typical nonlinear processes. In many situations, the control scheme is designed based on the requirement that the control directions are known a priori. However, the control direction in engineering systems might be time-varying and unavailable for control design. The Nussbaum-type gain design is efficient in dealing with the unknown control direction, which has been successfully developed for time-delay systems [1], robotic manipulators [2, 3], multiagent systems [4], and nonholonomic systems [5]. In this chapter, a novel Nussbaum-type gain design is developed while the fixed-time stability can be guaranteed.

To improve the infinite settling time for a conventional stabilizing design, Bhat and Bernstein presented the concept of finite-time stabilization [6]. Since then, finite-time stability analysis has been extended to stochastic dynamical systems. Furthermore, the finite-time stabilizing control design scheme has been combined with the event-triggered sampling technique to reduce the data transmission frequency [7]. Nonetheless, the finite-time performance depends on the beginning state, which may limit its viability. Polyakov developed a fixed-time stabilization strategy to solve this issue, in which the settling time may be changed independently of the original circumstances [8]. The fixed-time stability analysis has been further applied to strict feedback dynamic systems [9, 10]. However, fixed-time control synthesis is challenging owing to the existence of nonlinearity, such as unmodeled dynamics [11], backlash-like hysteresis [12], and input constraint [13]. To solve the fixed-time constrained tracking problem, the adaptive control strategies are presented [14, 15]. The fixed-time stability is further extended to the practical fixed-time stability, which lays the foundation for the boundedness analysis of signals in dynamical systems [16, 17]. This chapter studies the tracking control synthesis with practical fixed-time perfor-

Y. Yang et al., *Adaptive Nussbaum Design for Nonlinear Dynamical Systems*,
https://doi.org/10.1007/978-981-95-4997-9_4

mance. Furthermore, the settling time is shown to be solely determined by the design parameters.

Compared with the finite-time control design [18, 19], the control objective in this chapter can be achieved in a fixed time, which does not rely on the initial state. This is guaranteed by the novel design of the virtual control, adaptive Nussbaum-type gain, and filter for each subsystem. Compared with the recent results on the Nussbaum design [20–22], this chapter developed a novel design with fixed-time performance. With a fixed-time controller, both the tracking errors and the filter errors could converge to a small neighborhood of the origin within a fixed settling time, where the settling time does not depend on the initial condition.

The rest of this chapter is organized as follows. Section 4.2 provides the problem formulation. Section 4.3 establishes an adaptive controller design scheme under unknown control directions. In Sect. 4.4, the fixed-time stability for deterministic system is analyzed. Section 4.5 is the verification of the controller in a simulation example. Section 4.6 presents conclusion of this chapter.

4.2 Problem Formulation

In this chapter, the strict feedback nonlinear system evolving in continuous time is considered with the following dynamics as

$$\begin{aligned} \dot{x}_i(t) &= h_i(\bar{x}_i(t))\, x_{i+1}(t) + f_i(\bar{x}_i(t)), i = 1, 2, \ldots, n-1 \\ \dot{x}_n(t) &= h_n(\bar{x}_n(t))\, u(t) + f_n(\bar{x}_n(t)), \\ y &= x_1 \end{aligned} \tag{4.1}$$

where $u \in \mathbb{R}$ is the control signal, $y \in \mathbb{R}$ is the system output, $\bar{x}_i = [x_1 \ \cdots \ x_i]^{\mathrm{T}} \in \mathbb{R}^i$ is the system state, and $f_i(\cdot) : \mathbb{R}^i \to \mathbb{R}$ are the unknown smooth functions.

Lemma 4.1 *For the nonlinear dynamical system $\dot{x} = F(x)$ with Lipschitz function $F(x)$. Assume that there is a radially unbounded positive definite continuous function $V(x(t))$ that satisfies*

$$\begin{aligned} \dot{V}(x(t)) \leq\ & m(n \cdot h(t)\, N(\zeta(t)) + 1)\, \dot{\zeta}(t) \\ & -aV^p(x(t)) - bV^q(x(t)) + c, \end{aligned} \tag{4.2}$$

with unknown continuous bounded function $h(t)$, constants $m, n, a, b, c > 0$, $0 < p < 1$, and $q > 1$. Then, the equilibrium of the system is practically fixed-time stable with a settling time upper bounded as

$$T \leq \max\left\{\frac{1}{a\kappa(1-p)} + \frac{1}{b(q-1)}, \frac{1}{a(1-p)} + \frac{1}{b\kappa(q-1)}\right\}$$

with a positive constant $0 < \kappa < 1$.

Proof First, based on [1], one can obtain

$$m\left(n\cdot h\left(t\right)N\left(\zeta\left(t\right)\right)+1\right)\dot{\zeta}\left(t\right)\leq N_{\max}.$$

Denote $d=c+N_{\max}$, then one can obtain the following inequality from (4.2)

$$\dot{V}\leq -aV^{p}-bV^{q}+d. \tag{4.3}$$

Then, the fact in (4.3) further implies two cases as

$$\begin{aligned}\dot{V}&\leq -a\kappa V^{p}-a\left(1-\kappa\right)V^{p}-bV^{q}+d\\ \dot{V}&\leq -aV^{p}-b\kappa V^{q}-b\left(1-\kappa\right)V^{q}+d\end{aligned} \tag{4.4}$$

where $0<\kappa<1$.

As shown in the next, the parameter κ has a significant impact on the convergence set of the state $x(t)$ and the settling time. To show the effect of the parameter κ on the size of the convergence set and the rate of convergence, two cases are considered.

The First Case: For the case with (4.4), define

$$\begin{aligned}\Pi_a&=\left\{x|\,V\left(x\right)\leq\left(\frac{d}{a\left(1-\kappa\right)}\right)^{\frac{1}{p}}\right\},\\ \Pi_1&=\left\{x|\,V\left(x\right)\leq 1\right\}.\end{aligned}$$

When $x\notin\Pi_a$, one has $\dot{V}\leq -a\kappa V^{p}-bV^{q}$, which indicates that x moves toward the boundary of the set Π_a. At the boundary of the set Π_a, one has $\dot{V}=-a\kappa V^{p}-bV^{q}<0$. From LaSallel's invariance principle, the state x would remain in the set Π_a.

Suppose that $\left(\frac{d}{a(1-\kappa)}\right)^{\frac{1}{p}}\geq 1$, then $\dot{V}\leq -bV^{q}$. The traveling time from the initial condition to the boundary of the set Π_a can be calculated as

$$\begin{aligned}T_1=\int_0^{T_1}dt&\leq-\int_{V_0}^{\left(\frac{d}{a(1-\kappa)}\right)^{\frac{1}{p}}}\frac{dV}{bV^{q}}\\ &\leq\frac{\left(\frac{d}{a(1-\kappa)}\right)^{\frac{1-q}{p}}-{V_0}^{1-q}}{b\left(q-1\right)}\\ &\leq\frac{\left(\frac{d}{a(1-\kappa)}\right)^{\frac{1-q}{p}}}{b\left(q-1\right)},\end{aligned} \tag{4.5}$$

with the initial condition $V_0=V(t_0)$.

Suppose that $\left(\frac{d}{a(1-\kappa)}\right)^{\frac{1}{p}}<1$, then the state transition can be divided as two stages. The first stage corresponds to the process from the initial condition to the boundary of the set Π_1, which indicates that $\dot{V} \leq -bV^q$. The second stage corresponds to the process from the boundary of the set Π_1 to the boundary of the set Π_a, which implies that $\dot{V} \leq -a\kappa V^p$. Therefore, the traveling time from the initial condition V_0 to the boundary of the set Π_a can be calculated as

$$T_1 = \int_0^{T_1} dt \leq -\int_{V_0}^{1} \frac{dV}{bV^q} - \int_1^{\left(\frac{d}{a(1-\kappa)}\right)^{\frac{1}{p}}} \frac{dV}{a\kappa V^p}$$
$$\leq \frac{1}{a\kappa\,(1-p)} - \frac{\left(\frac{d}{a(1-\kappa)}\right)^{\frac{1-p}{p}}}{a\kappa\,(1-p)} + \frac{1}{b\,(q-1)}. \tag{4.6}$$

The Second Case: For the case with (4.4), define the set

$$\Pi_b = \left\{ x \mid V(x) \leq \left(\frac{d}{b\,(1-\kappa)}\right)^{\frac{1}{q}} \right\}.$$

Similar to the previous discussions, one can obtain

$$T_2 \leq \frac{1}{a\,(1-p)} - \frac{\left(\frac{d}{b(1-\kappa)}\right)^{\frac{1-p}{q}}}{a\,(1-p)} + \frac{1}{b\kappa\,(q-1)}, \tag{4.7}$$

when $\left(\frac{d}{b(1-\kappa)}\right)^{\frac{1}{q}} < 1$, and

$$T_2 \leq \frac{\left(\frac{d}{b(1-\kappa)}\right)^{\frac{1-q}{q}}}{b\kappa\,(q-1)}, \tag{4.8}$$

when $\left(\frac{d}{b(1-\kappa)}\right)^{\frac{1}{q}} \geq 1$.

Collecting the results in (4.5)–(4.8), the upper bounded of the settling time can be obtained as

$$T \leq \max\left\{ \frac{1}{a\kappa\,(1-p)} + \frac{1}{b\,(q-1)}, \frac{1}{a\,(1-p)} + \frac{1}{b\kappa\,(q-1)} \right\}. \tag{4.9}$$

This completes the proof. □

In light of the above notations and lemmas, the problem of interest in this chapter can be formulated as follows.

Problem: Given the reference $x_d(t)$ with bounded first-order and second-order derivatives $\dot{x}_d(t), \ddot{x}_d(t)$. This chapter aims to design an adaptive feedback scheme for system (4.1) such that

1. the system output signal $y(t)$ could track the reference $x_{\mathrm{d}}(t)$ with bounded error in fixed-settling time T_{max},
2. the closed-loop signals remain bounded.

4.3 Fixed-Time Nussbaum Design

In this section, a novel fixed-time adaptive controller design is developed in the framework of backstepping, where the virtual control input s_{i-1} is fed to the $i-1$th subsystem through the dynamic surface control (DSC) filter output α_{i-1}. For the nth subsystem, the control signal $\hat{u}$ is updated intermittently based on the sampling of u. Finally, the control signal $\hat{u}$ is applied to the system (4.1).

For ith subsystem, the tracking error is designed as

$$e_i = x_i - \alpha_{i-1} \tag{4.10}$$

where α_i with $i = 1, \ldots, n-1$ is the filter output for ith subsystem and $\alpha_0 = x_{\mathrm{d}}$ is the tracking reference signal.

For function approximation in control design problems, many techniques can be used, such as neural networks and fuzzy logic systems. The radial basis function neural network (RBFNN) is a popular choice for function approximation in nonlinear control problems due to its effectiveness in handling high-dimensional data, smoothness of function approximation, and adaptation to new data [23]. To deal with the unknown function f_i, the RBFNN is employed to represent the function as

$$f_i(x_i) = W_i^{*\mathrm{T}} \varphi_i(x_i) + \varepsilon_i^*(x_i), i = 1, \ldots, n \tag{4.11}$$

where W_i^* is the ideal network weight vector, $\varphi_i(x_i)$ is the RBFNN basis function, and $\varepsilon_i^*(x_i)$ is the approximation residual. Based on the universal approximation theorem [24], the approximation residual is upper bounded by a constant, i.e., $\left\|\varepsilon_1^*(x_i)\right\| \leq \varepsilon_i$ with $\varepsilon_i > 0$. The details of the fixed-time backstepping design are discussed as follows.

4.3.1 Step 1

For Step 1, the adaptive virtual control input is designed as

$$s_1 = \frac{1}{k_{g1}} N(\xi_1) \left(k_{11} e_1 + k_{12} e_1^3 + \frac{\hat{\theta}_1}{2\eta_1} e_1 \varphi_1^{\mathrm{T}} \varphi_1 - \dot{x}_{\mathrm{d}} \right), \tag{4.12}$$

where $\hat{\theta}_1(t)$ is the adaptive estimate of $\theta_1 = \|W_1^*\|^2$, designed as

$$\dot{\hat{\theta}}_1 = \frac{1}{2\eta_1} e_1^2 \varphi_1^{\mathrm{T}} \varphi_1 - \gamma_{11}\hat{\theta}_1 - \gamma_{12}\hat{\theta}_1^3, \tag{4.13}$$

with initial value $\hat{\theta}_1(0) = 0$, and positive constants η_1, γ_{11}, and γ_{12}.

The Nussbaum-type gain signal ξ_1 is designed as

$$\dot{\xi}_1 = k_{g1}\left(k_{11}e_1^2 + k_{12}e_1^4 + \frac{\hat{\theta}_1}{2\eta_1} e_1^2 \varphi_1^{\mathrm{T}} \varphi_1 - e_1 \dot{x}_{\mathrm{d}}\right). \tag{4.14}$$

To avoid complexity explosion in backstepping, a nonlinear differentiator is introduced:

$$\lambda_1 \dot{\alpha}_1 = (s_1 - \alpha_1) + (s_1 - \alpha_1)^3, \tag{4.15}$$

with filter error $\tilde{\alpha}_1 = \alpha_1 - s_1$ satisfying

$$\dot{\tilde{\alpha}}_1 = -\frac{1}{\lambda_1}\tilde{\alpha}_1 - \frac{1}{\lambda_1}\tilde{\alpha}_1^3 - \dot{s}_1 \tag{4.16}$$

with

$$\begin{aligned} \dot{s}_1 = & \frac{1}{k_{g1}} N(\xi_1)\left(k_{11}\dot{e}_1 + 3k_{12}e_1^2\dot{e}_1 - \ddot{\alpha}_{1-1}\right) \\ & + \frac{1}{k_{g1}} N(\xi_1)\left(\frac{\dot{\hat{\theta}}_1}{2\eta_1} e_1 \varphi_1^{\mathrm{T}} \varphi_1 + \frac{\hat{\theta}_1}{2\eta_1} \dot{e}_1 \varphi_1^{\mathrm{T}} \varphi_1 + \frac{\hat{\theta}_1}{\eta_1} e_1 \varphi_1^{\mathrm{T}} \dot{\varphi}_1\right) \\ & + \frac{1}{k_{gi}} \dot{N}(\xi_1)\dot{\xi}_1\left(k_{11}e_1 + k_{12}e_1^3 + \frac{\hat{\theta}_1}{2\eta_1} e_1 \varphi_1^{\mathrm{T}} \varphi_1 - \dot{\alpha}_{1-1}\right). \end{aligned} \tag{4.17}$$

The compact sets are defined as

$$\begin{aligned} \Omega_{x_{\mathrm{d}}} &= \left\{(x_{\mathrm{d}}, \dot{x}_{\mathrm{d}}, \ddot{x}_{\mathrm{d}}) \mid x_{\mathrm{d}}^2 + \dot{x}_{\mathrm{d}}^2 + \ddot{x}_{\mathrm{d}}^2 \leq R_{\mathrm{d}}\right\}, \\ \Omega_1 &= \left\{(e_1, \tilde{\theta}_1, \tilde{\alpha}_1) \mid \frac{1}{2}e_1^2 + \frac{1}{2}\tilde{\theta}_1^2 + \frac{1}{2}\tilde{\alpha}_1^2 \leq R_1\right\}. \end{aligned} \tag{4.18}$$

Since $\dot{s}_1$ is continuous on $\Omega_1 \times \Omega_{x_{\mathrm{d}}}$, it holds that

$$|\dot{s}_1| \leq \rho_1 \tag{4.19}$$

for positive constant ρ_1.

4.3.2 Step i $(2 \le i \le n-1)$

For Step i, the adaptive virtual control is recursively designed as

$$s_i = \frac{1}{k_{gi}} N(\xi_i) \left(k_{i1} e_i + k_{i2} e_i^3 + \frac{\hat{\theta}_i}{2\eta_i} e_i \varphi_i^{\mathrm{T}} \varphi_i - \dot{\alpha}_{i-1} \right), \tag{4.20}$$

where the adaptive law for $\hat{\theta}_i$ is

$$\dot{\hat{\theta}}_i = \frac{1}{2\eta_i} e_i^2 \varphi_i^{\mathrm{T}} \varphi_i - \gamma_{i1} \hat{\theta}_i - \gamma_{i2} \hat{\theta}_i^3. \tag{4.21}$$

The gain update rule for ξ_i is

$$\dot{\xi}_i = k_{gi} \left(k_{i1} e_i^2 + k_{i2} e_i^4 + \frac{\hat{\theta}_i}{2\eta_i} e_i^2 \varphi_i^{\mathrm{T}} \varphi_i - e_i \dot{\alpha}_{i-1} \right). \tag{4.22}$$

The corresponding filter dynamics is given by

$$\lambda_i \dot{\alpha}_i = (s_i - \alpha_i) + (s_i - \alpha_i)^3, \tag{4.23}$$

and the filter error dynamics becomes

$$\dot{\tilde{\alpha}}_i = -\frac{1}{\lambda_i} \tilde{\alpha}_i - \frac{1}{\lambda_i} \tilde{\alpha}_i^3 - \dot{s}_i, \tag{4.24}$$

with

$$\begin{aligned} \dot{s}_i = & \frac{1}{k_{gi}} N(\xi_i) \left(k_{i1} \dot{e}_i + 3 k_{i2} e_i^2 \dot{e}_i - \ddot{\alpha}_{i-1} \right) \\ & + \frac{1}{k_{gi}} N(\xi_i) \left(\frac{\dot{\hat{\theta}}_i}{2\eta_i} e_i \varphi_i^{\mathrm{T}} \varphi_i + \frac{\hat{\theta}_i}{2\eta_i} \dot{e}_i \varphi_i^{\mathrm{T}} \varphi_i + \frac{\hat{\theta}_i}{\eta_i} e_i \varphi_i^{\mathrm{T}} \dot{\varphi}_i \right) \\ & + \frac{1}{k_{gi}} \dot{N}(\xi_i) \dot{\xi}_i \left(k_{i1} e_i + k_{i2} e_i^3 + \frac{\hat{\theta}_i}{2\eta_i} e_i \varphi_i^{\mathrm{T}} \varphi_i - \dot{\alpha}_{i-1} \right). \end{aligned} \tag{4.25}$$

The compact sets are defined as

$$\begin{aligned} \Omega_{x_{\mathrm{d}}} &= \left\{ (x_{\mathrm{d}}, \dot{x}_{\mathrm{d}}, \ddot{x}_{\mathrm{d}}) \mid x_{\mathrm{d}}^2 + \dot{x}_{\mathrm{d}}^2 + \ddot{x}_{\mathrm{d}}^2 \le R_{\mathrm{d}} \right\}, \\ \Omega_i &= \left\{ (e_i, \tilde{\theta}_i, \tilde{\alpha}_i) \mid \frac{1}{2} e_i^2 + \frac{1}{2} \tilde{\theta}_i^2 + \frac{1}{2} \tilde{\alpha}_i^2 \le R_i \right\}. \end{aligned} \tag{4.26}$$

Since $\dot{s}_i$ is continuous on $\Omega_i \times \Omega_{x_d}$, it holds that

$$|\dot{s}_i| \le \rho_i, \tag{4.27}$$

for positive constant $\rho_i > 0$.

4.3.3 Step n

For the final subsystem, the adaptive Nussbaum-based continuous feedback control input is designed as

$$u(t) = \frac{N(\xi_n)}{k_{gn}} \left(k_{n1} e_n + k_{n2} e_n^3 + \frac{\hat{\theta}_n}{2\eta_n} e_n \varphi_n^{\mathrm{T}} \varphi_n - \dot{\alpha}_{n-1} \right). \tag{4.28}$$

Denote $\theta_n = \left\| W_n^* \right\|^2$. Then, the adaptive learning $\hat{\theta}_n(t)$ aims to compensate for the unknown weight norm θ_n with the learning error denoted as $\tilde{\theta}_n(t) = \theta_n - \hat{\theta}_n(t)$. The neural network learning is designed as

$$\dot{\hat{\theta}}_n = \frac{1}{2\eta_n} e_n^2 \varphi_n^{\mathrm{T}} \varphi_n - \gamma_{n1} \hat{\theta}_n - \gamma_{n2} \hat{\theta}_n^3, \tag{4.29}$$

with initial condition $\hat{\theta}_n(0) = 0$ and positive constants $\eta_n, \gamma_{n1}, \gamma_{n2}$.

The adaptive Nussbaum-type gain is designed as

$$\dot{\xi}_n = k_{gn} e_n \left(k_{n1} e_n + k_{n2} e_n^3 + \frac{\hat{\theta}_n}{2\eta_n} e_n \varphi_n^{\mathrm{T}} \varphi_n - \dot{\alpha}_{n-1} \right), \tag{4.30}$$

where k_{gn}, k_{n1}, and k_{n2} are positive constants.

4.4 Fixed-Time Stability Analysis

This section focuses on the stability analysis of the closed-loop system using the designs presented in the previous subsection.

Step 1

Substitute (4.1) and (4.11) into the time derivative of e_1 defined in (4.10), one has

$$\dot{e}_1 = W_1^{*\mathrm{T}} \varphi_1(x_1) + \varepsilon_1^*(x_1) + h_1 e_2 + h_1 s_1 + h_1 \tilde{\alpha}_1 - \dot{x}_{\mathrm{d}}. \tag{4.31}$$

Consider the Lyapunov candidate for the first subsystem as

$$V_1\left(e_1,\tilde{\alpha}_1,\tilde{\theta}_1\right)=\frac{1}{2}e_1^2+\frac{1}{2}\tilde{\alpha}_1^2+\frac{1}{2}\tilde{\theta}_1^2. \tag{4.32}$$

Differentiating V_1 yields

$$\begin{aligned}\dot{V}_1=&\,e_1\left[W_1^{*\mathrm{T}}\varphi_1\left(x_1\right)+\varepsilon_1^*\left(x_1\right)+h_1s_1+h_1e_2+h_1\tilde{\alpha}_1-\dot{x}_{\mathrm{d}}\right]\\&+\tilde{\alpha}_1\left(-\frac{1}{\lambda_i}\tilde{\alpha}_i-\frac{1}{\lambda_i}\tilde{\alpha}_i^3-\dot{s}_i\right)-\tilde{\theta}_1\dot{\hat{\theta}}_1.\end{aligned}$$

Applying Young's inquality, one has

$$e_1W_1^{*\mathrm{T}}\varphi_1\left(x_1\right)\leq\frac{\theta_1}{2\eta_1}e_1^2\varphi_1^{\mathrm{T}}\varphi_1+\frac{\eta_1}{2}. \tag{4.33}$$

Inserting (4.33) into $\dot{V}_1$, one has

$$\begin{aligned}\dot{V}_1\leq&\,\frac{\theta_1}{2\eta_1}e_1^2\varphi_1^T\varphi_1+\frac{\eta_1}{2}+h_1e_1s_1-e_1\dot{x}_{\mathrm{d}}+\tilde{\alpha}_1\left(-\frac{1}{\lambda_1}\tilde{\alpha}_1-\frac{1}{\lambda_1}\tilde{\alpha}_1^3-\dot{z}_1\right)\\&+h_1e_1e_2+h_1e_1\tilde{\alpha}_1+e_1\varepsilon_1-\tilde{\theta}_1\dot{\hat{\theta}}_1.\end{aligned}$$

Inserting the virtual control design (4.12) into above inequality yields

$$\begin{aligned}\dot{V}_1\leq&-k_{11}e_1^2-k_{12}e_1^4+\frac{1}{k_{g1}}\left[\frac{1}{k_{g1}}h_1N\left(\xi_1\right)+1\right]\dot{\xi}_1\\&+\frac{\tilde{\theta}_1}{2\eta_1}e_1^2\varphi_1^T\varphi_1-\frac{1}{\lambda_1}\tilde{\alpha}_1^2-\frac{1}{\lambda_1}\tilde{\alpha}_1^4-\tilde{\alpha}_1\dot{z}_1\\&-\tilde{\theta}_1\dot{\hat{\theta}}_1+\frac{\eta_1}{2}+h_1e_1e_2+h_1e_1\tilde{\alpha}_1+e_1\varepsilon_1.\end{aligned} \tag{4.34}$$

According to Young's inequality, one can further obtain

$$\begin{aligned}h_1e_1e_2&\leq\frac{1}{2}\left(h_1^2e_1^2+e_2^2\right)\leq\frac{1}{2}\left(\bar{h}_1^2e_1^2+e_2^2\right)\\h_1e_1\tilde{\alpha}_1&\leq\frac{1}{2}\left(h_1^2e_1^2+\tilde{\alpha}_1^2\right)\leq\frac{1}{2}\left(\bar{h}_1^2e_1^2+\tilde{\alpha}_1^2\right)\\-\tilde{\alpha}_1\dot{s}_1&\leq\left|\tilde{\alpha}_1\dot{s}_1\right|\leq\frac{1}{2}\left(\tilde{\alpha}_1^2+\dot{s}_1^2\right)\leq\frac{1}{2}\left(\tilde{\alpha}_1^2+\rho_1^2\right)\\e_1\varepsilon_1^*&\leq\frac{1}{2}\left(e_1^2+\varepsilon_1^2\right).\end{aligned} \tag{4.35}$$

Therefore, $\dot{V}_1$ further satisfies

$$\begin{aligned}
\dot{V}_1 \leq & -k_{11}e_1^2 - k_{12}e_1^4 - \frac{1}{\lambda_1}\tilde{\alpha}_1^2 - \frac{1}{\lambda_1}\tilde{\alpha}_1^4 + \frac{\bar{h}_1^2 e_1^2}{2} + \frac{\tilde{\alpha}_1^2}{2} + \frac{1}{2}\tilde{\alpha}_1^2 \\
& + \frac{1}{2}\rho_1^2 + \frac{e_1^2}{2} + \frac{\varepsilon_1^2}{2} + \frac{\eta_1}{2} + \frac{1}{k_{g1}}\left[\frac{1}{k_{g1}} h_1 N\left(\xi_1\right) + 1\right]\dot{\xi}_1 \\
& - \tilde{\theta}_1\left(\dot{\hat{\theta}}_1 - \frac{1}{2\eta_1} e_1^2 \varphi_1^T \varphi_1\right) + \frac{\bar{h}_1^2 e_1^2}{2} + \frac{e_2^2}{2}.
\end{aligned} \tag{4.36}$$

Taking the adaptive NN weight learning into above inequality results in

$$\begin{aligned}
\dot{V}_1 \leq & \frac{1}{k_{g1}}\left[\frac{1}{k_{g1}} h_1 N\left(\xi_1\right) + 1\right]\dot{\xi}_1 - \left(k_{11} - \bar{h}_1^2 - \frac{1}{2}\right) e_1^2 \\
& - k_{12}e_1^4 + \frac{e_2^2}{2} - \left(\frac{1}{\lambda_1} - 1\right)\tilde{\alpha}_1^2 - \frac{1}{\lambda_1}\tilde{\alpha}_1^4 \\
& + \frac{\varepsilon_1^2}{2} + \frac{\eta_1}{2} + \frac{1}{2}\rho_1^2 + \gamma_{11}\tilde{\theta}_1\hat{\theta}_1 + \gamma_{12}\tilde{\theta}_1\hat{\theta}_1^3.
\end{aligned} \tag{4.37}$$

Based on the Lemma 1.6, the following can be obtained

$$\begin{aligned}
-e_1^2 & \leq \frac{1}{3} - \frac{4}{3}\left(e_1^2\right)^{\frac{3}{4}}, \\
-\tilde{\alpha}_1^2 & \leq \frac{1}{3} - \frac{4}{3}\left(\tilde{\alpha}_1^2\right)^{\frac{3}{4}}, \\
-\tilde{\theta}_1^2 & \leq \frac{1}{3} - \frac{4}{3}\left(\tilde{\theta}_1^2\right)^{\frac{3}{4}}.
\end{aligned} \tag{4.38}$$

Similarly, one has

$$\begin{aligned}
\tilde{\theta}_1\hat{\theta}_1^3 & = \tilde{\theta}_1\theta_1^3 - 3\theta_1^2\tilde{\theta}_1^2 + 3\theta_1\tilde{\theta}_1^3 - \tilde{\theta}_1^4 \\
& \leq \gamma_{1,3}\frac{\tilde{\theta}_1^4}{4} + \gamma_{1,3}^{-\frac{1}{3}}\frac{3\theta_1^4}{4} + \gamma_{1,3}\frac{3\tilde{\theta}_1^4}{2} + \gamma_{1,3}^{-1}\frac{3\theta_1^4}{2} \\
& \quad + \gamma_{1,3}\frac{9\tilde{\theta}_1^4}{4} + \gamma_{1,3}^{-3}\frac{3\theta_1^4}{4} - \tilde{\theta}_1^4 \\
& \leq -\left(1 - 4\gamma_{1,3}\right)\tilde{\theta}_1^4 + \left(\frac{3}{4}\gamma_{1,3}^{-\frac{1}{3}} + \frac{3}{2}\gamma_{1,3}^{-1} + \frac{3}{4}\gamma_{1,3}^{-3}\right)\theta_1^4,
\end{aligned} \tag{4.39}$$

with constants $\gamma_{1,3} > 0$ and $1 - 4\gamma_{1,3} > 0$.

Collecting the results in (4.38) and (4.39), $\dot{V}_1$ further satisfies

$$\begin{aligned}\dot{V}_1 \le& \frac{1}{k_{g1}}\left[\frac{1}{k_{g1}}h_1 N(\xi_1)+1\right]\dot{\xi}_1 - k_{12}e_1^4 + \frac{e_2^2}{2}\\&-\left(k_{11}-\bar{h}_1^2-\frac{1}{2}\right)\left[\frac{4}{3}\left(e_1^2\right)^{\frac{3}{4}}-\frac{1}{3}\right]+\frac{\eta_1}{2}+\frac{1}{2}\rho_1^2\\&-\left(\frac{1}{\lambda_1}-1\right)\left[\frac{4}{3}\left(\tilde{\alpha}_1^2\right)^{\frac{3}{4}}-\frac{1}{3}\right]-\frac{1}{\lambda_1}\tilde{\alpha}_1^4+\frac{\varepsilon_1^2}{2}\\&+\gamma_{11}\left[\frac{1}{2}\theta_1^2+\frac{1}{6}-\frac{2}{3}\left(\tilde{\theta}_1^2\right)^{\frac{3}{4}}\right]\\&+\gamma_{12}\left[\left(\frac{3}{4}\gamma_{1,3}^{-\frac{1}{3}}+\frac{3}{2}\gamma_{1,3}^{-1}+\frac{3}{4}\gamma_{1,3}^{-3}\right)\theta_1^4-(1-4\gamma_{1,3})\tilde{\theta}_1^4\right].\end{aligned} \tag{4.40}$$

Step i ($2 \le i \le n-1$)

Substitute (4.1) and (4.11) into the time derivative of e_i defined in (4.10), one can obtain

$$\dot{e}_i = W_i^{*T}\varphi_i(\bar{x}_i) + \varepsilon_i^*(\bar{x}_i) + h_i e_{i+1} + h_i(\bar{x}_i(t))s_i + h_i\tilde{\alpha}_i - \dot{\alpha}_{i-1}. \tag{4.41}$$

The Lyapunov candidate for the ith subsystem is considered as

$$V_i\left(e_i,\tilde{\alpha}_i,\tilde{\theta}_i\right) = \frac{1}{2}e_i^2 + \frac{1}{2}\tilde{\alpha}_i^2 + \frac{1}{2}\tilde{\theta}_i^2. \tag{4.42}$$

Taking derivative of V_i, one has

$$\begin{aligned}\dot{V}_i =& e_i\dot{e}_i + \tilde{\alpha}_i\dot{\tilde{\alpha}}_i - \tilde{\theta}_i\dot{\hat{\theta}}_i\\=& e_i\left[W_i^{*T}\varphi_i(\bar{x}_i)+\varepsilon_i^*(\bar{x}_i)+h_i e_{i+1}+h_i s_i+h_i\tilde{\alpha}_i-\dot{\alpha}_{i-1}\right]\\&+\tilde{\alpha}_i\left(-\frac{1}{\lambda_i}\tilde{\alpha}_i-\frac{1}{\lambda_i}\tilde{\alpha}_i^3-\dot{z}_i\right)-\tilde{\theta}_i\dot{\hat{\theta}}_i.\end{aligned} \tag{4.43}$$

With Young's inequality, one has

$$e_i W_i^{*\mathrm{T}}\varphi_i(x_i) \le \frac{\theta_i}{2\eta_i}e_i^2\varphi_i^{\mathrm{T}}\varphi_i + \frac{\eta_i}{2}. \tag{4.44}$$

Substituting (4.44) into (4.43) yields

$$\begin{aligned}\dot{V}_i \le& \frac{\theta_i}{2\eta_i}e_i^2\varphi_i^T\varphi_i + \frac{\eta_i}{2} - h_i e_i s_i - e_i\dot{\alpha}_{i-1}\\&+h_i e_i e_{i+1} + h_i e_i\tilde{\alpha}_i + |e_i|\,\varepsilon_i - \tilde{\theta}_i\dot{\hat{\theta}}_i\\&+\tilde{\alpha}_i\left(-\frac{1}{\lambda_i}\tilde{\alpha}_i-\frac{1}{\lambda_i}\tilde{\alpha}_i^3-\dot{z}_i\right).\end{aligned} \tag{4.45}$$

From the virtual controller design (4.12), $\dot{V}_i$ satisfies

$$\begin{aligned}\dot{V}_i \leq & \frac{1}{k_{gi}}\left[\frac{1}{k_{gi}}h_i N(\xi_i)+1\right]\dot{\xi}_i+\tilde{\alpha}_i\left(-\frac{1}{\lambda_i}\tilde{\alpha}_i-\frac{1}{\lambda_i}\tilde{\alpha}_i^3-\dot{z}_i\right) \\ & -\tilde{\theta}_i\dot{\hat{\theta}}_i+\frac{\tilde{\theta}_i}{2\eta_i}e_i^2\varphi_i^T\varphi_i+\frac{\eta_i}{2}-k_{i1}e_i^2-k_{i2}e_i^4 \\ & +h_i e_i e_{i+1}+h_i e_i\tilde{\alpha}_i+e_i\varepsilon_i.\end{aligned} \tag{4.46}$$

Consider the following facts

$$\begin{aligned}h_i e_i e_{i+1} & \leq \frac{1}{2}\left(h_i^2e_i^2+e_{i+1}^2\right)\leq\frac{1}{2}\left(\bar{h}_i^2e_i^2+e_{i+1}^2\right) \\ h_i e_i\tilde{\alpha}_i & \leq \frac{1}{2}\left(h_i^2e_i^2+\tilde{\alpha}_i^2\right)\leq\frac{1}{2}\left(\bar{h}_i^2e_i^2+\tilde{\alpha}_i^2\right) \\ -\tilde{\alpha}_i\dot{z}_i & \leq |\tilde{\alpha}_i\dot{z}_i|\leq\frac{1}{2}\left(\tilde{\alpha}_i^2+\dot{z}_i^2\right)\leq\frac{1}{2}\left(\tilde{\alpha}_i^2+\rho_i^2\right) \\ e_i\varepsilon_i^* & \leq \frac{1}{2}\left(e_i^2+\varepsilon_i^2\right).\end{aligned} \tag{4.47}$$

Then, $\dot{V}_i$ satisfies

$$\begin{aligned}\dot{V}_i \leq & \frac{1}{k_{gi}}\left[\frac{1}{k_{gi}}h_i N(\xi_i)+1\right]\dot{\xi}_i-\tilde{\theta}_i\left(\dot{\hat{\theta}}_i-\frac{1}{2\eta_i}e_i^2\varphi_i^T\varphi_i\right)+\frac{\tilde{\theta}_i}{2\eta_i}e_i^2\varphi_i^T\varphi_i \\ & +\frac{\eta_i}{2}-k_{i1}e_i^2-k_{i2}e_i^4+\frac{\bar{h}_i^2e_i^2+e_{i+1}^2}{2}-\frac{1}{\lambda_i}\tilde{\alpha}_i^2-\frac{1}{\lambda_i}\tilde{\alpha}_i^4 \\ & +\frac{\bar{h}_i^2e_i^2+\tilde{\alpha}_i^2}{2}+\frac{e_i^2}{2}+\frac{\varepsilon_i^2}{2}+\frac{1}{2}\tilde{\alpha}_i^2+\frac{1}{2}\rho_i^2.\end{aligned} \tag{4.48}$$

Taking the adaptive NN weight design into above inequality yields

$$\begin{aligned}\dot{V}_i \leq & \frac{1}{k_{gi}}\left[\frac{1}{k_{gi}}h_i N(\xi_i)+1\right]\dot{\xi}_i \\ & -\left(k_{i1}-\bar{h}_i^2-\frac{1}{2}\right)e_i^2-k_{i2}e_i^4-\left(\frac{1}{\lambda_i}-1\right)\tilde{\alpha}_i^2-\frac{1}{\lambda_i}\tilde{\alpha}_i^4 \\ & +\frac{\eta_i}{2}+\frac{e_{i+1}^2}{2}+\frac{1}{2}\rho_i^2+\gamma_{i1}\tilde{\theta}_i\hat{\theta}_i+\gamma_{i2}\tilde{\theta}_i\hat{\theta}_i^3+\frac{\varepsilon_i^2}{2}.\end{aligned} \tag{4.49}$$

From Young's inequality, one has

$$
\begin{aligned}
-e_i^2 &\le \frac{1}{3}-\frac{4}{3}\left(e_i^2\right)^{\frac{3}{4}},\\
-\tilde{\alpha}_i^2 &\le \frac{1}{3}-\frac{4}{3}\left(\tilde{\alpha}_i^2\right)^{\frac{3}{4}}\\
\gamma_{i,1}\tilde{\theta}_i^{\mathrm{T}}\hat{\theta}_i+\gamma_{i,2}\tilde{\theta}_i\hat{\theta}_i^3 &\le -\frac{2}{3}\gamma_{i,1}\left(\tilde{\theta}_i^2\right)^{\frac{3}{4}}-\gamma_{i,2}\left(1-4\gamma_{i,3}\right)\tilde{\theta}_i^4\\
&\quad+\gamma_{i,1}\left(\frac{1}{2}\theta_i^2+\frac{1}{6}\right)+\gamma_{i,2}\left(\frac{3}{4}\gamma_{i,3}^{-\frac{1}{3}}+\frac{3}{2}\gamma_{i,3}^{-1}+\frac{3}{4}\gamma_{i,3}^{-3}\right)\theta_i^4,
\end{aligned}
$$

with $\gamma_{i,3}$ satisfying $\gamma_{i,3}>0$ and $1-4\gamma_{i,3}>0$. Based on above inequalities, the term $\dot{V}_i$ further satisfies

$$
\begin{aligned}
\dot{V}_i \le\ & \frac{1}{k_{gi}}\left[\frac{1}{k_{gi}}h_i N\left(\xi_i\right)+1\right]\dot{\xi}_i-k_{i2}e_i^4-\left(k_{i1}-\bar{h}_i^2-\frac{1}{2}\right)\left[\frac{4}{3}\left(e_i^2\right)^{\frac{3}{4}}-\frac{1}{3}\right]\\
&+\frac{e_{i+1}^2}{2}-\left(\frac{1}{\lambda_i}-1\right)\left[\frac{4}{3}\left(\tilde{\alpha}_i^2\right)^{\frac{3}{4}}-\frac{1}{3}\right]-\frac{1}{\lambda_i}\tilde{\alpha}_i^4\\
&-\frac{2}{3}\gamma_{i,1}\left(\tilde{\theta}_i^2\right)^{\frac{3}{4}}-\gamma_{i,2}\left(1-4\gamma_{i,3}\right)\tilde{\theta}_i^4+\gamma_{i,1}\left(\frac{1}{2}\theta_i^2+\frac{1}{6}\right)+\frac{\varepsilon_i^2}{2}\\
&+\frac{\eta_i}{2}+\frac{1}{2}\rho_i^2+\gamma_{i,2}\left(\frac{3}{4}\gamma_{i,3}^{-\frac{1}{3}}+\frac{3}{2}\gamma_{i,3}^{-1}+\frac{3}{4}\gamma_{i,3}^{-3}\right)\theta_i^4.
\end{aligned}
\tag{4.50}
$$

Step n

On the basis of (4.1), (4.10), (4.11), and (4.28), we have

$$
\dot{e}_n = W_n^{*\mathrm{T}}\varphi_n\left(x\right)+\varepsilon_n^*\left(x\right)+h_n u-\dot{\alpha}_{n-1}. \tag{4.51}
$$

Define the Lyapunov candidate for e_n and $\tilde{\theta}_n$ as $V_n\left(e_n,\tilde{\theta}_n\right)=\frac{1}{2}e_n^2+\frac{1}{2}\tilde{\theta}_n^2$. Differentiating V_n yields

$$
\begin{aligned}
\dot{V}_n &= e_n\left[W_n^{*\mathrm{T}}\varphi_n\left(x\right)+\varepsilon_n^*\left(x\right)+h_n u-\dot{\alpha}_{n-1}\right]-\tilde{\theta}_n\dot{\hat{\theta}}_n\\
&\le \frac{\theta_n}{2\eta_n}e_n^2\varphi_n^{\mathrm{T}}\varphi_n+\frac{\eta_n}{2}+e_n\varepsilon_n+h_n e_n u-e_n\dot{\alpha}_{n-1}-\tilde{\theta}_n\dot{\hat{\theta}}_n.
\end{aligned}
$$

Taking the control design (4.28) into above equation yields

$$
\begin{aligned}
\dot{V}_n \le\ & \frac{\theta_n}{2\eta_n}e_n^2\varphi_n^{\mathrm{T}}\varphi_n+\frac{\eta_n}{2}+e_n\varepsilon_n+\frac{1}{k_{gn}}\left[\frac{1}{k_{gn}}H_n N\left(\xi_n\right)+1\right]\dot{\xi}_n\\
&-e_n\left[k_{n1}e_n+k_{n2}e_n^3+\frac{\hat{\theta}_n}{2\eta_n}e_n\varphi_n^{\mathrm{T}}\varphi_n\right]-\tilde{\theta}_n\dot{\hat{\theta}}_n.
\end{aligned}
\tag{4.52}
$$

Recall Young's inequality and the fact $e_n\varepsilon_n \leq \frac{1}{2}\left(e_n^2+\varepsilon_n^2\right)$, one has

$$\begin{aligned}
\dot{V}_n \leq & \frac{1}{k_{gn}}\left[\frac{1}{k_{gn}}H_nN\left(\xi_n\right)+1\right]\dot{\xi}_n+\frac{\theta_n}{2\eta_n}e_n^2\varphi_n^{\mathrm{T}}\varphi_n+\frac{\eta_n}{2} \\
& +\frac{1}{2}\left(e_n^2+\varepsilon_n^2\right)-k_{n1}e_n^2-k_{n2}e_n^4-\frac{\hat{\theta}_n}{2\eta_n}e_n\varphi_n^{\mathrm{T}}\varphi_n-\tilde{\theta}_n\dot{\hat{\theta}}_n \\
\leq & \frac{1}{k_{gn}}\left[\frac{1}{k_{gn}}H_nN\left(\xi_n\right)+1\right]\dot{\xi}_n+\frac{\eta_n}{2}+\frac{1}{2}\left(e_n^2+\varepsilon_n^2\right) \\
& -k_{n1}e_n^2-k_{n2}e_n^4-\tilde{\theta}_n\left(\dot{\hat{\theta}}_n-\frac{1}{2\eta_n}e_n^2\varphi_n^T\varphi_n\right).
\end{aligned}$$

Inserting the adaptive NN design (4.29) into above inequality yields

$$\begin{aligned}
\dot{V}_n \leq & \frac{1}{k_{gn}}\left[\frac{1}{k_{gn}}H_nN\left(\xi_n\right)+1\right]\dot{\xi}_n-\left(k_{n1}-\frac{1}{2}\right)e_n^2-k_{n2}e_n^4 \\
& +\gamma_{n1}\tilde{\theta}_n\hat{\theta}_n+\gamma_{n2}\tilde{\theta}_n\hat{\theta}_n^3+\frac{\eta_n}{2}+\frac{1}{2}\varepsilon_n^2.
\end{aligned} \tag{4.53}$$

From Young's inequality, one has

$$\begin{aligned}
-e_n^2 \leq & \frac{1}{3}-\frac{4}{3}(e_n^2)^{\frac{3}{4}}, \\
\gamma_{n,1}\tilde{\theta}_n^{\mathrm{T}}\hat{\theta}_n+\gamma_{n,2}\tilde{\theta}_n^{\mathrm{T}}\hat{\theta}_n^3 \leq & -\frac{2}{3}\gamma_{n,1}\left|\tilde{\theta}_n\right|^{\frac{3}{2}}-\gamma_{n,2}\left(1-4\gamma_{n,3}\right)\tilde{\theta}_n^4 \\
& +\gamma_{n,1}\left(\frac{1}{2}\theta_n^2+\frac{1}{6}\right) \\
& +\gamma_{n,2}\left(\frac{3}{4}\gamma_{n,3}^{-\frac{1}{3}}+\frac{3}{2}\gamma_{n,3}^{-1}+\frac{3}{4}\gamma_{n,3}^{-3}\right)\theta_n^4.
\end{aligned}$$

From above inequalities, one has

$$\begin{aligned}
\dot{V}_n \leq & \frac{1}{k_{gn}}\left[\frac{1}{k_{gn}}H_nN\left(\xi_n\right)+1\right]\dot{\xi}_n-\left(k_{n1}-\frac{1}{2}\right)\left[\frac{4}{3}(e_n^2)^{\frac{3}{4}}-\frac{1}{3}\right] \\
& -k_{n2}e_n^4-\frac{2}{3}\gamma_{n,1}\left|\tilde{\theta}_n\right|^{\frac{3}{2}}-\gamma_{n,2}\left(1-4\gamma_{n,3}\right)\tilde{\theta}_n^4 \\
& +\gamma_{n,1}\left(\frac{1}{2}\theta_n^2+\frac{1}{6}\right)+\frac{\eta_n}{2}+\frac{1}{2}\varepsilon_n^2 \\
& +\gamma_{n,2}\left(\frac{3}{4}\gamma_{n,3}^{-\frac{1}{3}}+\frac{3}{2}\gamma_{n,3}^{-1}+\frac{3}{4}\gamma_{n,3}^{-3}\right)\theta_n^4.
\end{aligned} \tag{4.54}$$

Theorem 4.1 *Consider the system* (4.1) *with the presented design in Sect. 4.3. Then, there exists a positive constant T_m such that both the tracking error and closed-loop signals converge to a neighborhood of the origin in a finite time T with $T \leq T_m$.*

Proof From (4.40), one has

$$\dot{V}_1 \leq \frac{1}{k_{g1}}\left[\frac{1}{k_{g1}}h_1 N(\xi_1) + 1\right]\dot{\xi}_1 - a_1 V_1^{\frac{3}{4}} - b_1 V_1^2 + c_1 + \frac{e_2^2}{2} \tag{4.55}$$

with a_1, b_1, and c_1 defined as

$$\begin{aligned}
a_1 &= \min\left\{\frac{2^{\frac{11}{4}}}{3}\left(k_{11} - \frac{\bar{h}_1^2}{2} - \frac{1}{2}\right), \frac{2^{\frac{7}{4}}}{3}\gamma_{11}, \frac{2^{\frac{11}{4}}}{3}\left(\frac{1}{\lambda_1} - 1\right)\right\} \\
b_1 &= \frac{1}{3}\min\left\{4k_{12}, 4\gamma_{12}\left(1 - 4\gamma_{1,3}\right), \frac{4}{\lambda_1}\right\} \\
c_1 &= \frac{1}{3}\left(k_{11} - \bar{h}_1^2 - \frac{1}{2}\right) + \frac{1}{3}\left(\frac{1}{\lambda_1} - 1\right) + \frac{\varepsilon_1^2}{2} + \gamma_{11}\left(\frac{1}{2}\theta_1^2 + \frac{1}{6}\right) + \frac{\eta_1}{2} + \frac{1}{2}\rho_1^2 \\
&\quad + \gamma_{12}\left(\frac{3}{4}\gamma_{1,3}^{-\frac{1}{3}} + \frac{3}{2}\gamma_{1,3}^{-1} + \frac{3}{4}\gamma_{1,3}^{-3}\right)\theta_1^4.
\end{aligned} \tag{4.56}$$

From (4.50), one has

$$\dot{V}_i \leq \frac{1}{k_{gi}}\left[\frac{1}{k_{gi}}h_i N(\xi_i) + 1\right]\dot{\xi}_i - a_i V_i^{\frac{3}{4}} - b_i V_i^2 + c_i + \frac{e_{i+1}^2}{2}, \tag{4.57}$$

with

$$\begin{aligned}
a_i &= \min\left\{\frac{2^{\frac{11}{4}}}{3}\left(k_{i1} - \frac{\bar{h}_i^2}{2} - \frac{1}{2}\right), \frac{2^{\frac{7}{4}}}{3}\gamma_{1,1}, \frac{2^{\frac{11}{4}}}{3}\left(\frac{1}{\lambda_i} - 1\right)\right\} \\
b_i &= \frac{1}{3}\min\left\{4k_{i2}, 4\gamma_{i,2}\left(1 - 4\gamma_{i,3}\right), \frac{4}{\lambda_i}\right\} \\
c_i &= \frac{1}{3}\left(k_{i1} - \frac{\bar{h}_i^2}{2} - \frac{1}{2}\right) + \frac{1}{3}\left(\frac{1}{\lambda_i} - 1\right) + \frac{\varepsilon_i^2}{2} + \frac{\eta_i}{2} + \frac{1}{2}\rho_i^2 + \gamma_{i,1}\left(\frac{1}{2}\theta_i^2 + \frac{1}{6}\right) \\
&\quad + \gamma_{i,2}\left(\frac{3}{4}\gamma_{i,3}^{-\frac{1}{3}} + \frac{3}{2}\gamma_{i,3}^{-1} + \frac{3}{4}\gamma_{i,3}^{-3}\right)\theta_i^4.
\end{aligned} \tag{4.58}$$

From (4.54), one has

$$\dot{V}_n \leq \frac{1}{k_{gn}}\left[\frac{1}{k_{gn}}H_n N(\xi_n) + 1\right]\dot{\xi}_n - a_n V_n^{\frac{3}{4}} - b_n V_n^2 + c_n, \tag{4.59}$$

with

$$\begin{aligned} a_n &= \min\left\{\frac{2^{\frac{11}{4}}}{3}\left(k_{n1}-\frac{1}{2}\right), \frac{2^{\frac{7}{4}}}{3}\gamma_{n,1}\right\} \\ b_n &= \min\left\{2k_{n2}, 2\gamma_{n,2}\left(1-4\gamma_{n,3}\right)\right\} \\ c_n &= \frac{1}{3}\left(k_{n,1}-\frac{1}{2}\right)+\frac{\eta_n}{2}+\frac{1}{2}\varepsilon_n^2+\gamma_{n,1}\left(\frac{1}{2}\theta_n^2+\frac{1}{6}\right) \\ &+\gamma_{n,2}\left(\frac{3}{4}\gamma_{n,3}^{-\frac{1}{3}}+\frac{3}{2}\gamma_{n,3}^{-1}+\frac{3}{4}\gamma_{n,3}^{-3}\right)\theta_n^4. \end{aligned} \tag{4.60}$$

From the fact in (4.59) and Lemma 4.1, one can obtain V_n is bounded. Therefore, for nth subsystem, the boundedness of the terms e_n, $\tilde{\theta}_n$ and ξ_n can also be guaranteed. Based on the fact that e_n is bounded, from (4.57), one has

$$\dot{V}_{n-1} \leq \frac{1}{k_{g(n-1)}}\left[\frac{1}{k_{g(n-1)}}h_{n-1}N\left(\xi_{n-1}\right)+1\right]\dot{\xi}_{n-1}-a_{n-1}V_{n-1}^{\frac{3}{4}}-b_{n-1}V_{n-1}^2+\bar{c}_{n-1},$$

with $c_{n-1}+\left|\frac{e_n^2}{2}\right| \leq \bar{c}_{n-1}$. Applying again Lemma 4.1, one can get V_{n-1} is bounded. Recursively, one can obtain that V_i is bounded for $i = 1, \ldots, n-2$. Therefore, the terms e_i, $\tilde{\theta}_i$ and ξ_i are also bounded for $i = 1, \ldots, n-1$. Finally, based on Lemma 4.1, the settling time for fixed-time convergence can be obtained as $T \leq \max\limits_{1\leq i\leq n}\left\{\frac{4}{a_i\kappa}+\frac{1}{b_i}, \frac{4}{a_i}+\frac{1}{\kappa b_i}\right\}$ with a constant $0 < \kappa < 1$. This completes the proof. □

Remark 4.1 From the proof of Lemma 4.1 and Theorem 4.1, it can be observed that if a_i and b_i are big enough and $\bar{c}_i + N_{max}$ is small enough, then the size of residual sets $\Pi_a = \left\{x|\, V(x) \leq \left(\frac{d}{a(1-\kappa)}\right)^{\frac{1}{p}}\right\}$ and $\Pi_b = \left\{x|\, V(x) \leq \left(\frac{d}{a(1-\kappa)}\right)^{\frac{1}{q}}\right\}$ would shrink. Accordingly, the tracking error e_i and the parameter estimation error $\tilde{\theta}_i$ can be reduced.

4.5 Simulation Example

4.5.1 Example 1: Numerical Example

In this example, the following third-order nonlinear strict-feedback systems is considered as

$$\begin{aligned} \dot{x}_1 &= \left(1+x_1^2\right)x_2+x_1e^{-\frac{1}{2}x_1}, \\ \dot{x}_2 &= \left(3+\cos\left(x_1x_2\right)\right)x_3+x_1x_2^2, \\ \dot{x}_3 &= \left(1+\left|x_3\right|\right)u+0.2\cdot x_3\sin\left(x_1x_2\right), \\ y &= x_1 \end{aligned}$$

with the initial conditions $x_1(0) = 0.2$, $x_2(0) = 0.1$, $x_3(0) = 0.1$, and the control input u. The virtual control for the first and second subsystems are considered as

$$s_1 = \frac{1}{k_{g1}} N(\xi_1) \left(k_{11} e_1 + k_{12} e_1^3 + \frac{\hat{\theta}_1}{2\eta_1} e_1 \varphi_1^{\mathrm{T}} \varphi_1 - x_d \right),$$

$$s_2 = \frac{1}{k_{g2}} N(\xi_2) \left(k_{21} e_2 + k_{22} e_2^3 + \frac{\hat{\theta}_2}{2\eta_2} e_2 \varphi_2^{\mathrm{T}} \varphi_2 - \dot{\alpha}_1 \right).$$

The control input is designed as

$$u(t) = \frac{1}{k_{g3}} N(\xi_3) \left[k_{31} e_3 + k_{32} e_3^3 + \frac{\hat{\theta}_3}{2\eta_3} e_3 \varphi_3^{\mathrm{T}} \varphi_3 - e_2 \dot{\alpha}_2 \right].$$

The adaptive NN weights learning are designed as

$$\dot{\hat{\theta}}_i = \frac{1}{2\eta_i} e_i^2 \varphi_i^{\mathrm{T}} \varphi_i - \gamma_{i1} \hat{\theta}_i - \gamma_{i2} \hat{\theta}_i^3, i = 1, 2, 3$$

with the initial conditions $\hat{\theta}_1(0) = 1$, $\hat{\theta}_2(0) = 1.5$ and $\hat{\theta}_3(0) = 2$. The filters are designed as

$$\lambda_i \dot{\alpha}_i = (s_i - \alpha_i) + (s_i - \alpha_i)^3, i = 1, 2$$

with the initial conditions $\alpha_i(0) = 0$. The adaptive Nussbaum-type gains are designed as

$$\dot{\xi}_1 = k_{g1} \left(k_{11} e_1^2 + k_{12} e_1^4 + \frac{\hat{\theta}_1}{2\eta_1} e_1^2 \varphi_1^{\mathrm{T}} \varphi_1 - e_1 x_d \right),$$

$$\dot{\xi}_2 = k_{g2} e_2 \left(k_{21} e_2 + k_{22} e_2^3 + \frac{\hat{\theta}_2}{2\eta_2} e_2 \varphi_2^{\mathrm{T}} \varphi_2 - e_2 \dot{\alpha}_1 \right),$$

$$\dot{\xi}_3 = k_{g3} e_3 \left(k_{31} e_3 + k_{32} e_3^3 + \frac{\hat{\theta}_3}{2\eta_3} e_3 \varphi_3^{\mathrm{T}} \varphi_3 - e_3 \dot{\alpha}_2 \right),$$

with the initial conditions $\xi_1(0) = 2.3$, $\xi_2(0) = 2$ and $\xi_3(0) = 2$. The parameters in this example are selected $k_{11} = 15$, $k_{12} = 5$, $k_{21} = 30$, $k_{22} = 10$, $k_{31} = 30$, $k_{32} = 10$, $\lambda_1 = 0.3$, $\lambda_2 = 0.3$, $\gamma_{11} = 100$, $\gamma_{12} = 100$, $\gamma_{21} = 100$, $\gamma_{22} = 100$, $\gamma_{31} = 100$, $\gamma_{32} = 100$, $\eta_1 = 1$, $\eta_2 = 1$ and $\eta_3 = 1$. Based on the above parameters, one can know that $a_1 = 5.23$, $b_1 = 4.44$, $a_2 = 5.23$, $b_2 = 4.44$, $a_3 = 49.33$, and $b_3 = 20$, where a_1 and b_1 are defined in (4.56), a_2 and b_2 are defined in (4.58), and a_3 and b_3 are defined in (4.60). Therefore, from Lemma 4.1, the fixed-settling time for the third-order nonlinear strict-feedback systems can be determined as

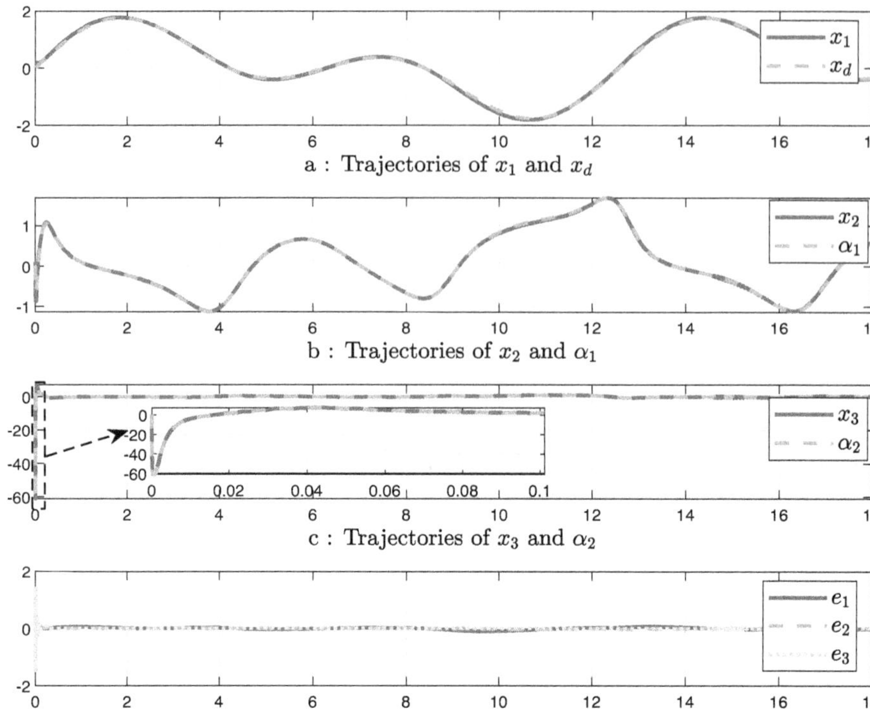

Fig. 4.1 System states, virtual control signal, and tracking errors for Example 1

$$T \leq \max_{1\leq i\leq 3}\left\{\frac{4}{a_i\kappa}+\frac{1}{b_i}, \frac{4}{a_i}+\frac{1}{\kappa b_i}\right\} = 1.5, \kappa = 0.70.$$

Figures 4.1, 4.2, and 4.3 present the simulation results for the proposed fixed-time control design. In Fig. 4.1, the trajectories of the system states x_1, x_2, and x_3 are shown to converge to the desired reference signal x_d and the filtered outputs α_1 and α_2, respectively, within a fixed time. Subfigure (a) illustrates that x_1 follows x_d with minimal tracking errors, confirming the fixed-time tracking performance. Subfigure (b) shows x_2 accurately tracking the virtual control α_1, while subfigure (c) demonstrates the convergence of x_3 to α_2 after a brief transient. Subfigure (d) shows the error signals e_1, e_2, and e_3, which converge to a neighborhood of zero within

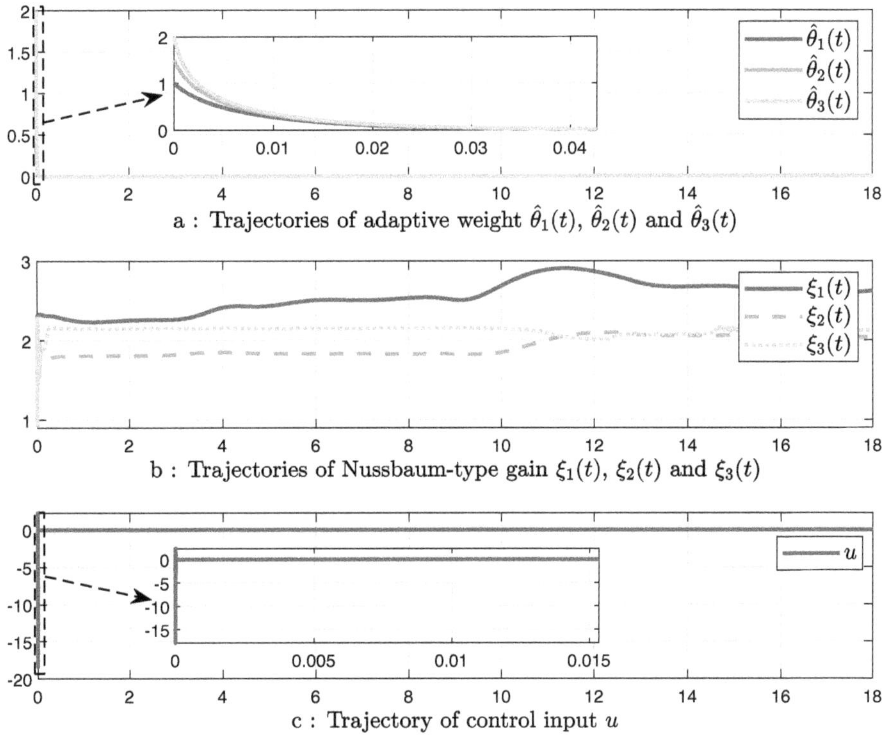

Fig. 4.2 Adaptive Nussbaum gains, parameter learning laws, and control signal for Example 1

the fixed time, verifying the practical stability of the control scheme. Figure 4.2 illustrates the evolution of the adaptive weights and Nussbaum-type gains. Subfigure (a) shows the adaptive weights $\hat{\theta}_1$, $\hat{\theta}_2$, and $\hat{\theta}_3$, which adjust smoothly and remain bounded. Subfigure (b) displays the trajectories of the Nussbaum-type gains ξ_1, ξ_2, and ξ_3, which also remain bounded, ensuring the stability of the overall system. Subfigure (c) depicts the control input u, which adapts dynamically and stabilizes as the system approaches the desired state. Figure 4.3 shows the performance of the dynamic surface filters. Subfigure (a) demonstrates that s_1 tracks α_1 with high accuracy, while subfigure (b) shows similar tracking performance for s_2 with respect to α_2. Subfigure (c) presents the outputs of the filters, $\tilde{\alpha}_1$ and $\tilde{\alpha}_2$, which converge to a neighborhood of zero within the fixed time, confirming the effectiveness of the filtering mechanism in avoiding differentiation errors.

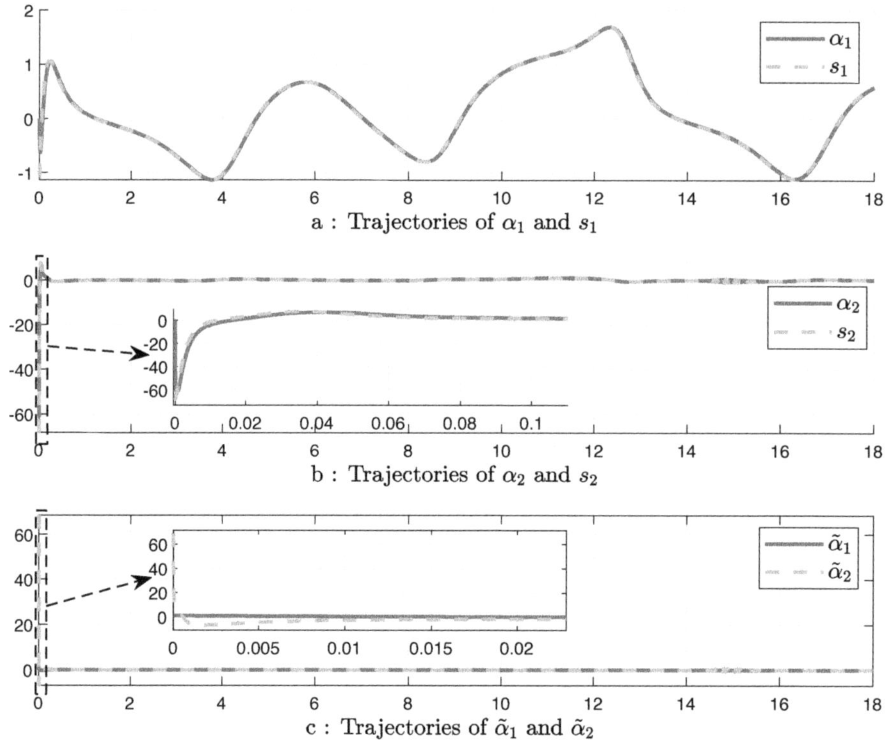

Fig. 4.3 Virtual control, DSC filter output, and filter errors for Example 1

4.5.2 Example 2: Single-Link Robot Arm

In this example, to illustrate the effectiveness of the proposed design, we consider the single-link robot arm in [25], whose dynamic can be described as

$$\dot{x}_1 = x_2 - \frac{0.12}{0.15}x_1$$
$$\dot{x}_2 = -\frac{9.8 * 0.25}{0.15}\sin(x_1) + \frac{1}{0.15}u$$

with x_1 and x_2 are the state of the system and u is the control input. The output tracking reference signal is selected as $x_d(t) = \sin(t) + \sin(0.5t)$.

The adaptive Nussbaum design for the first subsystem are developed as

$$\begin{aligned}
s_1 &= \frac{1}{k_{g1}} N(\xi_1) \left(k_{11} e_1 + k_{12} e_1^3 + \frac{\hat{\theta}_1}{2\eta_1} e_1 \varphi_1^{\mathrm{T}} \varphi_1 - x_d \right), \\
\dot{\xi}_1 &= k_{g1} \left(k_{11} e_1^2 + k_{12} e_1^4 + \frac{\hat{\theta}_1}{2\eta_1} e_1^2 \varphi_1^{\mathrm{T}} \varphi_1 - e_1 x_d \right), \\
\dot{\hat{\theta}}_1 &= \frac{1}{2\eta_1} e_1^2 \varphi_1^{\mathrm{T}} \varphi_1 - \gamma_{11} \hat{\theta}_1 - \gamma_{12} \hat{\theta}_1^3, \\
\lambda_1 \dot{\alpha}_1 &= (s_1 - \alpha_1) + (s_1 - \alpha_1)^3.
\end{aligned}$$

For the second subsystem, we consider the following design as

$$\begin{aligned}
u(t) &= \frac{1}{k_{g2}} N(\xi_2) \left[k_{21} e_2 + k_{22} e_2^3 + \frac{\hat{\theta}_2}{2\eta_2} e_2 \varphi_2^{\mathrm{T}} \varphi_2 - \dot{\alpha}_1 \right], \\
\dot{\xi}_2 &= k_{g2} e_2 \left[k_{21} e_2 + k_{22} e_2^3 + \frac{\hat{\theta}_2}{2\eta_2} e_2 \varphi_2^{\mathrm{T}} \varphi_2 - \dot{\alpha}_1 \right], \\
\dot{\hat{\theta}}_2 &= \frac{1}{2\eta_2} e_2^2 \varphi_2^{\mathrm{T}} \varphi_2 - \gamma_{21} \hat{\theta}_2 - \gamma_{22} \hat{\theta}_2^3.
\end{aligned}$$

The parameters are considered as $k_{11} = 40, k_{12} = 40, k_{21} = 40, k_{22} = 40, \tau_1 = 0.05$, $\gamma_{11} = 100$, $\gamma_{12} = 100$, $\gamma_{21} = 100$, $\gamma_{22} = 100$, $\eta_1 = 1$ and $\eta_2 = 1$.

Based on the above parameters, one can obtain that $a_1 = 21.30$, $b_1 = 36.67$, $a_2 = 82.96$ and $b_2 = 80$, where a_1 and b_1 are defined in (4.56) and a_2 and b_2 are defined in (4.60). From Lemma 4.1, the fixed-settling time for the systems can be determined as

$$T \le \max_{1 \le i \le 2} \left\{ \frac{4}{a_i \kappa} + \frac{1}{b_i}, \frac{4}{a_i} + \frac{1}{\kappa b_i} \right\} = 3.3, \kappa = 0.94.$$

The Nussbaum-type gains start with $\xi_1(0) = 2$ and $\xi_2(0) = 2$. The adaptive law is updated from $\hat{\theta}_1(0) = 1$ and $\hat{\theta}_2(0) = 1$.

Figures 4.4, 4.5, and 4.6 illustrate the simulation results for the proposed control design. In Fig. 4.4, the trajectories of the system states x_1 and x_2 are shown to converge to the reference signal x_{d} and the filtered output α_1, respectively. Subfigure (a) demonstrates that x_1 tracks x_{d} with minimal error, while subfigure (b) confirms the alignment of x_2 with α_1. Subfigure (c) further validates the effectiveness of the control design by showing the tracking errors e_1 and e_2 converging to a small neighborhood

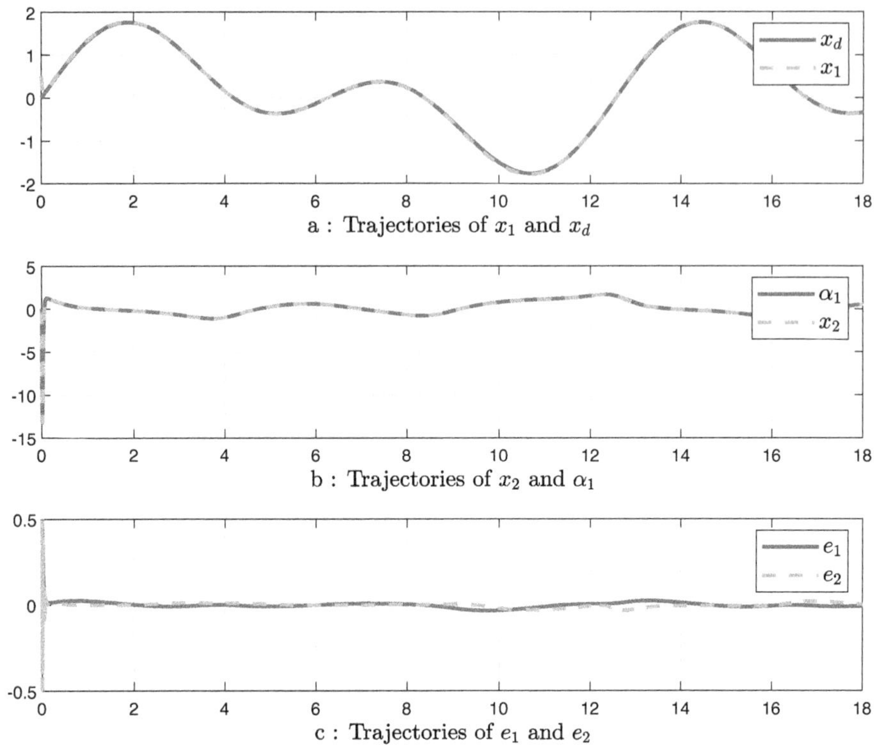

Fig. 4.4 System states, virtual control signal, and tracking errors for Example 2

near the origin, indicating satisfactory tracking performance. Figure 4.5 presents the evolutions of the Nussbaum-type gains and adaptive weights. Subfigure (a) shows the adaptive weights $\hat{\theta}_1$ and $\hat{\theta}_2$ converging smoothly to their steady values, while subfigure (b) illustrates the bounded behavior of the Nussbaum-type gains ξ_1 and ξ_2, ensuring system stability throughout the process. Figure 4.6 highlights the input and output behavior of the DSC filter. Subfigure (a) shows the convergence of s_1 to α_1 with high accuracy, and subfigure (b) depicts the diminishing of the filter error $\tilde{\alpha}_1$ within a fixed time. Subfigure (c) shows the control input u, which stabilizes over time, further validating the robustness of the proposed design.

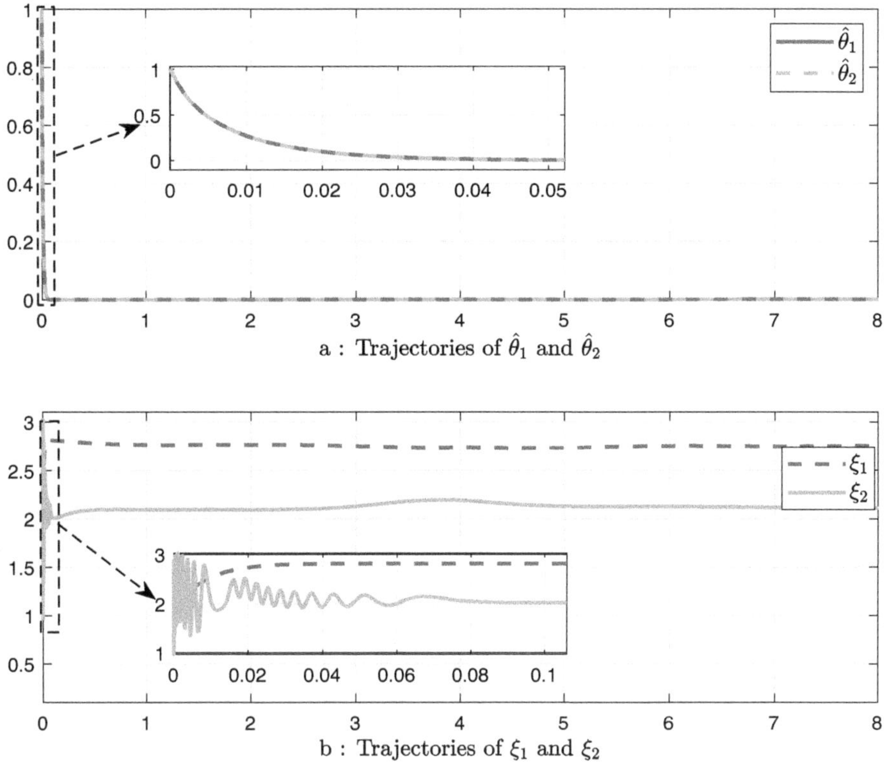

Fig. 4.5 Adaptive Nussbaum gains and parameter learning laws for Example 2

4.6 Conclusion

This chapter has investigated the tracking issue of nonlinear strict-feedback system with unknown control directions. The Nussbaum-type gain has been designed to handle multiple unknown control directions. Additionally, regardless of the initial state, the proposed solution could meet the fixed-time convergence requirement within the settling time. Finally, simulation experiments involving two examples have been carried out to verify the efficacy of the proposed design.

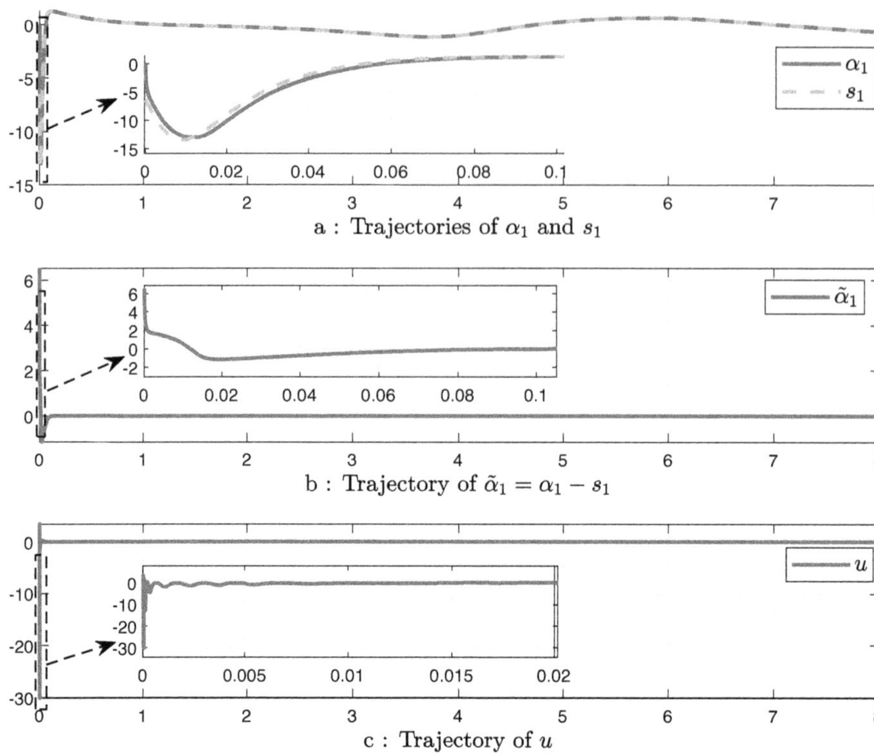

Fig. 4.6 Virtual control, DSC filter output, and filter errors for Example 2

References

1. Ge, S.S., Hong, F., Lee, T.H.: Adaptive neural control of nonlinear time-delay systems with unknown virtual control coefficients. IEEE Trans. Syst. Man Cybern. Part B (Cybernetics). **34**(1), 499–516 (2004)
2. Chen, C., Liu, Z., Xie, K., Liu, Y., Zhang, Y., Chen, C.L.P.: Adaptive fuzzy asymptotic control of MIMO systems with unknown input coefficients via a robust Nussbaum gain-based approach. IEEE Trans. Fuzzy Syst. **25**(5), 1252–1263 (2017)
3. Hua, C., Li, H., Li, K., Ding, W.: Low-computation tracking control of nonlinear systems with asymmetric full-state constraints and unknown control directions. IEEE Trans. Syst. Man Cybern.: Syst. **53**(2), 1051–1059 (2023)
4. Liu, G., Yang, Y., Zhao, X., Ahn, C.K.: Adaptive fuzzy practical bipartite synchronization for multi-agent systems with intermittent feedback under multiple unknown control directions. IEEE Trans. Fuzzy Syst. **32**(5), 3211–3225 (2024)
5. Liu, G., Yang, Y., Gao, W., Wunsch, D.C.: Adaptive Nussbaum design for nonholonomic systems with asymptotic stabilization against false data injection. IEEE Trans. Cybern. **55**(11), 5189–5201 (2025)
6. Bhat, S.P., Bernstein, D.S.: Continuous finite-time stabilization of the translational and rotational double integrators. IEEE Trans. Autom. Control **43**(5), 678–682 (1998)
7. Wang, A., Liu, L., Qiu, J., Feng, G.: Finite-time adaptive fuzzy control for nonstrict-feedback nonlinear systems via an event-triggered strategy. IEEE Trans. Fuzzy Syst. **28**(9), 2164–2174 (2020)

8. Polyakov, A.: Nonlinear feedback design for fixed-time stabilization of linear control systems. IEEE Trans. Autom. Control **57**(8), 2106–2110 (2012)
9. Wang, F., Liu, Z., Zhang, Y., Chen, C.L.P.: Adaptive quantized controller design via backstepping and stochastic small-gain approach. IEEE Trans. Fuzzy Syst. **24**(2), 330–343 (2016)
10. Liu, Z., Wang, F., Zhang, Y., Philip Chen, C.L.: Fuzzy adaptive quantized control for a class of stochastic nonlinear uncertain systems. IEEE Trans. Cybern. **46**(2), 524–534 (2016)
11. Sui, S., Chen, C.L.P., Tong, S.: A novel adaptive NN prescribed performance control for stochastic nonlinear systems. IEEE Trans. Neural Netw. Learn. Syst. **32**(7), 3196–3205 (2021)
12. Wang, J., Liu, Z., Zhang, Y., Chen, C.L.P.: A novel fuzzy control with filter-based event-triggered mechanism for nonlinear uncertain stochastic systems suffered input hysteresis. Fuzzy Sets Syst. **432**, 68–88 (2022)
13. Yang, Y., Tang, L., Zou, W., Guo, J., Ahn, C.K.: Dynamic event-triggered design with fixed-time performance and input dead-zone. IEEE Trans. Circuits Syst. II Express Briefs **69**(11), 4344–4348 (2022)
14. Bai, W., Liu, P.X., Wang, H.: Neural-network-based adaptive fixed-time control for nonlinear multiagent non-affine systems. IEEE Trans. Neural Netw. Learn. Syst. **35**(1), 570–583 (2024)
15. Lv, M., Li, Y., Wan, L., Dai, J., Chang, J.: Fast nonsingular fixed-time fuzzy fault-tolerant control for HFVs with guaranteed time-varying flight state constraints. IEEE Trans. Fuzzy Syst. **30**(11), 4555–4567 (2022)
16. Wang, F., Lai, G.: Fixed-time control design for nonlinear uncertain systems via adaptive method. Syst. & Control Lett. **140**, 104704 (2020)
17. Yang, Y., Liu, G., Li, Q., Ahn, C.K.: Multiple adaptive fuzzy Nussbaum-type functions design for stochastic nonlinear systems with fixed-time performance. Fuzzy Sets Syst. **476**, 108767 (2024)
18. Chen, W., Wen, C., Wu, J.: Global exponential/finite-time stability of nonlinear adaptive switching systems with applications in controlling systems with unknown control direction. IEEE Trans. Autom. Control **63**(8), 2738–2744 (2018)
19. Dong, G., Li, H., Ma, H., Lu, R.: Finite-time consensus tracking neural network FTC of multi-agent systems. IEEE Trans. Neural Netw. Learn. Syst. **32**(2), 653–662 (2021)
20. Zhang, J., Xiang, Z.: Event-triggered adaptive neural network sensor failure compensation for switched interconnected nonlinear systems with unknown control coefficients. IEEE Trans. Neural Netw. Learn. Syst. **33**(10), 5241–5252 (2022)
21. Wang, Y.W., Lei, Y., Bian, T., Guan, Z.H.: Distributed control of nonlinear multiagent systems with unknown and nonidentical control directions via event-triggered communication. IEEE Trans. Cybern. **50**(5), 1820–1832 (2020)
22. Li, B., Xia, J., Su, S.F., Sun, W., Zhang, H.: Event-triggered adaptive fuzzy tracking control for nonlinear systems with unknown control directions. IEEE Trans. Syst. Man Cybern.: Syst. **52**(7), 4648–4657 (2022)
23. Lewis, F.L.: Neural network control of robot manipulators. IEEE Expert. **11**(3), 64–75 (1996)
24. Tang, L., Yang, Y., Zou, W., Song, R.: Neuro-adaptive fixed-time control with novel command filter design for nonlinear systems with input dead-zone. Neurocomputing **511**, 308–318 (2022)
25. Wang, C., Wen, C., Hu, Q.: Event-triggered adaptive control for a class of nonlinear systems with unknown control direction and sensor faults. IEEE Trans. Autom. Control **65**(2), 763–770 (2020)

Chapter 5
Adaptive Fixed-Time Nussbaum Design for Stochastic Nonlinear Systems

5.1 Introduction

Engineering systems are commonly stochastic in nature owing to the existence of stochastic processes [1, 2]. Initiated by the stochastic integral, the mathematical foundation of the stochastic differential equation (SDE) has been established. With the equivalent integral formula about the SDE, one can obtain the SDE expression, including a Brownian motion [3]. Depending on how the Brownian motion term is treated, common ways to specify stochastic differential equations can be categorized as the Itô form [4] and the Stratonovich form [5]. In contrast to conventional integrals such as the Riemann integral, Stieltjes integral, or Lebesgue integral, the stochastic integral derived from the SDE is essentially the Itô integral. Different from the Itô integral, the Stratonovich integral is not a martingale and increases the difficulty of theoretical analysis. Therefore, the Itô formula is widely used and lays the foundation for the chain rule for stochastic processes. In addition, numerous academics have investigated stability analysis for stochastic dynamical systems [6–8]. Moreover, because of the imperfect modeling in complex systems, neural networks (NN) [9–11] and adaptive learning-based control [12, 13] have been successfully applied to deal with the uncertainty in nonlinear dynamical systems. The NN generates a black box model depending on the network topology. In contrast, Fuzzy Logic Systems (FLSs) formulate the system's behavior in a rule-based manner. As a function approximator, NN suffers from a lack of interpretability [14]. In contrast, the advantage of FLS lies in the linguistic representation for human understanding and the expert knowledge gained through rule-based modeling [15]. Due to its strong robustness, FLS is suitable for solving nonlinear, persistent dwell-time switching [16] and weighted Try-Once-Discard protocol [17] in process control. In this chapter, FLSs are exploited to tackle unknown control directions. The stochastic stabilizing design for stochastic multiagent systems has been developed in [18]. The FLS-based adaptive control design is combined with event-triggered feedback for a networked stochastic system to reduce the usage of communication resources in [7]. Further, for stochastic

Y. Yang et al., *Adaptive Nussbaum Design for Nonlinear Dynamical Systems*,
https://doi.org/10.1007/978-981-95-4997-9_5

nonlinear systems, the output synchronization problem has been investigated with multiple heterogeneous agents and the input dead zone in the agent's dynamics [19], and for stochastic uncertain nonlinear systems, the output tracking problem has been considered to take actuation saturation into account [12]. However, in these studies, stability and convergence could be achieved in an asymptotic manner, i.e., infinite time was required. In addition to steady-state performance, recent developments in engineering systems require the transient behavior to be sufficiently fast. As a result, it is required to improve the transient response of stochastic nonlinear systems [20].

With the existing Nussbaum-based design [21, 22], the system stability could be obtained indirectly for nonlinear systems with multiple unknown control coefficients by recursively applying the Lyapunov–Nussbaum stability theorem. In contrast, this chapter develops a novel multiple Nussbaum design to avoid the issue of mutual cancellation among multiple unknown control coefficients in stochastic systems, where the closed-loop stability could be obtained directly and the stability analysis could be greatly simplified. Though the Lyapunov–Nussbaum theorem [23, 24] and fixed-time Lyapunov theorem [25] have been extensively investigated, there is a gap between these two prevalent techniques. In our previous work [26], the Lyapunov–Nussbaum theorem is extended to the deterministic system with a single Nussbaum-type function and fixed-time performance. In this work, the gap between the Nussbaum design and fixed-time stability is bridged for the stochastic systems, which can efficiently deal with multiple unknown control coefficients in uncertain dynamics.

The remainder of this chapter is structured as follows: In Sect. 5.2, the problem statement is provided with the required assumptions and lemmas. In Sect. 5.3, the adaptive fuzzy backstepping design with multiple Nussbaum-type gains is presented to achieve fixed-time performance in stochastic nonlinear systems subject to multiple unknown control directions. The stability analysis for the presented Nussbaum design within the adaptive fuzzy backstepping framework is performed in Sects. 5.3 and 5.4, respectively. The simulation study is investigated in Sect. 5.5 to validate the effectiveness of the proposed design scheme. Finally, concluding remarks are made in Sect. 5.6.

5.2 Problem Formulation

Consider the nonlinear stochastic dynamical system evolving in continuous time as

$$\begin{aligned} \mathrm{d}x_i &= \left[h_i(\bar{x}_i)\, x_{i+1} + f_i(\bar{x}_i)\right] dt + g_i(\bar{x}_i)\, \mathrm{d}\omega, \\ \mathrm{d}x_n &= \left[h_n(\bar{x}_n)\, u + f_n(\bar{x}_n)\right] dt + g_n(\bar{x}_n)\, \mathrm{d}\omega, \\ y &= x_1, \end{aligned} \tag{5.1}$$

with the system state $\bar{x}_i = [x_1, x_2, \ldots, x_i]^{\mathrm{T}} \in \mathbb{R}^i$ and $x = [x_1, x_2, \ldots, x_n]^{\mathrm{T}} \in \mathbb{R}^n$, the control input signal $u \in \mathbb{R}$, the unknown smooth functions $f_i(\bar{x}_i(t)) : \mathbb{R}^i \rightarrow \mathbb{R}$, $g_i(\bar{x}_i(t)) : \mathbb{R}^i \rightarrow \mathbb{R}$, and $h_i(\bar{x}_i(t))$. Furthermore, $h_i(\bar{x}_i(t))$ represent the unknown control direction.

The following assumptions and lemmas are required for the subsequent discussions:

Assumption 5.1 The function $h_i(\bar{x}_i(t))$ has known upper and lower bounds, i.e., $0 < \underline{h}_i \leq |h_i(\bar{x}_i(t))| \leq \bar{h}_i$.

Assumption 5.2 The signals $\{x_\mathrm{d}(t), \dot{x}_\mathrm{d}(t), \ddot{x}_\mathrm{d}(t)\}$ are bounded.

Assumption 5.3 ([27]) The system (5.1) is input-to-state stable.

The conventional single Nussbaum-type function design suffers from mutual cancellation among multiple unknown control directions. In contrast, this chapter considers a novel class of Nussbaum-type functions design as

$$N_i(\xi) = 2\xi \sinh\left(\xi^2\right)\cos\left(2^{i-1}\xi\right) - 2^{i-1}\left(\cosh\left(\xi^2\right) - 1\right)\sin\left(2^{i-1}\xi\right). \quad (5.2)$$

As discussed later in Lemma 5.1, the novel Nussbaum-type function design (5.2) can effectively resolve the issue of mutual cancellation for the systems with multiple unknown control directions, which results in a direct analysis scheme in contrast to the conventional Nussbaum-based indirect analysis.

Lemma 5.1 *For the stochastic systems* (5.1)*, suppose that there exists a continuous positive definite function* $V(x(t))$*, which has the operator* $\mathcal{L}$ *satisfies*

$$\mathcal{L}V(x(t)) \leq \sum_{j=1}^{N}\left[h_j N_j\left(\xi_j\right) + 1\right]\dot{\xi}_j - aV^p(x(t)) - bV^q(x(t)) + c, \quad (5.3)$$

where a, b, c, m, and n are positive constants, $0 < p < 1$, $q > 1$, $h_j(\cdot)$ *is a smooth function satisfying* $0 < h_{j,\min} \leq \left|h_j\right| \leq h_{j,\max}$ *with unknown constant bounds* $h_{j,\min}$ *and* $h_{j,\max}$*, and* $N_j(\cdot)$ *is a Nussbaum-type function defined in* (5.2) *with* $\xi_j(t)$ *being the Nussbaum-type gain. Then, the following conclusions hold:*

- $V(t)$, $x(t)$, *and* $\left\{\xi_j(t)\right\}_{j=1}^{N}$ *are practical fixed-time stability in probability.*
- *There exists* $N_{max} > 0$ *such that* $\left|\sum_{j=1}^{N}\left[h_j N_j\left(\xi_j\right) + 1\right]\dot{\xi}_j\right| \leq N_{\max}$.
- *The expectation of the system state* $x(t)$ *converges to the set*

$$\left\{x \middle| V(x) \in \left(0, \min\left\{\left(\frac{l}{a(1-\kappa)}\right)^{\frac{1}{p}}, \left(\frac{l}{b(1-\kappa)}\right)^{\frac{1}{q}}\right\}\right]\right\}$$

with $l = c + N_{\max}$*. The settling time T satisfies* $E[T] \leq T_{\max}$,

$$T_{\max} = \max\left\{\frac{1}{a(1-p)} + \frac{1}{b\kappa(q-1)}, \frac{1}{a\kappa(1-p)} + \frac{1}{b(q-1)}\right\}$$

and $0 < \kappa < 1$.

Proof First, one can easily know that $-V^p(x(t)) - V^q(x(t)) \le -V(x(t))$ because $V(\cdot)$ is positive definite. Then, one has

$$\mathcal{L}V \le \sum_{j=1}^{N} \left[h_j(t) N_j(\xi_j) + 1\right]\dot{\xi}_j - \vartheta V + c,$$

where $\vartheta = \min\{a, b\}$. Define $W(t) = V(t)e^{\vartheta t}$, one has

$$\begin{aligned} E[W(t)]\big|_0^t &= E\left[\int_0^t \mathcal{L}W(\sigma)\,\mathrm{d}\sigma\right] E[V(t)] \\ &= E\left[\int_0^t \mathcal{L}V(\sigma)\,\mathrm{d}\sigma\right] \end{aligned} \tag{5.4}$$

for all t. It can be noted that $\mathcal{L}W(t) = \vartheta V(t)e^{\vartheta t} + \mathcal{L}V(t)e^{\vartheta t}$, from which one has

$$\begin{aligned} E[W(t)]|_0^t &= \int_0^t \left[\vartheta V(\sigma)e^{\vartheta\sigma} + \mathcal{L}V(\sigma)e^{\vartheta\sigma}\right]\mathrm{d}\sigma \\ &\le \int_0^t \left\{\sum_{j=1}^{N}\left[h_j(\sigma)N_j(\xi_j(\sigma)) + 1\right]\dot{\xi}_j(\sigma) + c\right\} e^{\vartheta\sigma}\mathrm{d}\sigma. \end{aligned} \tag{5.5}$$

In addition, one also has

$$\begin{aligned} &\int_0^t \left[\sum_{j=1}^{N}\left[h_j(\sigma)N_j(\xi_j(\sigma)) + 1\right]\dot{\xi}_j(\sigma) + c\right] e^{\vartheta\sigma}\mathrm{d}\sigma \\ &\le \int_0^t \sum_{j=1}^{N}\left[h_j(\sigma)N_j(\xi_j(\sigma)) + 1\right]\dot{\xi}_j(\sigma)e^{\vartheta\sigma}\mathrm{d}\sigma + \frac{c}{\vartheta}\left(e^{\vartheta t} - 1\right). \end{aligned} \tag{5.6}$$

Inserting the above inequality into (5.5) yields

$$\begin{aligned} &E[W(t)] - E[W(0)] \\ &\le \frac{c}{\vartheta}\left(e^{\vartheta t} - 1\right) + \int_0^t \sum_{j=1}^{N}\left[h_j(\sigma)N_j(\xi_j(\sigma)) + 1\right]\dot{\xi}_j(\sigma)e^{\vartheta\sigma}\mathrm{d}\sigma, \end{aligned}$$

which implies that

$$E\left[V\left(t\right)\right] \leq \frac{c}{\vartheta}\left(1-e^{-\vartheta t}\right)+\int_0^t \sum_{j=1}^{N} h_j\left(\sigma\right) N_j\left(\xi_j\left(\sigma\right)\right)\dot{\xi}_j\left(\sigma\right) e^{-\vartheta(t-\sigma)}\mathrm{d}\sigma$$
$$+\int_0^t \sum_{j=1}^{N} \dot{\xi}_j\left(\sigma\right) e^{-\vartheta(t-\sigma)}\mathrm{d}\sigma + E\left[V\left(0\right)\right]\cdot e^{-\vartheta t}.$$

In the following, we prove the boundedness of $\left\{\xi_j(t)\right\}_{j=1}^{N}$ by contradiction. Suppose that $\xi_1, \ldots, \xi_\delta$ are unbounded and $\xi_{\delta+1}, \ldots, \xi_N$ are bounded, then the above inequality implies

$$E\left[V\left(t\right)\right] \leq E\left[V\left(0\right)\right]\cdot e^{-\vartheta t}+\frac{c}{\vartheta}\left(1-e^{-\vartheta t}\right)+Y\left(t\right)+c_1$$

with

$$c_1 = \sum_{j=\delta+1}^{N}\int_0^t h_j\left(\sigma\right) N_j\left(\xi_j\left(\sigma\right)\right)\dot{\xi}_j\left(\sigma\right) e^{-\vartheta(t-\sigma)}\mathrm{d}\sigma$$
$$+\sum_{j=\delta+1}^{N}\int_0^t e^{-\vartheta(t-\sigma)}\dot{\xi}_j\left(\sigma\right)\mathrm{d}\sigma + c_0 \tag{5.7}$$

being bounded and

$$Y\left(t\right) = \int_0^t \sum_{j=1}^{\delta}\left|h_j\right| \operatorname{sign}\left(h_j\right)\left(\sigma\right) N_j\left(\xi_j\left(\sigma\right)\right)\dot{\xi}_j\left(\sigma\right) e^{-\vartheta(t-\sigma)}\mathrm{d}\sigma$$
$$+\int_0^t \sum_{j=1}^{\delta}\dot{\xi}_j\left(\sigma\right) e^{-\vartheta(t-\sigma)}\mathrm{d}\sigma \tag{5.8}$$

from which one can easily observe that the boundedness (or unboundedness) of ξ_j implies the boundedness (or unboundedness) of $Y(t)$. Accordingly, we investigate the following two cases.

1. $\xi_j\left(t\right)$ has no upper bound for $j = 1, \ldots, \delta$.
   First, consider the case that $\xi_1, \ldots, \xi_\delta$ do not have upper bound on $\left[0, t_f\right)$. To discuss the boundedness of ξ_j, define three monotonically increasing sequences $\left\{\Xi_{1,j,r}\right\}_{r=1}^{\infty}$, $\left\{\Xi_{2,j,r}\right\}_{r=1}^{\infty}$, and $\left\{\Xi_{3,j,r}\right\}_{r=1}^{\infty}$ as

$$\Xi_{1,j,r} = \frac{(4r-1)\pi}{2^j}+v_j\pi$$
$$\Xi_{2,j,r} = \frac{(4r-1)\pi}{2^j}+v_j\pi+\frac{\theta_j}{2^{j-1}}$$
$$\Xi_{3,j,r} = \frac{(4r+1)\pi}{2^j}+v_j\pi-\frac{\theta_j}{2^{j-1}}$$

where $0 \leq \theta_j \leq \frac{\pi}{2}$ and $v_j = \sum_{k=1}^{j} \left(2^{-k} + 2^{-k}\mathrm{sign}\left(h_k\right)\right) \geq 0$, which satisfies

$$
\begin{aligned}
2^{j-1}v_j &= 2^{j-1}\sum_{k}^{j} \left(2^{-k} + 2^{-k}\mathrm{sign}\left(h_k\right)\right) \\
&= 2^{j-1}\sum_{k=1}^{j-1} \left[2^{-k} + 2^{-k}\mathrm{sign}\left(h_k\right)\right] + \frac{1}{2}\left[1 + \mathrm{sign}\left(h_j\right)\right].
\end{aligned} \tag{5.9}
$$

Based on (5.9), one knows that $2^{j-1}v_j$ is a positive odd integer when $\mathrm{sign}\left(h_j\right) = 1$, and $2^{j-1}v_j$ is a positive even integer when $\mathrm{sign}\left(h_j\right) = -1$. Accordingly, define the increasing time sequences $\left\{t_{1,j,r}\right\}_{r=1}^{\infty}$, $\left\{t_{2,j,r}\right\}_{r=1}^{\infty}$, and $\left\{t_{3,j,r}\right\}_{r=1}^{\infty}$ as $\xi_j\left(t_{1,j,r}\right) = \Xi_{1,j,r}$, $\xi_j\left(t_{2,j,r}\right) = \Xi_{2,j,r}$, and $\xi_j\left(t_{3,j,r}\right) = \Xi_{3,j,r}$, respectively. From the above definitions, one can ensure that $\lim_{r\to\infty} t_{1,j,r} = \lim_{r\to\infty} t_{2,j,r} = \lim_{r\to\infty} t_{3,j,r} = t_f$ and $\lim_{r\to\infty} \Xi_{1,j,r} = \lim_{r\to\infty} \Xi_{2,j,r} = \lim_{r\to\infty} \Xi_{3,j,r} = \infty$ holds.

Next, one can obtain

$$
\begin{aligned}
Y(t) \leq& \sum_{j=1}^{\delta}\int_{t_{2,j,r}}^{t_{3,j,r}} \left|h_j\right| \mathrm{sign}\left(h_j\right) N_j\left(\xi_j\right)\dot{\xi}_j e^{-\vartheta\left(t_{3,j,r}-\sigma\right)}\mathrm{d}\sigma \\
&+\sum_{j=1}^{\delta}\int_{t_{1,j,r}}^{t_{2,j,r}} \left|h_j\right| h_j N_j\left(\xi_j\right)\dot{\xi}_j e^{-\vartheta\left(t_{3,j,r}-\sigma\right)}\mathrm{d}\sigma \\
&+\sum_{j=1}^{\delta}\int_{0}^{t_{1,j,r}} \left|h_j\right| \mathrm{sign}\left(h_j\right) N_j\left(\xi_j\right)\dot{\xi}_j e^{-\vartheta\left(t_{3,j,r}-\sigma\right)}\mathrm{d}\sigma \\
&+\sum_{j=1}^{\delta}\int_{0}^{t_{3,j,r}} e^{-\vartheta\left(t_{3,j,r}-\sigma\right)}\dot{\xi}_j\mathrm{d}\sigma + c_1.
\end{aligned} \tag{5.10}
$$

By integrating over the interval $\xi_j \in \left[0, \Xi_{1,j,r}\right]$, we can get

$$
\begin{aligned}
&\int_0^{\Xi_1} \mathrm{sign}\left(h_j\right) N_j\left(\xi_j\right)\mathrm{d}\xi_j \\
&= \mathrm{sign}\left(h_j\right)\left(\cosh\left(\Xi_{1,j,r}^2\right) - 1\right)\cos\left(2r\pi - \frac{1}{2}\pi + 2^{j-1}v_j\pi\right),
\end{aligned}
$$

where $2^{j-1}v_j$ is a positive integer from the discussion of (5.9). Then, one can obtain $\int_0^{\Xi_1} \mathrm{sign}\left(h_j\right) N_j\left(\xi_j\right)\mathrm{d}\xi_j = 0$. Similar to the above inequality, one has $\int_0^{\frac{(4r+1)\pi}{2^i}+v_j\pi} \mathrm{sign}\left(h_j\right) N_j\left(\xi_j\right)\mathrm{d}\xi_j = 0$ and

$$
\begin{aligned}
&\int_0^{\frac{r\pi}{2^{j-2}}+v_j\pi} \operatorname{sign}\left(h_j\right) N_j\left(\xi_j\right) \mathrm{d}\xi_j \\
&= \operatorname{sign}\left(h_j\right) \cos\left(2^{j-1} v_j \pi\right)\left(\cosh\left(\left(\frac{r\pi}{2^{j-2}}+v_j\pi\right)^2\right)-1\right) \\
&< 0.
\end{aligned}
$$

Define the intervals $\Omega_{j,0}$, $\Omega_{j,1}$, $\Omega_{j,2}$, …, $\Omega_{j,2r}$ as

$$
\begin{aligned}
\Omega_{j,0} &= \left[v_j\pi, \frac{\pi}{2^j}+v_j\pi\right], \\
\Omega_{j,1} &= \left[\frac{\pi}{2^j}+v_j\pi, \frac{3\pi}{2^j}+v_j\pi\right], \\
\Omega_{j,2} &= \left[\frac{3\pi}{2^j}+v_j\pi, \frac{5\pi}{2^j}+v_j\pi\right], \\
&\vdots \\
\Omega_{j,2r} &= \left[\frac{(4r-1)\pi}{2^j}+v_j\pi, \frac{(4r+1)\pi}{2^i}+v_j\pi\right].
\end{aligned}
\tag{5.11}
$$

Define $\underline{\iota}_0$, $\underline{\iota}$, $\iota_{j,0}, \iota_{j,1}, \iota_{j,2}, \ldots, \iota_{j,2r}$ as the instants satisfying $\xi_j\left(\underline{\iota}_0\right)=\frac{\pi}{2}$, $\xi_j\left(\underline{\iota}\right)=v_j\pi$, $\xi_j\left(\iota_{j,0}\right)=\frac{\pi}{2^j}+v_j\pi$, $\xi_j\left(\iota_{j,1}\right)=\frac{3\pi}{2^j}+v_j\pi$, $\xi_j\left(\iota_{j,2}\right)=\frac{5\pi}{2^j}+v_j\pi$, $\cdots$, $\xi_j\left(\iota_{j,2r}\right)=\frac{(4r+1)\pi}{2^i}+v_j\pi$. Based on the properties of $cos(\cdot)$, one can know that in the interval $\Omega_{j,2r}$, the term $\int_0^{\xi_j} \operatorname{sign}\left(h_j\right) N_j(\sigma)\,\mathrm{d}\sigma$ should be non-positive. Therefore, the first term of $Y(t)$ can also be represented as

$$
\begin{aligned}
&\sum_{j=1}^{\delta} \int_{t_{2,j,r}}^{t_{3,j,r}} h_j N_j\left(\xi_j\right) \dot{\xi}_j e^{-\delta\left(t_{3,j,r}-\sigma\right)} \mathrm{d}\sigma \\
&\leq \sum_{j=1}^{\delta} e^{-\vartheta\left(t_{3,j,r}-t_{2,j,r}\right)} \int_{\Xi_{2,j,r}}^{\Xi_{3,j,r}} \left|h_j\right| \operatorname{sign}\left(h_j\right) N_j\left(\xi_j\right) \mathrm{d}\xi_j \\
&\leq \sum_{j=1}^{\delta} H_{\min} e^{-\vartheta\left(t_{3,j,r}-t_{2,j,r}\right)} \int_{\Xi_{2,j,r}}^{\Xi_{3,j,r}} \operatorname{sign}\left(h_j\right) N_j\left(\xi_j\right) \mathrm{d}\xi_j \\
&= -\sum_{j=1}^{\delta} H_{\min} e^{-\vartheta\left(t_{3,j,r}-t_{2,j,r}\right)} \sin\left(\theta_j\right) \\
&\quad \cdot \left(\cosh\left(\Xi_{3,j,r}^2\right)+\cosh\left(\Xi_{2,j,r}^2\right)\right)
\end{aligned}
\tag{5.12}
$$

with $H_{\min}=\min_j h_{j,\min}$.

Furthermore, the second part of $Y(t)$ is

$$\begin{aligned}
&\sum_{j=1}^{\delta}\int_{t_{1,j,r}}^{t_{2,j,r}} h_j N_j\left(\xi_j\right)\dot{\xi}_j e^{-\vartheta\left(t_{3,j,r}-\sigma\right)}\mathrm{d}\sigma \\
&\leq H_{\min}\sum_{j=1}^{\delta} e^{-\vartheta\left(t_{2,j,r}-t_{1,j,r}\right)}\int_{\Xi_{1,j,r}}^{\Xi_{2,j,r}} \operatorname{sign}\left(h_j\right) N_j\left(\xi_j\right)\mathrm{d}\xi_j \\
&\leq 0.
\end{aligned} \tag{5.13}$$

Similar to discussion on the interval $\Omega_{j,2r}$, one can get

$$\int_{\iota_{j,w-1}}^{\iota_{j,w}} \operatorname{sign}\left(h_j\right) N_j\left(\xi_j\right)\dot{\xi}_j e^{-\vartheta\left(\iota_{j,w}-\sigma\right)}\mathrm{d}\sigma \leq 0 \tag{5.14}$$

for $w = 1, 2, \ldots, 2r-1$ on the interval $\left[\iota_{j,w-1}, \iota_{j,w}\right]$. Additionally, one has $\int_0^{\frac{\pi}{2}} \operatorname{sign}\left(h_j\right) N_j\left(\xi_j\right)\mathrm{d}\xi_j = 0$, which implies

$$\begin{aligned}
&\int_0^{\underline{\iota}_0} \operatorname{sign}\left(h_j\right) N_j\left(\xi_j\right)\dot{\xi}_j e^{-\vartheta\left(\underline{\iota}-\sigma\right)}\mathrm{d}\sigma \\
&\leq \int_0^{\underline{\iota}_0} \operatorname{sign}\left(h_j\right) N_j\left(\xi_j\right)\dot{\xi}_j\mathrm{d}\sigma \\
&= 0
\end{aligned} \tag{5.15}$$

if $h_j > 0$ and

$$\begin{aligned}
&\int_0^{\underline{\iota}_0} \operatorname{sign}\left(h_j\right) N_j\left(\xi_j\right)\dot{\xi}_j e^{-\vartheta\left(\underline{\iota}-\sigma\right)}\mathrm{d}\sigma \\
&\leq e^{-\vartheta\left(\underline{\iota}-\underline{\iota}_0\right)}\int_0^{\underline{\iota}_0} \operatorname{sign}\left(h_j\right) N_j\left(\xi_j\right)\dot{\xi}_j\mathrm{d}\sigma \\
&= 0
\end{aligned}$$

if $h_j < 0$. With these facts, one has

$$\int_0^{\underline{\iota}} \operatorname{sign}\left(h_j\right) N_j\left(\xi_j\right)\dot{\xi}_j e^{-\vartheta\left(\underline{\iota}-\sigma\right)}\mathrm{d}\sigma \leq 0.$$

Therefore, the third part of $Y(t)$ satisfies

$$\begin{aligned}
&\sum_{j=1}^{\delta}\int_{0}^{t_{1,j,r}} h_j N_j\left(\xi_j\right)\dot{\xi}_j e^{-\vartheta\left(t_{3,j,r}-\sigma\right)}\mathrm{d}\sigma \\
&\leq \sum_{w=1}^{2r-1}\sum_{j=1}^{\delta} h_{j,\min}\int_{\iota_{j,w-1}}^{\iota_{j,w}} \operatorname{sign}\left(h_j\right) N_j\left(\xi_j\right)\dot{\xi}_j e^{-\vartheta\left(\iota_{j,w}-\sigma\right)}\mathrm{d}\sigma \\
&\quad + \sum_{j=1}^{\delta} h_{j,\min}\int_{0}^{\underline{\iota}} \operatorname{sign}\left(h_j\right) N_j\left(\xi_j\right) e^{-\vartheta\left(v_j\pi-\sigma\right)}\mathrm{d}\sigma .
\end{aligned} \tag{5.16}$$

Based on the above analysis, one has

$$\sum_{j=1}^{\delta}\int_{0}^{t_{1,j,r}} \left|h_j\right| \operatorname{sign}\left(h_j\right) N_j\left(\xi_j\right)\dot{\xi}_j e^{-\vartheta\left(t_{3,j,r}-\sigma\right)}\mathrm{d}\sigma \leq 0.$$

Collecting the results in (5.12), (5.13), and

$$\begin{aligned}
&\sum_{j=1}^{\delta}\int_{0}^{t_{1,j,r}} \left|h_j\right| \operatorname{sign}\left(h_j\right) N_j\left(\xi_j\right)\dot{\xi}_j e^{-\vartheta\left(t_{3,j,r}-\sigma\right)}\mathrm{d}\sigma \leq 0, \\
&\sum_{j=1}^{\delta}\int_{0}^{t_{3,j,r}} e^{-\vartheta\left(t_{3,j,r}-\sigma\right)}\dot{\xi}_j\mathrm{d}\sigma \leq \sum_{j=1}^{\delta}\Xi_{3,j,r} - \sum_{j=1}^{\delta}\xi_j(0)
\end{aligned}$$

the inequality in (5.10) further implies

$$\begin{aligned}
Y(t) \leq & -H_{\min}\sum_{j=1}^{\delta}\sin\left(\theta_j\right)\left(\cosh\left(\Xi_{3,j,r}^2\right)+\cosh\left(\Xi_{2,j,r}^2\right)\right) \\
& + \sum_{j=1}^{\delta}\Xi_{3,j,r} - \sum_{j=1}^{\delta}\xi_j(0) + c_1
\end{aligned}$$

which is equivalent to

$$Y(t) \leq -\sum_{j=1}^{\delta} c_3\cosh\left(\Xi_{3,j,r}^2\right)\left(1+\frac{\cosh\left(\Xi_{2,j,r}^2\right)}{\cosh\left(\Xi_{3,j,r}^2\right)} - \frac{\Xi_{3,j,r}}{c_3\cosh\left(\Xi_{3,j,r}^2\right)}\right) + c_2 \tag{5.17}$$

with $c_2 = -\sum_{i=1}^{\delta} \xi_i(0) + c_1$ and $c_3 = H_{\min} \sin(\theta_j)$. It can be observed that $\lim_{t \to t_f} Y(t) = -\infty$, which contradicts with the fact that $Y(t) \geq 0$. Therefore, the terms $\xi_1, \ldots, \xi_\delta$ are upper-bounded on $[0, t_f)$.

2. $\xi_j(t)$ has no lower bound for $j = 1, \ldots, \delta$.
   Consider the case that $\xi_1, \ldots, \xi_\delta$ do not have lower bound on $[0, t_f)$, which is equivalent to $-\xi_1, \ldots, -\xi_\delta$ do not have upper bound on $[0, t_f)$. Since $\int_0^{\xi_j} \operatorname{sign}(h_j) N_j(\tau) d\tau$ is a even function, the term $Y(t)$ defined in (5.8) further satisfies

$$Y(t) \leq -\sum_{j=1}^{\delta} d_3 \cosh\left(\Xi_{3,j,r}^2\right) \left(1 + \frac{\cosh\left(\Xi_{2,j,r}^2\right)}{\cosh\left(\Xi_{3,j,r}^2\right)} + \frac{\Xi_{3,j,r}}{c_3 \cosh\left(\Xi_{3,j,r}^2\right)}\right) + d_2 \tag{5.18}$$

with

$$d_1 = \sum_{j=\delta+1}^{M} \int_0^t h_j(\sigma) N_j(\xi_j(\sigma)) \dot{\xi}_j(\sigma) e^{-\vartheta(t-\sigma)} \mathrm{d}\sigma + c_0 - \sum_{j=\delta+1}^{M} \int_0^t e^{-\vartheta(t-\sigma)} \dot{\xi}_j(\sigma) \, \mathrm{d}\sigma \tag{5.19}$$

$d_2 = \sum_{j=1}^{\delta} \xi_j(0) + d_1$, and $d_3 = H_{\min} e^{-\vartheta(t_{3,j,r} - t_{2,j,r})} \sin(\theta_j)$. From (5.18), one can observe that $\lim_{t \to t_f} Y(t) = -\infty$, which contradicts with the fact that $Y(t) \geq 0$. Therefore, the terms $\xi_1, \ldots, \xi_\delta$ are lower-bounded on $[0, t_f)$.
   Collecting the above discussions, one can conclude that the signal $\xi_j(t)$ is bounded on the interval $[0, T)$ for $j = 1, \ldots, \delta$. Accordingly, the term $V(t)$ is also bounded on the interval $[0, T)$.

Consider the fact that the term $\sum_{j=1}^{N} \left[h_j(t) N_j(\xi_j) + 1\right] \dot{\xi}_j$ is integrable, which implies that $\left|\sum_{j=1}^{N} \left[h_j(t) N_j(\xi_j) + 1\right] \dot{\xi}_j\right| \leq N_{\max}$ with the constant N_{max}. Furthermore, we can rewrite (5.3) as $\mathcal{L}V(t) \leq -aV^p(t) - bV^q(t) + l$. In what follows, we investigate the settling time and convergence set through two cases.

1. The inequality $\mathcal{L}V(t) \leq -aV^p(t) - bV^q(t) + l$ is equivalent to

$$\mathcal{L}V(t) \leq -a\kappa V^p(t) - a(1-\kappa) V^p(t) - bV^q(t) + l.$$

Denote

$$\begin{aligned}\Omega_a &= \{x|V(x)=\phi\},\\ \Omega_{a+} &= \{x|V(x)>\phi\},\\ \Omega_{a-} &= \{x|V(x)<\phi\},\\ \Omega_s &= \{x|V(x)=1\},\end{aligned}$$

with $\phi=\left(\frac{l}{a(1-\kappa)}\right)^{\frac{1}{p}}$. Then, traveling time from the initial time to the set Ω_a and Ω_s can be denoted as

$$\begin{aligned}T_\phi &= \{E(t)|V(x(t))=\phi\}\\ T_s &= \{E(t)|V(x(t))=1\}.\end{aligned}$$

First, when $x\in\Omega_{a+}$, one has

$$\mathcal{L}V(t)\le -a\kappa V^p(t)-bV^q(t) \tag{5.20}$$

which indicates that the state x in Ω_{a+} would move toward Ω_a. In addition, considering the case when $x\in\Omega_a$, one has

$$\mathcal{L}V(t)\le -a\kappa V^p(t)-bV^q(t)<0. \tag{5.21}$$

Based on LaSalle's invariance principle, $x(t)$ would move toward the interior of set Ω_{a-}. Note that Ω_{a-} is a compact set and $\mathcal{L}V(x(t))<0$ on its boundary. Then, the set Ω_{a-} is forward invariant.

Next, we discuss the settling time for the state x converging to the set Ω_a. First, when $\phi=\left(\frac{l}{a(1-\kappa)}\right)^{\frac{1}{p}}\ge 1$, one has

$$\mathcal{L}V(t)\le -bV^q(t). \tag{5.22}$$

The settling time for the state traveling from the set Ω_s to the set Ω_a is

$$E[T]=T_\phi\le\frac{\phi^{1-q}}{b(q-1)}. \tag{5.23}$$

When $\phi=\left(\frac{l}{a(1-\kappa)}\right)^{\frac{1}{p}}<1$, from the above result, the traveling time T_s from the initial condition to the set Ω_s is calculated as $\frac{1}{b(q-1)}$. However, when the state travels from the set Ω_s to the set Ω_a, one has $-bV^q<0$ and $\mathcal{L}V\le -a\kappa V^p$. Therefore, the settling time satisfies

$$E\left[T\right]=T_{\phi}\leq\frac{1}{b\left(q-1\right)}+\frac{1}{a\kappa\left(1-p\right)}-\frac{\phi^{1-p}}{a\kappa\left(1-p\right)}. \tag{5.24}$$

2. The inequality $\mathcal{L}V\left(t\right)\leq-aV^{p}\left(t\right)-bV^{q}\left(t\right)+l$ is equivalent to

$$\mathcal{L}V\left(t\right)\leq-aV^{p}\left(t\right)-b\kappa V^{q}\left(t\right)-b\left(1-\kappa\right)V^{q}\left(t\right)+l.$$

Denote

$$\begin{aligned}\Omega_{b}&=\left\{x|V\left(x\right)=\Lambda\right\},\\ \Omega_{b+}&=\left\{x|V\left(x\right)>\Lambda\right\},\\ \Omega_{b-}&=\left\{x|V\left(x\right)<\Lambda\right\},\end{aligned}$$

with $\Lambda=\left(\frac{l}{b(1-\kappa)}\right)^{\frac{1}{q}}$. Then, the traveling time T_{Λ} from the initial time to Ω_{Λ} is

$$T_{\Lambda}=\left\{E\left(t\right)|\,V\left(x\left(t\right)\right)=\Lambda\right\}.$$

Similar to the previous case, the set Ω_{b-} is forward invariant and the trajectory starting outside Ω_{b-} will converge to the set Ω_{b-} in fixed time, with the settling time satisfying

$$E\left[T\right]=T_{\Lambda}\leq\frac{\Lambda^{1-q}}{b\kappa\left(q-1\right)}$$

for $\Lambda\geq1$, and

$$E\left[T\right]=T_{\Lambda}\leq\frac{1}{a\left(1-p\right)}-\frac{\Lambda^{1-p}}{a\left(1-p\right)}+\frac{1}{b\kappa\left(q-1\right)}$$

for $\Lambda<1$.

Accordingly, by collecting the results in the above two cases and Definitions 1.6–1.8, the solution to the stochastic system is globally stochastically finite-time stable. The state x can be guaranteed to converge to the set Ω_{ab} within fixed settling time $E\left[T\right]\leq T_{\max}$, where

$$\begin{aligned}\Omega_{ab}&=\left\{x|\,V\left(x\right)\in\left(0,\min\left\{\left(\frac{l}{a\left(1-\kappa\right)}\right)^{\frac{1}{p}},\left(\frac{l}{b\left(1-\kappa\right)}\right)^{\frac{1}{q}}\right\}\right]\right\},\\ T_{\max}&=\max\left\{\frac{1}{a\kappa\left(1-p\right)}+\frac{1}{b\left(q-1\right)},\frac{1}{a\left(1-p\right)}+\frac{1}{b\kappa\left(q-1\right)}\right\}.\end{aligned}$$

This completes the proof. □

Remark 5.1 It is worthwhile to highlight the notable differences between Lemma 5.1 and Lemma 1 in [26]. First, the work in [26] focuses on deterministic systems, whereas the current work aims to deal with stochastic systems. Second, the Nussbaum design in [26] is essentially the single Nussbaum-type gain, whereas the Nussbaum method in this chapter is a multiple Nussbaum-type gain design, as shown in the summation term in Lemma 5.1. Moreover, the conventional Nussbaum-type functions could not be directly added together because of the issue of mutual cancellation. In contrast, with the multiple Nussbaum design in this chapter, the Nussbaum-type functions could be added together for the Lyapunov stability analysis while avoiding mutual cancellation.

5.3 Fixed-Time Nussbaum Design

5.3.1 Step 1

For the first subsystem, consider the tracking error $e_1 = x_1 - x_d$. Then, one has

$$de_1 = \left[h_1(x_1)e_2 + h_1(x_1)s_1 + h_1(x_1)\tilde{\alpha}_1 + f_1(x_1) - \dot{x}_d\right]dt + g_1(x_1)\,d\omega. \tag{5.25}$$

The virtual control input is designed as

$$s_1 = \frac{1}{k_{g1}}N_1(\xi_1)\left(k_{1,1}e_1 + k_{1,2}e_1^5 + \frac{\hat{\theta}_1}{2\rho_1}e_1^3\varphi_1^{\mathrm{T}}(x_1)\varphi_1(x_1) - \dot{x}_d\right) \tag{5.26}$$

with the adaptive Nussbaum-type gain

$$\dot{\xi}_1 = k_{g1}\left(k_{1,1}e_1^4 + k_{1,2}e_1^8 + \frac{\hat{\theta}_1}{2\rho_1}e_1^6\varphi_1^{\mathrm{T}}(x_1)\varphi_1(x_1) - e_1^3\dot{x}_d\right),$$

and k_{g1} being a positive constant. The conventional DSC filter can efficiently deal with the issue of repeated differentiation of virtual control. However, the filtered error cannot converge in a fixed settling time. To ensure the filter design convergence in a fixed settling time, a novel nonlinear DSC filter is considered as

$$\lambda_1\dot{\alpha}_1 = (s_1 - \alpha_1) + (s_1 - \alpha_1)^3, \tag{5.27}$$

where $\tilde{\alpha}_1 = \alpha_1 - s_1$ denotes the filter error. Define

$$F_1(x_1, e_1) = \frac{9}{4}e_1 + \frac{3}{4\eta_1}e_1\|g_1\|^4 + f_1(x_1), \tag{5.28}$$

one has $F_1(x_1, e_1) = W_1^{*\mathrm{T}}\varphi_1(x_1) + \varepsilon_1(x_1, e_1)$, which can be approximated by the FLS $\hat{F}_1(x_1, e_1) = W_1^{\mathrm{T}}(t)\varphi_1(x_1, e_1)$ with residual $\varepsilon_1(x_1, e_1)$. Furthermore, denote

$\theta_1 = \left\| W_1^* \right\|^2$, which is not available owing to the unknown functions $f_1(\cdot)$ and $g_1(\cdot)$. Then, we design the adaptive law to compensate for the unknown functions as

$$\dot{\hat{\theta}}_1 = \Gamma_1 \left[\frac{1}{2\rho_1} e_1^6 \varphi_1^{\mathrm{T}}(x_1)\varphi_1(x_1) - \gamma_{1,1}\hat{\theta}_1 - \gamma_{1,2}\hat{\theta}_1^3 \right], \tag{5.29}$$

with $\tilde{\theta}_1 = \theta_1 - \hat{\theta}_1$.

Consider the Lyapunov candidate for (5.25) as $V_1(t) = \frac{1}{4}e_1(t)^4 + \frac{1}{2}\tilde{\alpha}_1(t)^2 + \frac{1}{2}\tilde{\theta}_1(t)^{\mathrm{T}}\Gamma_1^{-1}\tilde{\theta}_1(t)$. Accordingly, one has

$$\begin{aligned} \mathcal{L}V_1(t) = {} & e_1^3\left[h_1(x_1)e_2 + h_1(x_1)s_1 + h_1(x_1)\tilde{\alpha}_1 + f_1(x_1) - \dot{x}_d\right] \\ & + \tilde{\alpha}_1\dot{\tilde{\alpha}}_1 - \tilde{\theta}_1^{\mathrm{T}}\Gamma_1^{-1}\dot{\hat{\theta}}_1 + \frac{3}{2}e_1^2 g_1^{\mathrm{T}} g_1. \end{aligned}$$

From Lemma 1.6, one has $e_1^2 g_1^{\mathrm{T}} g_1 \le \frac{1}{2\eta_1}e_1^4\|g_1\|^4 + \frac{\eta_1}{2}$ with a positive constant η_1. Inserting this relation into $\mathcal{L}V_1(t)$ yields

$$\begin{aligned} \mathcal{L}V_1(t) \le {} & e_1^3 h_1(x_1)e_2 + e_1^3 h_1(x_1)s_1 + e_1^3 h_1(x_1)\tilde{\alpha}_1 + e_1^3 f_1(x_1) - e_1^3\dot{x}_d \\ & - \tilde{\theta}_1^{\mathrm{T}}\Gamma_1^{-1}\dot{\hat{\theta}}_1 + \frac{3}{4\eta_1}e_1^4\|g_1\|^4 + \frac{3\eta_1}{4} - \frac{1}{\lambda_1}\tilde{\alpha}_1^2 - \frac{1}{\lambda_1}\tilde{\alpha}_1^4 - \tilde{\alpha}_1\dot{s}_1. \end{aligned}$$

According to Lemma 1.6 and Assumption 5.1, one has

$$\begin{aligned} e_1^3 h_1(x_1)\tilde{\alpha}_1 &\le \frac{3}{4}e_1^4 + \frac{1}{4}h_1^4(x_1)\tilde{\alpha}_1^4 \le \frac{3}{4}e_1^4 + \frac{1}{4}\bar{h}_1^4\tilde{\alpha}_1^4, \\ -\tilde{\alpha}_1\dot{s}_1 &\le |\tilde{\alpha}_1\dot{s}_1| \le \frac{1}{2}\left(\tilde{\alpha}_1^2 + \dot{s}_1^2\right) \le \frac{\tilde{\alpha}_1^4}{4} + \frac{\dot{s}_1^2}{2} + \frac{1}{4}, \\ e_1^3 h_1(x_1)e_2 &\le \frac{3}{4}e_1^4 + \frac{1}{4}h_1^4(x_1)e_2^4 \le \frac{3}{4}e_1^4 + \frac{1}{4}\bar{h}_1^4 e_2^4. \end{aligned}$$

Inserting the above into $\mathcal{L}V_1(t)$ yields

$$\begin{aligned} \mathcal{L}V_1(t) \le {} & \frac{1}{k_{g1}^2}h_1(x_1)N_1(\xi_1)\dot{\xi}_1 + \frac{3}{4}e_1^4 + \frac{1}{4}\bar{h}_1^4 e_2^4 + \frac{3}{4}e_1^4 + \frac{1}{4}\bar{h}_1^4\tilde{\alpha}_1^4 \\ & - \frac{9}{4}e_1^4 - e_1^3\dot{x}_d + e_1^3\left[f_1(x_1) + \frac{3}{4\eta_1}e_1\|g_1\|^4 + \frac{9}{4}e_1\right] \\ & - \tilde{\theta}_1^{\mathrm{T}}\Gamma_1^{-1}\dot{\hat{\theta}}_1 + \frac{3\eta_1}{4} - \frac{1}{\lambda_1}\tilde{\alpha}_1^2 - \frac{1}{\lambda_1}\tilde{\alpha}_1^4 + \frac{\tilde{\alpha}_1^4}{4} + \frac{\dot{s}_1^2}{2} + \frac{1}{4}. \end{aligned} \tag{5.30}$$

Consider the function approximation using FLS as in (5.28). Then, $\mathcal{L}V_1(t)$ further satisfies

$$\begin{aligned}\mathcal{L}V_1(t) \leq& \frac{1}{k_{g1}^2} h_1(x_1) N_1(\xi_1) \dot{\xi}_1 + e_1^3 W_1^{*\mathrm{T}} \varphi_1(x_1) \\ &+ e_1^3 \varepsilon_1 + \frac{1}{4} \bar{h}_1^4 e_2^4 - \frac{3}{4} e_1^4 - e_1^3 \dot{x}_d - \tilde{\theta}_1^{\mathrm{T}} \Gamma_1^{-1} \dot{\hat{\theta}}_1 \\ &+ \frac{3\eta_1}{4} - \frac{1}{\lambda_1} \tilde{\alpha}_1^2 + \frac{1}{4} \bar{h}_1^4 \tilde{\alpha}_1^4 - \frac{1}{\lambda_1} \tilde{\alpha}_1^4 + \frac{\tilde{\alpha}_1^4}{4} + \frac{\dot{s}_1^2}{2} + \frac{1}{4}. \end{aligned} \tag{5.31}$$

Moreover, one can obtain the following based on Lemma 1.6 as

$$\begin{aligned} e_1^3 W_1^{*\mathrm{T}} \varphi_1(x_1) &\leq \frac{1}{2\rho_1} e_1^6 \theta_1 \varphi_1^{\mathrm{T}}(x_1) \varphi_1(x_1) + \frac{\rho_1}{2}, \\ e_1^3 \varepsilon_1 &\leq \frac{3}{4} e_1^4 + \frac{1}{4} \varepsilon_1^4. \end{aligned}$$

Inserting the above into (5.31) yields

$$\begin{aligned}\mathcal{L}V_1(t) \leq& \frac{1}{k_{g1}^2} h_1(x_1) N_1(\xi_1) \dot{\xi}_1 + \frac{1}{2\rho_1} e_1^6 \hat{\theta}_1 \varphi_1^{\mathrm{T}}(x_1) \varphi_1(x_1) - e_1^3 \dot{x}_d \\ &+ \frac{1}{4} \bar{h}_1^4 \tilde{\alpha}_1^4 + \frac{\dot{s}_1^2}{2} + \frac{1}{2\rho_1} e_1^6 \tilde{\theta}_1 \varphi_1^{\mathrm{T}}(x_1) \varphi_1(x_1) + \frac{1}{4} \bar{h}_1^4 e_2^4 - \tilde{\theta}_1^{\mathrm{T}} \Gamma_1^{-1} \dot{\hat{\theta}}_1 \\ &+ \frac{3\eta_1}{4} - \frac{1}{\lambda_1} \tilde{\alpha}_1^2 + \frac{\tilde{\alpha}_1^4}{4} + \frac{1}{4} - \frac{1}{\lambda_1} \tilde{\alpha}_1^4 + \frac{\rho_1}{2} + \frac{1}{4} \varepsilon_1^4. \end{aligned}$$

Recalling the virtual control design s_1 in (5.26), one has

$$\begin{aligned}\mathcal{L}V_1(t) \leq& \frac{1}{k_{g1}} \left[\frac{1}{k_{g1}} h_1(x_1) N_1(\xi_1) + 1 \right] \dot{\xi}_1 - k_{1,1} e_1^4 - k_{1,2} e_1^8 \\ &+ \tilde{\theta}_1 \left[\frac{1}{2\rho_1} e_1^6 \varphi_1^{\mathrm{T}}(x_1) \varphi_1(x_1) - \Gamma_1^{-1} \dot{\hat{\theta}}_1 \right] + \frac{1}{4} \bar{h}_1^4 \tilde{\alpha}_1^4 - \frac{1}{\lambda_1} \tilde{\alpha}_1^4 \\ &+ \frac{\tilde{\alpha}_1^4}{4} + \frac{\dot{s}_1^2}{2} + \frac{1}{4} \bar{h}_1^4 e_2^4 - \frac{1}{\lambda_1} \tilde{\alpha}_1^2 + \frac{1}{4} + \frac{\rho_1}{2} + \frac{1}{4} \varepsilon_1^4 + \frac{3\eta_1}{4}. \end{aligned} \tag{5.32}$$

Substituting the adaptive law into the above inequality, one has

$$\begin{aligned}\mathcal{L}V_1(t) \leq& \frac{1}{k_{g1}} \left[\frac{1}{k_{g1}} h_1(x_1) N_1(\xi_1) + 1 \right] \dot{\xi}_1 - k_{1,1} e_1^4 - k_{1,2} e_1^8 \\ &+ \gamma_{1,1} \tilde{\theta}_1^{\mathrm{T}} \hat{\theta}_1 + \gamma_{1,2} \tilde{\theta}_1^{\mathrm{T}} \hat{\theta}_1^3 + \frac{1}{4} \bar{h}_1^4 e_2^4 - \frac{1}{\lambda_1} \tilde{\alpha}_1^2 \\ &+ \frac{1}{4} \bar{h}_1^4 \tilde{\alpha}_1^4 - \frac{1}{\lambda_1} \tilde{\alpha}_1^4 + \frac{\tilde{\alpha}_1^4}{4} + \frac{\dot{s}_1^2}{2} + \frac{1}{4} + \frac{\rho_1}{2} + \frac{1}{4} \varepsilon_1^4 + \frac{3\eta_1}{4}. \end{aligned} \tag{5.33}$$

5.3.2 Step $i\,(2 \le i \le n-1)$

Consider the tracking error for the ith subsystem as $e_i = x_i - \alpha_{i-1}$. Accordingly, one has

$$\begin{aligned} \mathrm{d}e_i &= \mathrm{d}x_i - \mathrm{d}\alpha_{i-1} \\ &= \left[h_i\left(\bar{x}_i\right)e_{i+1} + f_i\left(\bar{x}_i\right) - \dot{\alpha}_{i-1} + h_i\left(\bar{x}_i\right)\tilde{\alpha}_i + h_i\left(\bar{x}_i\right)s_i\right]dt + g_i\left(\bar{x}_i\right)\mathrm{d}\omega. \end{aligned}$$

The recursive virtual control input s_i is designed as

$$s_i = \frac{1}{k_{gi}}N_i\left(\xi_i\right)\left[k_{i,1}e_i + k_{i,2}e_i^5 + \frac{\hat{\theta}_i}{2\rho_i}e_i^3\varphi_i^{\mathrm{T}}\left(\bar{x}_i\right)\varphi_i\left(\bar{x}_i\right) - \dot{\alpha}_{i-1}\right], \tag{5.34}$$

with the adaptive Nussbaum-type gain as

$$\dot{\xi}_i = k_{gi}\left(k_{i,1}e_i^4 + k_{i,2}e_i^8 + \frac{\hat{\theta}_i}{2\rho_i}e_i^6\varphi_i^{\mathrm{T}}\left(\bar{x}_i\right)\varphi_i\left(\bar{x}_i\right) - e_i^3\dot{\alpha}_{i-1}\right). \tag{5.35}$$

To avoid the repeated differentiation, we design a novel DSC as

$$\lambda_i\dot{\alpha}_i = \left(s_i - \alpha_i\right) + \left(s_i - \alpha_i\right)^3,$$

with the filter error $\tilde{\alpha}_i = \alpha_i - s_i$ satisfying

$$\dot{\tilde{\alpha}}_i = \dot{\alpha}_i - \dot{s}_i = -\frac{1}{\lambda_i}\tilde{\alpha}_i - \frac{1}{\lambda_i}\tilde{\alpha}_i^3 - \dot{s}_i.$$

Consider the Lyapunov candidate $V_i\left(t\right) = \frac{1}{4}e_i^4 + \frac{1}{2}\tilde{\alpha}_i^2 + \frac{1}{2}\tilde{\theta}_i^{\mathrm{T}}\Gamma_i^{-1}\tilde{\theta}_i$. Then,

$$\begin{aligned} \mathcal{L}V_i\left(t\right) &= e_i^3\left[h_i\left(\bar{x}_i\right)e_{i+1} + f_i\left(\bar{x}_i\right) + h_i\left(\bar{x}_i\right)\tilde{\alpha}_i + h_i\left(\bar{x}_i\right)s_i - \dot{s}_{i-1}\right] \\ &\quad + \tilde{\alpha}_i\dot{\tilde{\alpha}}_i - \tilde{\theta}_i^{\mathrm{T}}\Gamma_i^{-1}\dot{\hat{\theta}}_i + \frac{3}{2}e_i^2 g_i^{\mathrm{T}}g_i. \end{aligned}$$

From Young's inequality, $e_i^2 g_i^{\mathrm{T}}g_i \le \frac{1}{2\eta_i}e_i^4\|g_i\|^4 + \frac{\eta_i}{2}$ with $\eta_i > 0$, one obtains

$$\begin{aligned} \mathcal{L}V_i\left(t\right) &\le e_i^3h_i\left(\bar{x}_i\right)e_{i+1} + e_i^3h_i\left(\bar{x}_i\right)s_i + e_i^3f_i\left(\bar{x}_i\right) \\ &\quad + e_i^3h_i\left(\bar{x}_i\right)\tilde{\alpha}_i - e_i^3\dot{\alpha}_{i-1} - \tilde{\theta}_i^{\mathrm{T}}\Gamma_i^{-1}\dot{\hat{\theta}}_i + \frac{3}{4\eta_i}e_i^4\|g_i\|^4 \\ &\quad + \frac{3\eta_i}{4} - \frac{1}{\lambda_i}\tilde{\alpha}_i^2 - \frac{1}{\lambda_i}\tilde{\alpha}_i^4 - \tilde{\alpha}_i\dot{s}_i. \end{aligned}$$

From Assumption 5.1 and Lemma 1.6, one has

$$
\begin{aligned}
e_i^3 h_i(\bar{x}_i)\tilde{\alpha}_i &\le \frac{3}{4}e_i^4 + \frac{1}{4}h_i^4(\bar{x}_i)\tilde{\alpha}_i^4 \le \frac{3}{4}e_i^4 + \frac{1}{4}\bar{h}_i^4\tilde{\alpha}_i^4,\\
-\tilde{\alpha}_i\dot{s}_i &\le |\tilde{\alpha}_i\dot{s}_i| \le \frac{1}{2}\left(\tilde{\alpha}_i^2 + \dot{s}_i^2\right) \le \frac{\tilde{\alpha}_i^4}{4} + \frac{\dot{s}_i^2}{2} + \frac{1}{4},\\
e_i^3 h_i(\bar{x}_i)e_{i+1} &\le \frac{3}{4}e_i^4 + \frac{1}{4}h_i^4(\bar{x}_i)e_{i+1}^4 \le \frac{3}{4}e_i^4 + \frac{1}{4}\bar{h}_i^4 e_{i+1}^4.
\end{aligned}
$$

Inserting the above into $\mathcal{L}V_i(t)$ yields

$$
\begin{aligned}
\mathcal{L}V_i(t) \le{}& \frac{1}{k_{gi}^2}h_i(\bar{x}_i)N_i(\xi_i)\dot{\xi}_i + e_i^3\left[f_i(\bar{x}_i) + \frac{3}{4\eta_i}e_i\|g_i\|^4 + \frac{9}{4}e_i\right]\\
&-\frac{3}{4}e_i^4 - e_i^3\dot{\alpha}_{i-1} + \frac{1}{4}\bar{h}_i^4 e_{i+1}^4 - \tilde{\theta}_i^{\mathrm{T}}\Gamma_i^{-1}\dot{\hat{\theta}}_i + \frac{3\eta_i}{4}\\
&-\frac{1}{\lambda_i}\tilde{\alpha}_i^2 - \frac{1}{\lambda_i}\tilde{\alpha}_i^4 + \frac{1}{4}\bar{h}_i^4\tilde{\alpha}_i^4 + \frac{\tilde{\alpha}_i^4}{4} + \frac{\dot{s}_i^2}{2} + \frac{1}{4}.
\end{aligned}
\tag{5.36}
$$

Define $F_i(\bar{x}_i, e_i) = \frac{3}{4\eta_i}e_i\|g_i\|^4 + f_i(\bar{x}_i) + \frac{9}{4}e_i$, one has $F_i(\bar{x}_i, e_i) = W_i^{*\mathrm{T}}\varphi_i(\bar{x}_i, e_i) + \varepsilon_i(\bar{x}_i, e_i)$ with $\theta_i = \left\|W_i^*\right\|^2$, which is not available because of the unknown functions $f_i(\cdot)$ and $g_i(\cdot)$. Then, one has

$$
\begin{aligned}
\mathcal{L}V_i(t) \le{}& \frac{1}{k_{gi}^2}h_i(\bar{x}_i)N_i(\xi_i)\dot{\xi}_i + e_i^3\left[W_i^{*\mathrm{T}}\varphi_i(\bar{x}_i) + \varepsilon_i\right]\\
&-\frac{3}{4}e_i^4 - e_i^3\dot{\alpha}_{i-1} + \frac{1}{4}\bar{h}_i^4 e_{i+1}^4 - \tilde{\theta}_i^{\mathrm{T}}\Gamma_i^{-1}\dot{\hat{\theta}}_i + \frac{3\eta_i}{4}\\
&-\frac{1}{\lambda_i}\tilde{\alpha}_i^2 - \frac{1}{\lambda_i}\tilde{\alpha}_i^4 + \frac{1}{4}\bar{h}_i^4\tilde{\alpha}_i^4 + \frac{\tilde{\alpha}_i^4}{4} + \frac{\dot{s}_i^2}{2} + \frac{1}{4}.
\end{aligned}
$$

To ensure the fixed-time stability for the adaptive fuzzy control design, we should compensate for the unknown function $F_i(\bar{x}_i, e_i)$ and eliminate the term $-\tilde{\theta}_i^{\mathrm{T}}\Gamma_i^{-1}\dot{\hat{\theta}}_i$ in (5.36). Therefore, we consider adaptive fuzzy learning as

$$
\dot{\hat{\theta}}_i = \Gamma_i\left[\frac{1}{2\rho_i}e_i^6\varphi_i^{\mathrm{T}}(\bar{x}_i)\varphi_i(\bar{x}_i) - \gamma_{i,1}\hat{\theta}_i - \gamma_{i,2}\hat{\theta}_i^3\right],
\tag{5.37}
$$

with the parameter error $\tilde{\theta}_i = \theta_i - \hat{\theta}_i$. Note that

$$
\begin{aligned}
e_i^3 W_i^{*\mathrm{T}}\varphi_i(\bar{x}_i) &\le \frac{1}{2\rho_i}e_i^6\theta_i\varphi_i^{\mathrm{T}}(\bar{x}_i)\varphi_i(\bar{x}_i) + \frac{\rho_i}{2},\\
e_i^3\varepsilon_i &\le \frac{3}{4}e_i^4 + \frac{1}{4}\varepsilon_i^4,
\end{aligned}
\tag{5.38}
$$

which can be obtained using Young's inequality. Inserting the above into $\mathcal{L}V_i(t)$ yields

$$\begin{aligned}\mathcal{L}V_i(t) \leq & \frac{1}{k_{gi}^2} h_i(\bar{x}_i) N_i(\xi_i)\dot{\xi}_i + \frac{1}{2\rho_i} e_i^6 \hat{\theta}_i \varphi_i^{\mathrm{T}}(\bar{x}_i)\varphi_i(\bar{x}_i) - e_i^3 \dot{\alpha}_{i-1} \\ & + \frac{1}{4}\bar{h}_i^4 e_{i+1}^4 - \tilde{\theta}_i^{\mathrm{T}} \Gamma_i^{-1} \dot{\hat{\theta}}_i - \frac{1}{\lambda_i}\tilde{\alpha}_i^2 - \frac{1}{\lambda_i}\tilde{\alpha}_i^4 + \frac{1}{4}\bar{h}_i^4 \tilde{\alpha}_i^4 + \frac{\tilde{\alpha}_i^4}{4} \\ & + \frac{1}{2\rho_i} e_i^6 \tilde{\theta}_i \varphi_i^{\mathrm{T}}(\bar{x}_i)\varphi_i(\bar{x}_i) + \frac{\dot{s}_i^2}{2} + \frac{1}{4} + \frac{3\eta_i}{4} + \frac{\rho_i}{2} + \frac{1}{4}\varepsilon_i^4. \end{aligned} \quad (5.39)$$

By recalling the virtual control input design for s_i in (5.34), one has

$$\begin{aligned}\mathcal{L}V_i(t) \leq & \frac{1}{k_{gi}}\left[\frac{1}{k_{gi}} h_i(\bar{x}_i) N_i(\xi_i) + 1\right]\dot{\xi}_i - k_{i,1}e_i^4 - k_{i,2}e_i^8 + \frac{1}{4}\bar{h}_i^4 e_{i+1}^4 \\ & - \frac{1}{\lambda_i}\tilde{\alpha}_i^2 - \frac{1}{\lambda_i}\tilde{\alpha}_i^4 + \frac{1}{4}\bar{h}_i^4\tilde{\alpha}_i^4 + \frac{\tilde{\alpha}_i^4}{4} + \frac{\dot{s}_i^2}{2} + \frac{1}{4} + \frac{3\eta_i}{4} + \frac{\rho_i}{2} + \frac{1}{4}\varepsilon_i^4 \\ & + \tilde{\theta}_i\left(\frac{1}{2\rho_i} e_i^6 \varphi_i^{\mathrm{T}}(\bar{x}_i)\varphi_i(\bar{x}_i) - \Gamma_i^{-1}\dot{\hat{\theta}}_i\right). \end{aligned}$$

Substituting the adaptive law design (5.37) into the above result yields

$$\begin{aligned}\mathcal{L}V_i(t) \leq & \frac{1}{k_{gi}}\left[\frac{1}{k_{gi}} h_i(\bar{x}_i) N_i(\xi_i) + 1\right]\dot{\xi}_i - k_{i,1}e_i^4 - k_{i,2}e_i^8 \\ & + \gamma_{i,1}\tilde{\theta}_i^{\mathrm{T}}\hat{\theta}_i + \gamma_{i,2}\tilde{\theta}_i^{\mathrm{T}}\hat{\theta}_i^3 + \frac{1}{4}\bar{h}_i^4 e_{i+1}^4 - \frac{1}{\lambda_i}\tilde{\alpha}_i^2 - \frac{1}{\lambda_i}\tilde{\alpha}_i^4 + \frac{1}{4}\bar{h}_i^4\tilde{\alpha}_i^4 \\ & + \frac{\tilde{\alpha}_i^4}{4} + \frac{\dot{s}_i^2}{2} + \frac{1}{4} + \frac{3\eta_i}{4} + \frac{\rho_i}{2} + \frac{1}{4}\varepsilon_i^4. \end{aligned}$$

5.3.3 Step n

For the nth subsystem, we consider the tracking error as $e_n = x_n - \alpha_{n-1}$, with the dynamics

$$\mathrm{d}e_n = \mathrm{d}x_n - \mathrm{d}\alpha_{n-1} = \left[h_n(\bar{x}_n)u + f_n(\bar{x}_n) - \dot{\alpha}_{n-1}\right]dt + g_n(\bar{x}_n)\mathrm{d}\omega.$$

Design the control input signal u as

$$u = \frac{1}{k_{gn}} N_n(\xi_n)\left(k_{n,1}e_n + k_{n,2}e_n^5 + \frac{\hat{\theta}_n}{2\rho_n} e_n^3 \varphi_n^{\mathrm{T}}(\bar{x}_n)\varphi_n(\bar{x}_n) - \dot{\alpha}_{n-1}\right), \quad (5.40)$$

with the adaptive Nussbaum-type gain

$$\dot{\xi}_n = k_{gn}\left(k_{n,1}e_n^4 + k_{n,2}e_n^8 + \frac{\hat{\theta}_n}{2\rho_n}e_n^6\varphi_n^{\mathrm{T}}(\bar{x}_n)\varphi_n(\bar{x}_n) - e_n^3\dot{\alpha}_{n-1}\right). \tag{5.41}$$

Consider the Lyapunov candidate $V_n(t) = \frac{1}{4}e_n^4 + \frac{1}{2}\tilde{\theta}_n^{\mathrm{T}}\Gamma_n^{-1}\tilde{\theta}_n$. Differentiating V_n yields

$$\mathcal{L}V_n(t) = e_n^3\left[h_n(\bar{x}_n)u + f_n(\bar{x}_i) - \dot{\alpha}_{n-1}\right] - \tilde{\theta}_n^{\mathrm{T}}\Gamma_n^{-1}\dot{\hat{\theta}}_n + \frac{3}{2}e_n^2 g_n^{\mathrm{T}} g_n.$$

Further, considering Young's inequality $e_n^2 g_n^{\mathrm{T}} g_n \leq \frac{1}{2\eta_n}e_n^4\|g_n\|^4 + \frac{\eta_n}{2}$ with a positive constant η_n, one obtains

$$\mathcal{L}V_n(t) \leq e_n^3 h_n(\bar{x}_n)u - \tilde{\theta}_n^{\mathrm{T}}\Gamma_n^{-1}\dot{\hat{\theta}}_n + e_n^3 f_n(\bar{x}_n) + \frac{3}{4\eta_n}e_n^4\|g_n\|^4 + \frac{3\eta_n}{4} - e_n^3\dot{\alpha}_{n-1}.$$

Substituting u into the above equation yields

$$\begin{aligned}
\mathcal{L}V_n(t) \leq\ & \frac{1}{k_{gn}^2}h_n(\bar{x}_n)N_n(\xi_n)\dot{\xi}_n + e_n^3 f_n(\bar{x}_n) - e_n^3\dot{\alpha}_{n-1} \\
& -\tilde{\theta}_n^{\mathrm{T}}\Gamma_n^{-1}\dot{\hat{\theta}}_n + \frac{3}{4\eta_n}e_n^4\|g_n\|^4 + \frac{3\eta_n}{4} \\
\leq\ & h_n(\bar{x}_n)N_n(\xi_n)\dot{\xi}_n + e_n^3\left[f_n(\bar{x}_n) + \frac{3}{4\eta_n}e_n^4\|g_n\|^4 + \frac{3}{4}e_n\right] \\
& -\frac{3}{4}e_n^4 - e_n^3\dot{\alpha}_{n-1} - \tilde{\theta}_n^{\mathrm{T}}\Gamma_n^{-1}\dot{\hat{\theta}}_n + \frac{3\eta_n}{4}.
\end{aligned} \tag{5.42}$$

Define $F_n(\bar{x}_n, e_n) = f_n(\bar{x}_n) + \frac{3}{4\eta_n}e_n\|g_n\|^4 + \frac{3}{4}e_n$. With the FLS as the function approximator, one has $F_n(\bar{x}_n, e_n) = W_n^{*\mathrm{T}}\varphi_n(\bar{x}_n, e_n) + \varepsilon_n(\bar{x}_n, e_n)$ with $\theta_n = \|W_n^*\|^2$, which is not available owing to the unknown functions $f_n(\cdot)$ and $g_n(\cdot)$. Then, one obtains

$$\begin{aligned}
\mathcal{L}V_n(t) \leq\ & e_n^3 W_i^{*\mathrm{T}}\varphi_n(\bar{x}_i) + e_n^3\varepsilon_n - \frac{3}{4}e_n^4 - e_n^3\dot{\alpha}_{n-1} - \tilde{\theta}_n^{\mathrm{T}}\Gamma_n^{-1}\dot{\hat{\theta}}_n \\
& + \frac{3\eta_n}{4} + \frac{1}{k_{gn}^2}h_n(\bar{x}_n)N_n(\xi_n)\dot{\xi}_n.
\end{aligned}$$

To resolve this issue, we consider the adaptive law to compensate for the unknown functions as

$$\dot{\hat{\theta}}_n = \Gamma_n\left[\frac{1}{2\rho_n}e_n^6\varphi_n^{\mathrm{T}}(\bar{x}_n)\varphi_n(\bar{x}_n) - \gamma_{n,1}\hat{\theta}_n - \gamma_{n,2}\hat{\theta}_n^3\right], \tag{5.43}$$

with $\tilde{\theta}_n = \theta_n - \hat{\theta}_n$. From Young's inequalities $W_n^{*\mathrm{T}} \cdot \varphi_n \cdot e_n^3 \leq \frac{1}{2\rho_n}\varphi_n^{\mathrm{T}} \cdot \varphi_n \cdot e_n^6 \cdot \theta_n + \frac{1}{2}\rho_n$ and $e_n^3\varepsilon_n \leq \frac{3}{4}e_n^4 + \frac{1}{4}\varepsilon_n^4$, one has

$$\begin{aligned}\mathcal{L}V_n(t) \leq{}& \frac{1}{2\rho_n}e_n^6\hat{\theta}_n\varphi_n^{\mathrm{T}}(\bar{x}_n)\varphi_n(\bar{x}_n) - e_n^3\dot{\alpha}_{n-1}\\&+\frac{1}{k_{gn}^2}h_n(\bar{x}_n)N_n(\xi_n)\dot{\xi}_n + \frac{3\eta_n}{4} + \frac{\rho_n}{2}\\&+\frac{1}{2\rho_n}e_n^6\tilde{\theta}_n\varphi_n^{\mathrm{T}}(\bar{x}_n)\varphi_n(\bar{x}_n) - \tilde{\theta}_n^{\mathrm{T}}\Gamma_n^{-1}\dot{\hat{\theta}}_n + \frac{1}{4}\varepsilon_n^4.\end{aligned}$$

Collecting the control input design (5.40) and the adaptive law design (5.43), the term $\mathcal{L}V_n(t)$ further satisfies

$$\begin{aligned}\mathcal{L}V_n(t) \leq{}& \frac{1}{k_{gn}}\left[\frac{1}{k_{gn}}h_n(\bar{x}_n)N_n(\xi_n) + 1\right]\dot{\xi}_n\\&-k_{n,1}e_n^4 - k_{n,2}e_n^8 + \frac{1}{4}\varepsilon_n^4 + \frac{3\eta_n}{4} + \frac{\rho_n}{2}\\&+\gamma_{n,1}\tilde{\theta}_n^{\mathrm{T}}\hat{\theta}_n + \gamma_{n,2}\tilde{\theta}_n^{\mathrm{T}}\hat{\theta}_n^3.\end{aligned} \tag{5.44}$$

5.4 Fixed-Time Stability Analysis

We investigate the following fundamental Nussbaum-based Lyapunov analysis for a stochastic system.

Theorem 5.1 *Under Assumptions 5.1–5.3, consider the adaptive robust backstepping design from Sect. 5.3. Then, the error signals e_i, adaptive learning errors $\tilde{\theta}_i$, Nussbaum-type gain $\xi_i(t)$, and DSC filter errors $\tilde{\alpha}_i$ remain the solution to the stochastic system is semi-globally uniformly ultimately bounded.*

Proof First, applying Lemma 1.6 yields

$$\begin{aligned}-\tilde{\theta}_1^2 &\leq \frac{1}{3} - \frac{4}{3}\left(\tilde{\theta}_1^2\right)^{\frac{3}{4}},\\-\tilde{\alpha}_1^2 &\leq \frac{1}{3} - \frac{4}{3}\left(\tilde{\alpha}_1^2\right)^{\frac{3}{4}},\\-e_1^4 &\leq \frac{1}{3} - \frac{4}{3}\left(e_1^4\right)^{\frac{3}{4}},\end{aligned} \tag{5.45}$$

and

$$
\begin{aligned}
\tilde{\theta}_1\hat{\theta}_1 &\le \frac{1}{2}\theta_1^2 - \frac{1}{2}\tilde{\theta}_1^2 \\
&\le \frac{1}{2}\theta_1^2 + \frac{1}{6} - \frac{2}{3}\left|\tilde{\theta}_1\right|^{\frac{3}{2}}, \\
\tilde{\theta}_1\hat{\theta}_1^3 &= \tilde{\theta}_1\theta_1^3 - 3\theta_1^2\tilde{\theta}_1^2 + 3\theta_1\tilde{\theta}_1^3 - \tilde{\theta}_1^4 \\
&\le \gamma_{1,3}\frac{\tilde{\theta}_1^4}{4} + \gamma_{1,3}^{-\frac{1}{3}}\frac{3\theta_1^4}{4} + \gamma_{1,3}\frac{3\tilde{\theta}_1^4}{2} + \gamma_{1,3}^{-1}\frac{3\theta_1^4}{2} + \gamma_{1,3}\frac{9\tilde{\theta}_1^4}{4} + \gamma_{1,3}^{-3}\frac{3\theta_1^4}{4} - \tilde{\theta}_1^4 \\
&\le -\left(1 - 4\gamma_{1,3}\right)\tilde{\theta}_1^4 + \left(\frac{3}{4}\gamma_{1,3}^{-\frac{1}{3}} + \frac{3}{2}\gamma_{1,3}^{-1} + \frac{3}{4}\gamma_{1,3}^{-3}\right)\theta_1^4
\end{aligned}
$$

with $\gamma_{1,3} > 0$, $1 - 4\gamma_{1,3} > 0$. From the above results, we have

$$
\begin{aligned}
\gamma_{1,1}\tilde{\theta}_1^{\mathrm{T}}\hat{\theta}_1 + \gamma_{1,2}\tilde{\theta}_1^{\mathrm{T}}\hat{\theta}_1^3 \le\ & \gamma_{1,2}\left[-\left(1 - 4\gamma_{1,3}\right)\tilde{\theta}_1^4 + \left(\frac{3}{4}\gamma_{1,3}^{-\frac{1}{3}} + \frac{3}{2}\gamma_{1,3}^{-1} + \frac{3}{4}\gamma_{1,3}^{-3}\right)\theta_1^4\right] \\
& + \gamma_{1,1}\left(\frac{1}{2}\theta_1^2 + \frac{1}{6} - \frac{2}{3}\left|\tilde{\theta}_1\right|^{\frac{3}{2}}\right) \\
\le\ & -\gamma_{1,2}\left(1 - 4\gamma_{1,3}\right)\tilde{\theta}_1^4 + \gamma_{1,2}\left(\frac{3}{4}\gamma_{1,3}^{-\frac{1}{3}} + \frac{3}{2}\gamma_{1,3}^{-1} + \frac{3}{4}\gamma_{1,3}^{-3}\right)\theta_1^4 \\
& - \frac{2}{3}\gamma_{1,1}\left|\tilde{\theta}_1\right|^{\frac{3}{2}} + \gamma_{1,1}\left(\frac{1}{2}\theta_1^2 + \frac{1}{6}\right).
\end{aligned} \tag{5.46}
$$

Inserting (5.45) and (5.46) into (5.33), one has

$$
\begin{aligned}
\mathcal{L}V_1(t) \le\ & \frac{1}{k_{g1}}\left[\frac{1}{k_{g1}}h_1(x_1)N_1(\xi_1) + 1\right]\dot{\xi}_1 - k_{1,2}e_1^8 + \frac{1}{4}\bar{h}_1^4 e_2^4 \\
& + k_{1,1}\left(\frac{1}{3} - \frac{4}{3}\left(e_1^4\right)^{\frac{3}{4}}\right) - \frac{4}{3\lambda_1}\left(\tilde{\alpha}_1^2\right)^{\frac{3}{4}} - \left(\frac{1}{\lambda_1} - \frac{1}{4} - \frac{\bar{h}_1^4}{4}\right)\tilde{\alpha}_1^4 \\
& - \frac{2}{3}\gamma_{1,1}\left|\tilde{\theta}_1\right|^{\frac{3}{2}} - \gamma_{1,2}\left(1 - 4\gamma_{1,3}\right)\tilde{\theta}_1^4 + \gamma_{1,1}\left(\frac{1}{2}\theta_1^2 + \frac{1}{6}\right) \\
& + \gamma_{1,2}\left(\frac{3}{4}\gamma_{1,3}^{-\frac{1}{3}} + \frac{3}{2}\gamma_{1,3}^{-1} + \frac{3}{4}\gamma_{1,3}^{-3}\right)\theta_1^4 + \frac{1}{4}\varepsilon_1^4 \\
& + \frac{3\eta_1}{4} + \frac{1}{2}\dot{s}_1^2 + \frac{1}{4} + \frac{\rho_1}{2} + \frac{1}{3\lambda_1}.
\end{aligned} \tag{5.47}
$$

Differentiating the virtual control s_1 yields

$$
\begin{aligned}
\dot{s}_1 =\ & \frac{1}{k_{g1}}\dot{N}_1(\xi_1)\dot{\xi}_1\left[k_{1,1}e_1 + k_{1,2}e_1^5 + \frac{\hat{\theta}_1}{2\rho_1}e_1^3\varphi_1^{\mathrm{T}}(x_1)\varphi_1(x_1) - \dot{x}_d\right] \\
& + \frac{1}{2\rho_1 k_{g1}}N_1(\xi_1)\left[\dot{\hat{\theta}}_1 e_1^3\varphi_1^{\mathrm{T}}(\bar{x}_1)\varphi_1(\bar{x}_1) + 3\hat{\theta}_1 e_1^2\dot{e}_1\varphi_1^{\mathrm{T}}(x_1)\varphi_1(x_1)\right. \\
& \left. + 2\hat{\theta}_1 e_1^3\varphi_1^{\mathrm{T}}(x_1)\cdot\dot{\varphi}_1(x_1)\right] + \frac{1}{k_{g1}}N_1(\xi_1)\left(k_{1,1}\dot{e}_1 + 5k_{1,2}e_1^4\dot{e}_1 - \ddot{x}_d\right).
\end{aligned}
$$

Define the sets

$$\Omega_{x_d} = \left\{ (x_d, \dot{x}_d, \ddot{x}_d) |\, x_d^2 + \dot{x}_d^2 + \ddot{x}_d^2 \le R_0 \right\},$$
$$\Omega_1 = \left\{ \left(e_1, \tilde{\theta}_1, \tilde{\alpha}_1\right) \Big| \, \frac{1}{4}e_1^4 + \frac{1}{2}\tilde{\theta}_1^2 + \frac{1}{2}\tilde{\alpha}_1^2 \le R_1 \right\},$$

where $R_0 > 0$ and $R_1 > 0$. On the compact set $\Omega_1 \times \Omega_{x_d}$, $\dot{s}_1$ is continuous and $|\dot{s}_1| \le \chi_1$ with $\chi_1 > 0$. Therefore, (5.47) implies that

$$\begin{aligned}
\mathcal{L}V_1(t) \le{}& \frac{1}{k_{g1}}\left[\frac{1}{k_{g1}}h_1(x_1)N_1(\xi_1)+1\right]\dot{\xi}_1 - k_{1,1}\left(\frac{1}{3}-\frac{4}{3}\left(e_1^4\right)^{\frac{3}{4}}\right)\\
&-\frac{4}{3\lambda_1}\left(\tilde{\alpha}_1^2\right)^{\frac{3}{4}} - k_{1,2}e_1^8 - \left(\frac{1}{\lambda_1}-\frac{1}{4}-\frac{\bar{h}_1^4}{4}\right)\tilde{\alpha}_1^4 + \frac{1}{4}\bar{h}_1^4 e_2^4\\
&-\gamma_{1,2}\left(1-4\gamma_{1,3}\right)\tilde{\theta}_1^4 - \frac{2}{3}\gamma_{1,1}\left|\tilde{\theta}_1\right|^{\frac{3}{2}} + \gamma_{1,1}\left(\frac{1}{2}\theta_1^2+\frac{1}{6}\right)\\
&+\gamma_{1,2}\left(\frac{3}{4}\gamma_{1,3}^{-\frac{1}{3}}+\frac{3}{2}\gamma_{1,3}^{-1}+\frac{3}{4}\gamma_{1,3}^{-3}\right)\theta_1^4\\
&+\frac{1}{4}\varepsilon_1^4+\frac{3\eta_1}{4}+\frac{1}{2}\chi_1^2+\frac{1}{4}+\frac{\rho_1}{2}+\frac{1}{3\lambda_1}.
\end{aligned}$$

Furthermore, $\mathcal{L}V_1(t)$ further satisfies

$$\begin{aligned}
\mathcal{L}V_1(t) \le{}& \frac{1}{k_{g1}}\left[\frac{1}{k_{g1}}h_1(\bar{x}_1)N_1(\xi_1)+1\right]\dot{\xi}_1 - k_{1,2}\cdot 2^2\left(\frac{1}{2}e_1^4\right)^2\\
&+\frac{1}{4}\bar{h}_1^4e_2^4+\frac{1}{4}\varepsilon_1^4+\frac{3\eta_1}{4}-\frac{4^{\frac{7}{4}}}{3}k_{1,1}\left(\frac{1}{4}e_1^4\right)^{\frac{3}{4}}-\frac{4}{3\lambda_1}2^{\frac{3}{4}}\left(\frac{1}{2}\tilde{\alpha}_1^2\right)^{\frac{3}{4}}\\
&-\left(\frac{4}{\lambda_1}-1-\bar{h}_1^4\right)\left(\frac{1}{2}\tilde{\alpha}_1^2\right)^2+\frac{1}{2}\chi_1^2+\frac{\rho_1}{2}-\gamma_{1,1}\Gamma_1\frac{2^{\frac{7}{4}}}{3}\left(\frac{1}{2}\tilde{\theta}_1^{\mathrm{T}}\Gamma_1^{-1}\tilde{\theta}_1\right)^{\frac{3}{4}}\\
&-4\gamma_{1,2}\left(1-4\gamma_{1,3}\right)\Gamma_1\left(\frac{1}{2}\tilde{\theta}_1^{\mathrm{T}}\Gamma_1^{-1}\tilde{\theta}_1\right)^2+\frac{1}{4}+\frac{1}{3\lambda_1}\\
&+\gamma_{1,1}\left(\frac{1}{2}\theta_1^2+\frac{1}{6}\right)+\gamma_{1,2}\left(\frac{3}{4}\gamma_{1,3}^{-\frac{1}{3}}+\frac{3}{2}\gamma_{1,3}^{-1}+\frac{3}{4}\gamma_{1,3}^{-3}\right)\theta_1^4+\frac{1}{3}k_{1,1}.
\end{aligned} \tag{5.48}$$

Considering the virtual control design (5.34), differentiating s_i yields

$$\begin{aligned}
\dot{s}_i ={}& \frac{1}{k_{gi}}\dot{N}_i(\xi_i)\dot{\xi}_i\left[k_{i,1}e_i+k_{i,2}e_i^5+\frac{\hat{\theta}_i}{2\rho_i}e_i^3\varphi_i^{\mathrm{T}}(\bar{x}_i)\varphi_i(\bar{x}_i)-\dot{\alpha}_{i-1}\right]\\
&+\frac{1}{2\rho_ik_{gi}}N_i(\xi_i)\left[\dot{\hat{\theta}}_ie_i^3\varphi_i^{\mathrm{T}}(\bar{x}_i)\varphi_i(\bar{x}_i)+3\hat{\theta}_ie_i^2\dot{e}_i\varphi_i^{\mathrm{T}}(\bar{x}_i)\varphi_i(\bar{x}_i)\right.\\
&\left.+2\hat{\theta}_ie_i^3\varphi_i^{\mathrm{T}}(\bar{x}_i)\cdot\dot{\varphi}_i(\bar{x}_i)\right]+\frac{1}{k_{gi}}N_i(\xi_i)\left(k_{i,1}\dot{e}_i+5k_{i,2}e_i^4\dot{e}_i-\ddot{\alpha}_{i-1}\right).
\end{aligned}$$

Consider the sets

$$\Omega_{x_d} = \left\{ (x_d, \dot{x}_d, \ddot{x}_d) \middle| \, x_d^2 + \dot{x}_d^2 + \ddot{x}_d^2 \le R_0 \right\},$$
$$\Omega_i = \left\{ \left(e_i, \tilde{\theta}_i, \tilde{\alpha}_i\right) \middle| \, \frac{1}{4}e_i^4 + \frac{1}{2}\tilde{\theta}_i^2 + \frac{1}{2}\tilde{\alpha}_i^2 \le R_i \right\},$$

with positive constants $R_0 > 0$ and $R_i > 0$. On the compact set $\Omega_i \times \Omega_{x_d}$, $\dot{s}_i$ is continuous and $|\dot{s}_i| \le \chi_i$ with $\chi_i > 0$. Then, $\mathcal{L}V_i(t)$ further satisfies

$$\begin{aligned}\mathcal{L}V_i(t) \le &\frac{1}{k_{gi}}\left[\frac{1}{k_{gi}}h_i(\bar{x}_i)N_i(\xi_i) + 1\right]\dot{\xi}_i - k_{i,1}e_i^4 - k_{i,2}e_i^8 \\ &+\gamma_{i,1}\tilde{\theta}_i^{\mathrm{T}}\hat{\theta}_i + \gamma_{i,2}\tilde{\theta}_i^{\mathrm{T}}\hat{\theta}_i^3 + \frac{1}{4}\bar{h}_i^4 e_{i+1}^4 - \frac{1}{\lambda_i}\tilde{\alpha}_i^2 - \frac{1}{\lambda_i}\tilde{\alpha}_i^4 + \frac{1}{4}\bar{h}_i^4\tilde{\alpha}_i^4 \\ &+\frac{\tilde{\alpha}_i^4}{4} + \frac{\chi_i^2}{2} + \frac{1}{4} + \frac{3\eta_i}{4} + \frac{\rho_i}{2} + \frac{1}{4}\varepsilon_i^4. \end{aligned} \tag{5.49}$$

From Young's inequality, one can obtain $-\tilde{\alpha}_i^2 \le \frac{1}{3} - \frac{4}{3}(\tilde{\alpha}_i^2)^{\frac{3}{4}}$, $-e_i^4 \le \frac{1}{3} - \frac{4}{3}(e_i^4)^{\frac{3}{4}}$, and

$$\begin{aligned}\gamma_{i,1}\tilde{\theta}_i^{\mathrm{T}}\hat{\theta}_i + \gamma_{i,2}\tilde{\theta}_i\hat{\theta}_i^3 \le &-\frac{2}{3}\gamma_{i,1}\left|\tilde{\theta}_i\right|^{\frac{3}{2}} - \gamma_{i,2}\left(1 - 4\gamma_{i,3}\right)\tilde{\theta}_i^4 + \gamma_{i,1}\left(\frac{1}{2}\theta_i^2 + \frac{1}{6}\right) \\ &+\gamma_{i,2}\left(\frac{3}{4}\gamma_{i,3}^{-\frac{1}{3}} + \frac{3}{2}\gamma_{i,3}^{-1} + \frac{3}{4}\gamma_{i,3}^{-3}\right)\theta_i^4,\end{aligned}$$

with $\gamma_{i,3} > 0$ and $1 - 4\gamma_{i,3} > 0$. Substituting the above two inequalities into (5.49), one obtains

$$\begin{aligned}\mathcal{L}V_i(t) \le &\frac{1}{k_{gi}}\left[\frac{1}{k_{gi}}h_i(\bar{x}_i)N_i(\xi_i) + 1\right]\dot{\xi}_i - \frac{4^{\frac{7}{4}}}{3}k_{i,1}\left(\frac{1}{4}e_i^4\right)^{\frac{3}{4}} \\ &-\frac{2^{\frac{7}{4}}}{3}\gamma_{i,1}\Gamma_i\left(\frac{1}{2}\tilde{\theta}_i^{\mathrm{T}}\Gamma_i^{-1}\tilde{\theta}_i\right)^{\frac{3}{4}} - \frac{2^{\frac{11}{4}}}{3\lambda_i}\left(\frac{1}{2}\tilde{\alpha}_i^2\right)^{\frac{3}{4}} - 4k_{i,2}\left(\frac{1}{2}e_i^4\right)^2 \\ &-\left(\frac{4}{\lambda_i} - \bar{h}_i^4 - 1\right)\left(\frac{1}{2}\tilde{\alpha}_i^2\right)^2 + \frac{1}{4}\bar{h}_i^4 e_{i+1}^4 + \frac{\chi_i^2}{2} + \frac{1}{4}\varepsilon_i^4 + \frac{1}{3\lambda_i} + \frac{1}{3}k_{i,1} \\ &-4\gamma_{i,2}\left(1 - 4\gamma_{i,3}\right)\Gamma_i\left(\frac{1}{2}\tilde{\theta}_i^{\mathrm{T}}\Gamma_i^{-1}\tilde{\theta}_i\right)^2 + \frac{1}{4} + \frac{3\eta_i}{4} + \frac{\rho_i}{2} \\ &+\gamma_{i,1}\left(\frac{1}{2}\theta_i^2 + \frac{1}{6}\right) + \gamma_{i,2}\left(\frac{3}{4}\gamma_{i,3}^{-\frac{1}{3}} + \frac{3}{2}\gamma_{i,3}^{-1} + \frac{3}{4}\gamma_{i,3}^{-3}\right)\theta_i^4.\end{aligned} \tag{5.50}$$

Moreover, note that

$$\begin{aligned}\gamma_{n,1}\tilde{\theta}_n^{\mathrm{T}}\hat{\theta}_n+\gamma_{n,2}\tilde{\theta}_n\hat{\theta}_n^3\le&-\frac{2}{3}\gamma_{n,1}\left|\tilde{\theta}_n\right|^{\frac{3}{2}}-\gamma_{n,2}\left(1-4\gamma_{n,3}\right)\tilde{\theta}_n^4\\&+\gamma_{n,1}\left(\frac{1}{2}\theta_n^2+\frac{1}{6}\right)\\&+\gamma_{n,2}\left(\frac{3}{4}\gamma_{n,3}^{-\frac{1}{3}}+\frac{3}{2}\gamma_{n,3}^{-1}+\frac{3}{4}\gamma_{n,3}^{-3}\right)\theta_n^4,\end{aligned}\tag{5.51}$$

with constants $\gamma_{n,3}>0$ and $1-4\gamma_{n,3}>0$. Then, one can get

$$\begin{aligned}\mathcal{L}V_n(t)\le&\frac{1}{k_{gn}}\left(\frac{1}{k_{gn}}h_n(\bar{x}_n)N_n(\xi_n)+1\right)\dot{\xi}_n-\frac{4^{\frac{7}{4}}}{3}k_{n,1}\left(\frac{1}{4}e_n^4\right)^{\frac{3}{4}}\\&-4k_{n,2}\left(\frac{1}{4}e_n^4\right)^2-\frac{2}{3}\gamma_{n,1}\Gamma_n\left(\frac{1}{2}\tilde{\theta}_n^{\mathrm{T}}\Gamma_n^{-1}\tilde{\theta}_n\right)^{\frac{3}{4}}+\frac{1}{3}k_{n,1}\\&-\gamma_{n,2}\left(1-4\gamma_{n,3}\right)\Gamma_n\left(\frac{1}{2}\tilde{\theta}_n^{\mathrm{T}}\Gamma_n^{-1}\tilde{\theta}_n\right)^2+\gamma_{n,1}\left(\frac{1}{2}\theta_n^2+\frac{1}{6}\right)\\&+\gamma_{n,2}\left(\frac{3}{4}\gamma_{n,3}^{-\frac{1}{3}}+\frac{3}{2}\gamma_{n,3}^{-1}+\frac{3}{4}\gamma_{n,3}^{-3}\right)\theta_n^4+\frac{1}{4}\varepsilon_n^4+\frac{3\eta_n}{4}+\frac{\rho_n}{2}.\end{aligned}\tag{5.52}$$

Consider $V=\frac{1}{4}\sum_{i=1}^{n}e_i^4+\frac{1}{2}\sum_{i=1}^{n-1}\tilde{\alpha}_i^2+\frac{1}{2}\sum_{i=1}^{n}\tilde{\theta}_i^{\mathrm{T}}\Gamma_i^{-1}\tilde{\theta}_i$. Then, one gets $\mathcal{L}V=\sum_{i=1}^{n}\mathcal{L}V_i(t)$. From (5.48), (5.50), (5.52), and Lemma 1.7, one has

$$\mathcal{L}V\le\sum_{i=1}^{n}\frac{1}{k_{gi}}\left[\frac{1}{k_{gi}}h_i(\bar{x}_i)N_i(\xi_i)+1\right]\dot{\xi}_i-aV^{\frac{3}{4}}-bV^2+c\tag{5.53}$$

where

$$\begin{aligned}a=&\min_{1\le i\le n}\{a_i\},\\b=&\min_{1\le i\le n}\{b_i\},\\c=&\sum_{i=1}^{n}\left[\gamma_{i,1}\left(\frac{1}{2}\theta_i^2+\frac{1}{6}\right)+\gamma_{i,2}\left(\frac{3}{4}\gamma_{i,3}^{-\frac{1}{3}}+\frac{3}{2}\gamma_{i,3}^{-1}+\frac{3}{4}\gamma_{i,3}^{-3}\right)\theta_i^4+\frac{\rho_i}{2}+\frac{3\eta_i}{4}+\frac{1}{4}\varepsilon_i^4\right]\\&+\sum_{i=1}^{n-1}\left(\frac{\chi_i^2}{2}+\frac{1}{4}+\frac{1}{3\lambda_i}\right)+\sum_{i=2}^{n}\frac{1}{3}\left(k_{i,1}-\frac{1}{4}\bar{h}_{i-1}^4\right)+\frac{1}{3}k_{1,1}\end{aligned}$$

with

$$\begin{aligned}
a_1 &= \min\left\{\frac{4^{\frac{7}{4}}}{3}k_{1,1}, \frac{4}{3\lambda_1}\cdot 2^{\frac{3}{4}}, \frac{2^{\frac{7}{4}}}{3}\gamma_{1,1}\cdot\Gamma_1\right\},\\
a_i &= \min\left\{\frac{4^{\frac{7}{4}}}{3}\left(k_{i,1}-\frac{1}{4}\bar{h}_{i-1}^4\right), \frac{2^{\frac{11}{4}}}{3\lambda_i}, \frac{2^{\frac{7}{4}}}{3}\gamma_{i,1}\cdot\Gamma_i\right\},\\
a_n &= \min\left\{\frac{4^{\frac{7}{4}}}{3}k_{n,1}-\frac{1}{4}\bar{h}_{n-1}^4, \frac{2}{3}\gamma_{n,1}\Gamma_n\right\},\\
b_1 &= \frac{1}{3}\min\left\{4k_{1,2}, \frac{4}{3\lambda_1}-1-\bar{h}_1^4, 4\gamma_{1,2}\left(1-4\gamma_{1,3}\right)\cdot\Gamma_1\right\},\\
b_i &= \frac{1}{3}\min\left\{4k_{i,2}, \frac{4}{3\lambda_i}-\bar{h}_i^4-1, 4\gamma_{i,2}\left(1-4\gamma_{i,3}\right)\cdot\Gamma_i\right\},\\
b_n &= \frac{1}{2}\min\left\{4k_{n,2}, \gamma_{n,2}\left(1-4\gamma_{n,3}\right)\Gamma_n\right\}.
\end{aligned} \tag{5.54}$$

Through Lemma 5.1, one can ensure that the error signals e_i, adaptive learning errors $\tilde{\theta}_i$, Nussbaum-type gain $\xi_i(t)$ for $i=1,\ldots,n$ and DSC filter errors $\tilde{\alpha}_i$ for $i=1,\ldots,n-1$ are all bounded. The proof has been completed. □

Remark 5.2 With the linear term $(s_i-\alpha_i)$ and the nonlinear term $(s_i-\alpha_i)^3$ in the presented novel nonlinear DSC design, one can finally obtain that $\mathcal{L}V(t)$ satisfies the inequality (5.53) containing the fractional-order term $V^{\frac{3}{4}}(t)$ and integer-order term $V^2(t)$. Finally, from Lemma 5.1, one can conclude the fixed-time performance of the closed-loop signals including the DSC filter error.

Remark 5.3 The effect of the standard Brownian motion on the closed-loop system is captured by the diffusion term $g_i(\cdot)$ for $i=1,2,\ldots,n$. From Lemma 5.1, the closed-loop stability can be ensured by the inequality (5.3), where the operator $\mathcal{L}$ is defined in (1.3) containing the diffusion term. Therefore, the virtual control input design (5.26) and (5.34); the controller (5.40); the adaptive fuzzy learning (5.29), (5.37), and (5.43); and the multiple Nussbaum-type gains design (5.27), (5.35), and (5.41) take the effect of the diffusion term $g_i(\cdot)$ into consideration such that the fact in (5.53) holds, which further ensures the boundedness of signals $\left\{e_i, \tilde{\alpha}_i, \tilde{\theta}_i\right\}$.

5.5 Simulation Example

In this section, the proposed control scheme is verified through the short-period dynamics of a missile in [28] as

$$\begin{aligned}
dx_1 &= \left(x_2-1.02x_1-1.28x_1^3\right)dt,\\
dx_2 &= \left[\left(-70.12-360.27x_1^2\right)u-57.17x_1-322.16x_1^3\right]dt\\
&\quad+\left(-70.12-360.27x_1^2\right)\mathrm{d}\omega,
\end{aligned}$$

Table 5.1 Fuzzy sets of the application example

l	1	2	3	4	5
F_1^l	NL	NS	ZO	PS	PL
F_2^l	NL	NS	ZO	PS	PL

with the angle x_1, the pitch rate x_2, and the fin position u.

The fuzzy sets in this subsection are selected as listed in Table 5.1. The corresponding fuzzy membership functions are considered as $\mu_{F_i^l}(x_i) = \exp\left[-\frac{(x_i-9+3l)^2}{5}\right]$ with $i = 1, 2$ and $l = 1, 2, 3, 4, 5$. With the above design, we choose the parameters and initial conditions as $k_{11} = 80$, $k_{12} = 40$, $k_{21} = 100$, $k_{22} = 100$, $\tau_1 = 0.1$, $\gamma_{11} = 500, \gamma_{12} = 500, \gamma_{21} = 10, \gamma_{22} = 10, \Gamma_1 = 500, \Gamma_2 = 500, \rho_1 = 0.1, \rho_2 = 0.1$, $x_1(0) = 0.5$, $x_2(0) = 0.5$, $\xi_1(0) = -1$, and $\xi_2(0) = 0.2$.

In the simulation, we compare the results of using the different Nussbaum-type functions. Figure 5.1a shows that the result of the system output x_1 could track the desired signal with a satisfactory level. It can be observed that, with the traditional

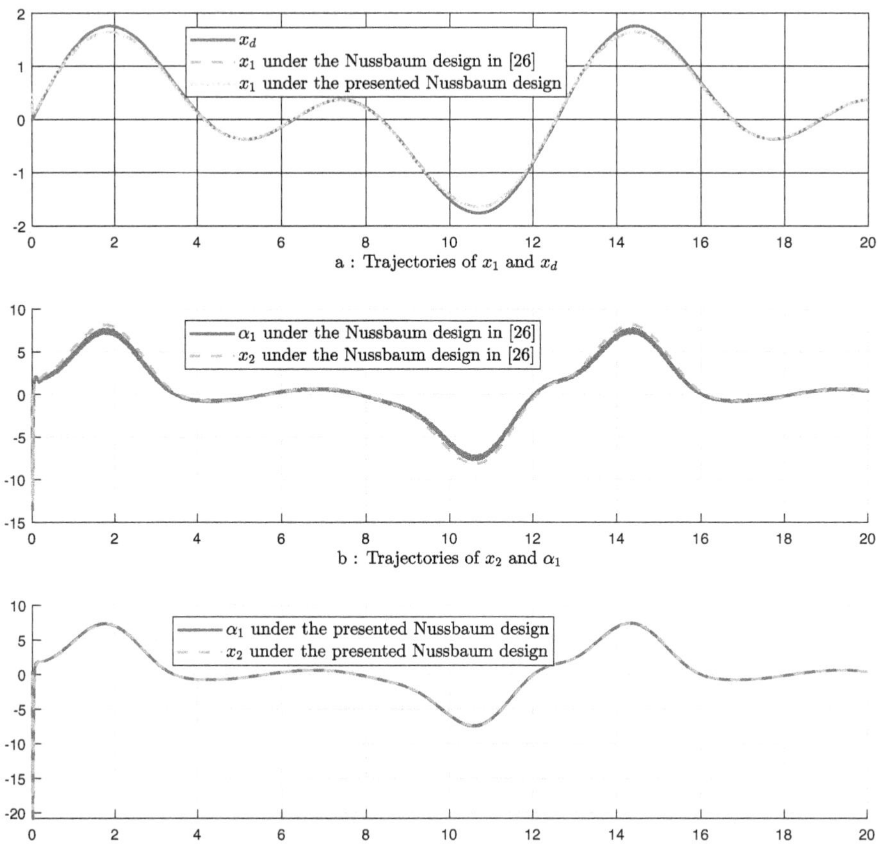

Fig. 5.1 The comparison of state variables and DSC filter output

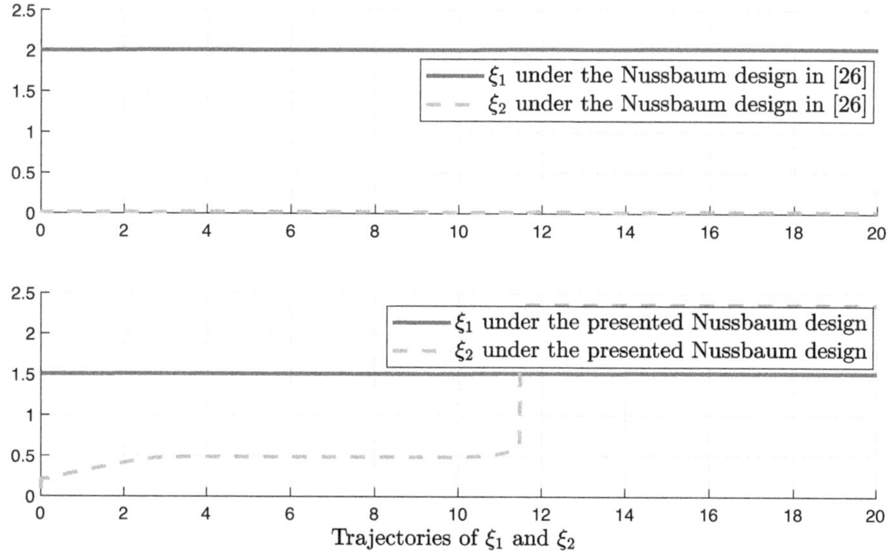

Fig. 5.2 The comparison of Nussbaum-type gain

Nussbaum-type function design, a large deviation exists in the early stage, which is dissatisfactory and should be improved for the stability concern. With the presented Nussbaum design scheme, the tracking error converges to the origin with negligible overshoot. The evolution of filter output α_1 and system state x_2 with the proposed Nussbaum design and the traditional Nussbaum-type function are shown in Fig. 5.1b and c, respectively. It is shown that the Nussbaum-type function designed in this chapter requires less control effort compared to the traditional Nussbaum design, as shown by the signals norm in Fig. 5.1b and c, respectively. The evolution of the signals ξ_1 and ξ_2 is shown in Fig. 5.2, which verifies the boundedness of Nussbaum-type gains designed in this chapter. The tracking errors e_1 and e_2 for the first and second subsystems are illustrated in Fig. 5.3a and b, respectively. The tracking performance with the Nussbaum design in this chapter is superior to the case with the method in [26], as shown in Fig. 5.3a and b. Figure 5.4a shows the convergence performance of the fuzzy logic parameters of the system. Figure 5.4b shows the control signal u of the system, which can ensure practical fixed-time stability in probability for the system with stochastic interference.

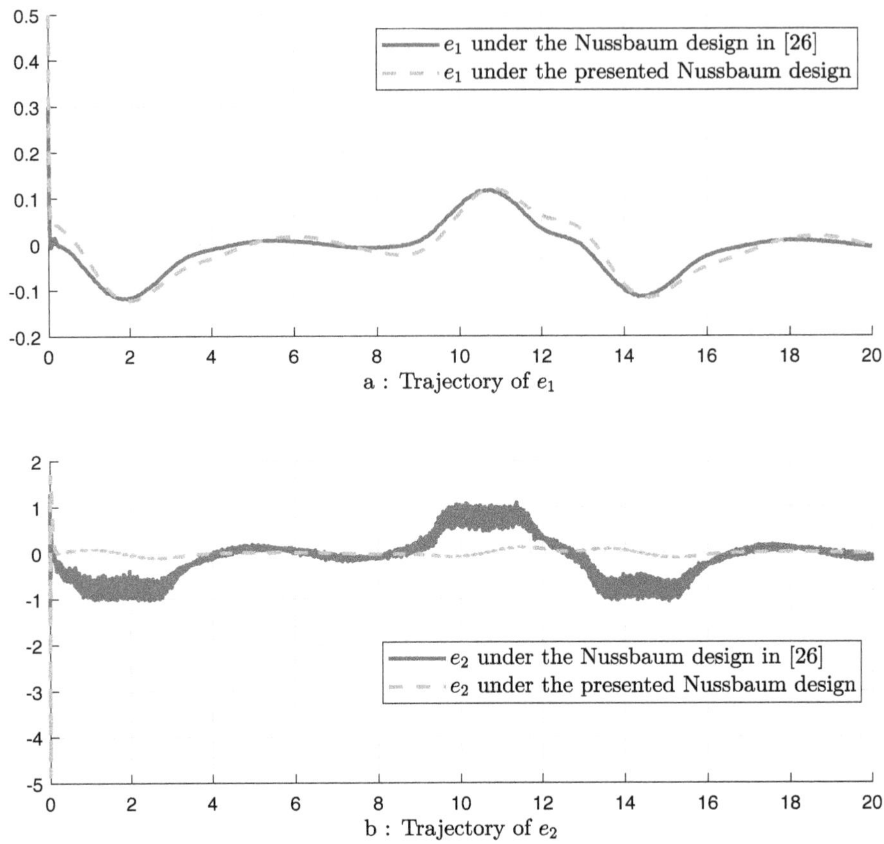

Fig. 5.3 The comparison of tracking errors

5.6 Conclusion

The tracking problem for a class of stochastic strict-feedback nonlinear systems with multiple unknown control directions is considered by adaptive fuzzy backstepping design in this work. An adaptive law is combined with FLSs to handle unknown nonlinear functions in the system dynamics. To bridge the gap between the Nussbaum design and the fixed-time stability, the multiple Nussbaum-type functions are designed to compensate for the multiple unknown control coefficients separately and efficiently. Theoretical discussions ensure the fixed-time control performance. Simulation studies validate the presented design in the end.

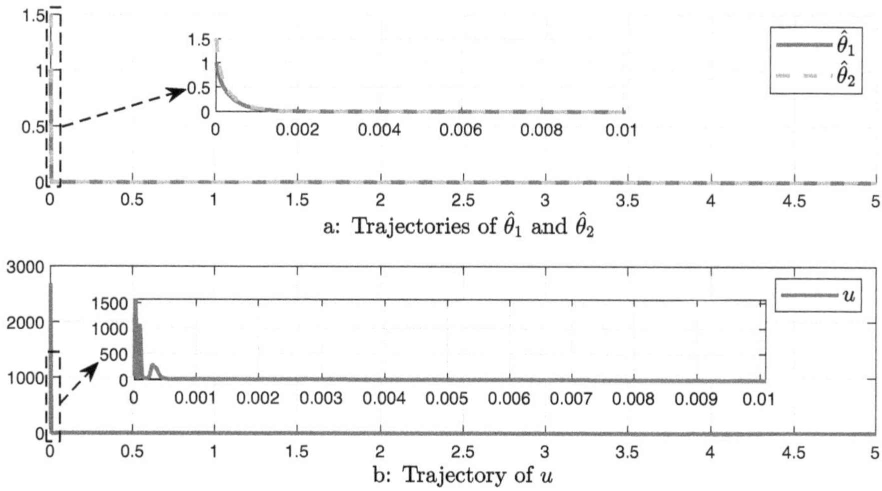

Fig. 5.4 Adaptive parameters and control signal

References

1. Florchinger, P.: Lyapunov-like techniques for stochastic stability. SIAM J. Control. Optim. **33**(4), 1151–1169 (1995)
2. Khasminskii, R.: Stochastic Stability of Differential Equations. Springer, London (2011)
3. Särkkä, S., Solin, A.: Applied Stochastic Differential Equations. Cambridge University Press, Cambridge (2019)
4. Itô, K.: Multiple wiener integral. J. Math. Soc. Jpn. **3**(1), 157–169 (1951)
5. Stratonovich, R.L.: Conditional Markov Processes and Their Application to the Theory of Optimal Control. American Elsevier Pub.Co, New York (1968)
6. Ji, H.B., Xi, H.S.: Adaptive output-feedback tracking of stochastic nonlinear systems. IEEE Trans. Autom. Control **51**(2), 355–360 (2006)
7. Wang, J., Liu, Z., Zhang, Y., Philip Chen, C.L.: A novel fuzzy control with filter-based event-triggered mechanism for nonlinear uncertain stochastic systems suffered input hysteresis. Fuzzy Sets Syst. **432**, 68–88 (2022)
8. Liu, Z., Wang, J., Chen, C.L.P., Zhang, Y.: Event trigger fuzzy adaptive compensation control of uncertain stochastic nonlinear systems with actuator failures. IEEE Trans. Fuzzy Syst. **26**(6), 3770–3781 (2018)
9. Wang, H., Liu, K., Liu, X., Chen, B., Lin, C.: Neural-based adaptive output-feedback control for a class of nonstrict-feedback stochastic nonlinear systems. IEEE Trans. Cybern. **45**(9), 1977–1987 (2015)
10. Yang, Y., Modares, H., Vamvoudakis, K.G., Lewis, F.L.: Cooperative finitely excited learning for dynamical games. IEEE Trans. Cybern. **54**(2), 797–810 (2024)
11. Yang, Y., Wunsch, D., Yin, Y.: Hamiltonian-driven adaptive dynamic programming for continuous nonlinear dynamical systems. IEEE Trans. Neural Netw. Learn. Syst. **28**(8), 1929–1940 (2017)
12. Gao, Y.F., Sun, X.M., Wen, C., Wang, W.: Adaptive tracking control for a class of stochastic uncertain nonlinear systems with input saturation. IEEE Trans. Autom. Control **62**(5), 2498–2504 (2017)

13. Yang, Y., Pan, Y., Xu, C.Z., Wunsch, D.C.: Hamiltonian-driven adaptive dynamic programming with efficient experience replay. IEEE Trans. Neural Netw. Learn. Syst. **35**(3), 3278–3290 (2024)
14. Singh, G.: Think positive: an interpretable neural network for image recognition. Neural Netw. **151**, 178–189 (2022)
15. Liu, G., Yang, Y., Zhao, X., Ahn, C.K.: Adaptive fuzzy practical bipartite synchronization for multi-agent systems with intermittent feedback under multiple unknown control directions. IEEE Trans. Fuzzy Syst. **32**(5), 3211–3225 (2024)
16. Wang, J., Xia, J., Shen, H., Xing, M., Park, J.H.: $\mathcal{H}_\infty$ synchronization for fuzzy Markov jump chaotic systems with piecewise-constant transition probabilities subject to PDT switching rule. IEEE Trans. Fuzzy Syst. **29**(10), 3082–3092 (2021)
17. Wang, J., Yang, C., Xia, J., Wu, Z.G., Shen, H.: Observer-based sliding mode control for networked fuzzy singularly perturbed systems under weighted try-once-discard protocol. IEEE Trans. Fuzzy Syst. **30**(6), 1889–1899 (2022)
18. Wang, F., Chen, B., Lin, C., Li, X.: Distributed adaptive neural control for stochastic nonlinear multiagent systems. IEEE Trans. Cybern. **47**(7), 1795–1803 (2017)
19. Hua, C., Zhang, L., Guan, X.: Distributed adaptive neural network output tracking of leader-following high-order stochastic nonlinear multiagent systems with unknown dead-zone input. IEEE Trans. Cybern. **47**(1), 177–185 (2017)
20. You, Z., Wang, F.: Adaptive fast finite-time fuzzy control of stochastic nonlinear systems. IEEE Trans. Fuzzy Syst. **30**(7), 2279–2288 (2022)
21. Ge, S.S., Hong, F., Lee, T.H.: Adaptive neural control of nonlinear time-delay systems with unknown virtual control coefficients. IEEE Trans. Syst. Man Cybern. Part B (Cybernetics). **34**(1), 499–516 (2004)
22. Shojaei, F., Arefi, M.M., Khayatian, A., Karimi, H.R.: Observer-based fuzzy adaptive dynamic surface control of uncertain nonstrict feedback systems with unknown control direction and unknown dead-zone. IEEE Trans. Syst. Man Cybern.: Syst. **49**(11), 2340–2351 (2019)
23. Huang, J., Wang, W., Wen, C., Zhou, J.: Adaptive control of a class of strict-feedback time-varying nonlinear systems with unknown control coefficients. Automatica **93**, 98–105 (2018)
24. Su, Y.: Cooperative global output regulation of second-order nonlinear multi-agent systems with unknown control direction. IEEE Trans. Autom. Control **60**(12), 3275–3280 (2015)
25. Jin, X.: Adaptive fixed-time control for MIMO nonlinear systems with asymmetric output constraints using universal barrier functions. IEEE Trans. Autom. Control **64**(7), 3046–3053 (2019)
26. Liu, G., Yang, Y., Ding, D.W., Li, Q.: Adaptive event-based fixed-time tracking design for strict-feedback nonlinear systems with unknown control coefficients. Neurocomputing **541**, 126278 (2023)
27. Wen, C., Zhou, J., Liu, Z., Su, H.: Robust adaptive control of uncertain nonlinear systems in the presence of input saturation and external disturbance. IEEE Trans. Autom. Control **56**(7), 1672–1678 (2011)
28. Rajagopal, K., Balakrishnan, S.N., Busemeyer, J.R.: Neural network-based solutions for stochastic optimal control using path integrals. IEEE Trans. Neural Netw. Learn. Syst. **28**(3), 534–545 (2017)

Part III
Advanced Topics with Nussbaum Design

Chapter 6
Adaptive Nussbaum Design for Multiagent Systems with Constraints

6.1 Introduction

In many engineering applications, the complete measurement of system states is impossible or at a high cost. For example, the measurement of joint velocity and acceleration for wheeled mobile robot is more challenging compared to the joint position measurement technique [1]. Therefore, when the state cannot be directly measured, output feedback controller based on the observer design is desired. Typical observer designs for nonlinear systems are Luenberger observer [2–4], K-filter [5–7], MT-filter [8], etc. These methods are based on the requirement that the system dynamics is Lipschitz captured by the Lipschitz constant. However, the Lipschitz constant is usually difficult to be obtained, which makes the observer design difficult. In contrast, high-gain observer design developed in [9–11] can efficiently solve this issue by replacing the Lipschitz constant with a bounded time-varying function of the system output. In this chapter, we extend the high-gain observer design for single-agent system to MAS to estimate the unmeasured states for each follower.

The controller design for dynamical systems has to be implemented on digital platforms and driven by power systems. Therefore, the control input signals have to be guaranteed to be within a certain level. This requirement can be viewed as the input saturation constraint and has been extensively focused recently. The optimal control theory has been applied to the input constrained controller design, where the nonquadratic penalty on the control input is developed to guarantee that the control input does not exceed the predefined saturation level [12]. This method has also been extended to tracking control problem [13], differential game [14], and the case with intermittent feedback [15]. However, the input constraint considered in these results is essentially the type of smooth saturation. Another type of non-smooth input constraint, usually captured by the saturation, is more challenging. Adaptive control with projection technique has been used to design the control with bounded norm [16]. However, this method only guarantees the boundedness of the control input signal and the user-defined input saturation level still cannot be directly

Y. Yang et al., *Adaptive Nussbaum Design for Nonlinear Dynamical Systems*,
https://doi.org/10.1007/978-981-95-4997-9_6

achieved. This issue is tackled by the auxiliary system-based method [17], where an additional dynamical system is combined with the backstepping design framework.

Compared to the distributed control method [18, 19], we develop a novel observer with time-varying feedback gain for each followers to estimate the unmeasured states, which can relax the Lipschitz continuous requirement about the agent's dynamics. Both the input saturation and output constraint are considered simultaneously within the presented framework for heterogeneous MAS. Also, the method developed in this chapter can be easily extended to the case of homogeneous MAS.

The remainder of this chapter is as follows. In Sect. 5.2, the leader–follower synchronization problem for a group of MASs with nonlinear strict-feedback follower dynamics is formulated. In Sect. 6.3, the adaptive Nussbaum controller is developed within the framework of the backstepping method. The stability analysis for the adaptive Nussbaum design is shown in Sect. 6.3. Simulation examples are conducted in Sect. 6.5 to validate the presented scheme. Finally, the concluding remarks are made in Sect. 5.6.

6.2 Problem Formulation

We consider the heterogeneous MASs with n followers and a leader. The dynamics of each follower is given by

$$\begin{aligned} \dot{x}_{i,j} &= f_{i,j}\left(\bar{x}_{i,j}\right) + x_{i,j+1}, \quad j = 1, ..., n-1 \\ \dot{x}_{i,n} &= f_{i,n}\left(\bar{x}_{i,n}\right) + \Gamma\left(u_i\right), \\ y_i &= x_{i,1}, \end{aligned} \tag{6.1}$$

where $x_{i,j} \in \mathbb{R}$ is the state with $\bar{x}_{i,j} = \left[x_{i,1}^{\mathrm{T}} \cdots x_{i,j}^{\mathrm{T}}\right]^{\mathrm{T}} \in \mathbb{R}^i$ and $x_i = \bar{x}_{i,n} \in \mathbb{R}^n$, $u_i \in \mathbb{R}$ is the control input, y_i is the output, and $f_{i,j}(\cdot) : \mathbb{R}^i \to \mathbb{R}$ is the unknown function for the ith follower for $j = 1, ..., n$. Note that only y_i is available for the ith follower and its neighbors to synthesize the synchronization controller.

Assumption 6.1 For each follower, there exists at least a directed path from the leader to the follower.

Assumption 6.1 is a common condition about the communication graph in the synchronization protocol design for MAS. Based on Assumption 6.1, one can obtain

$$b_i + \sum_{k \in \mathcal{N}_i} a_{i,k} \geq 1, \quad \forall i \in \mathcal{F}.$$

This condition ensures that every follower is influenced either directly by the leader or indirectly through other followers, guaranteeing connectivity within the network.

Assumption 6.2 The follower system (6.1) is input-to-state stable.

Assumption 6.2 is reasonable since an unstable plant cannot be globally stabilized in the presence, which has been explained in [17].

Assumption 6.3 ([9]) For each follower, there exists a locally non-negative function $\iota_i(y_i)$ such that

$$\left|f_{i,j}\left(\bar{x}_{i,j}\right)-f_{i,j}\left(\hat{\bar{x}}_i\right)\right| \leq \iota_i\left(y_i\right)\left(\left|x_{i,2}-\hat{x}_{i,2}\right|+\cdots+\left|x_{i,j}-\hat{x}_{i,j}\right|\right), \quad j=2, \ldots, n$$

where $\iota_i(\cdot)$ is a strictly positive function.

Remark 6.1 Existing results are usually based on the Lipschitz continuity of the unknown function $f_{i,j}(\cdot)$ with a Lipschitz constant [4, 20]. However, the Lipschitz constant is usually difficult to be obtained for uncertain nonlinear functions. To tackle this issue, where the non-negative function $L(y)$ is presented as an alternative to the Lipschitz constant.

In this chapter, we consider the non-smooth input saturation constraints for each follower,

$$\Gamma_i\left(u_i\right)=\begin{cases} sign\left(u_i\right)\bar{u}_i, & |u| \geq \bar{u}_i \\ u_i, & |u| \geq \bar{u}_i \end{cases} \tag{6.2}$$

with the upper bound $\bar{u}_i>0$. The controller design and stability analysis is difficult because the saturation function (6.2) is non-smooth. Therefore, we adopt the following smooth approximation:

$$\begin{aligned} \hat{\Gamma}_i\left(u_i\right) &= \bar{u}_i \tanh\left(\frac{u_i(t)}{\bar{u}_i}\right) \\ &= \bar{u}_i \frac{e^{u_i/\bar{u}_i}-e^{-u_i/\bar{u}_i}}{e^{u_i/\bar{u}_i}+e^{-u_i/\bar{u}_i}}. \end{aligned} \tag{6.3}$$

Remark 6.2 The relation between the saturation function $\Gamma(u_i)$ and the smooth approximation $\hat{\Gamma}(u_i)$ is

$$\Gamma_i\left(u_i\right)=\hat{\Gamma}_i\left(u_i\right)+h_i\left(u_i\right) \tag{6.4}$$

where $h_i(u_i)$ is the approximation residual. The smooth approximation function $\hat{\Gamma}_i(\cdot)$ has the following properties:

1. The approximation residual $h_i(u_i)$ satisfies [17]

$$h_i\left(u_i\right) \leq h_i^*=\bar{u}_i(1-\tanh(1)). \tag{6.5}$$

2. The smooth approximation function $\hat{\Gamma}_i(u_i)$ has upper-bounded gradient, i.e.:

$$\begin{aligned} \varpi_i &= \left| \frac{\partial \hat{\Gamma}_i(u_i)}{\partial u_i} \right| \\ &= \left| \frac{4}{\left(e^{u_i/\bar{u}_i} + e^{-u_i/\bar{u}_i}\right)^2} \right| \leq 1. \end{aligned} \tag{6.6}$$

3. The smooth approximation function $\hat{\Gamma}_i(u_i)$ and the function $\bar{\Gamma}(u_i) = \frac{\partial \hat{\Gamma}_i(u_i)}{\partial u_i} u_i$ are upper-bounded, i.e.:

$$\begin{aligned} \left| \hat{\Gamma}_i(u_i) \right| &= \bar{u}_i \left| \tanh\left(\frac{u_i}{\bar{u}_i}\right) \right| \leq \bar{u}_i \\ \left\| \bar{\Gamma}_i(u_i) \right\| &= \left| \frac{4u_i}{\left(e^{u_i/\bar{u}_i} + e^{-u_i/\bar{u}_i}\right)^2} \right| \leq \frac{\bar{u}_i}{2}. \end{aligned} \tag{6.7}$$

With the saturation approximation (6.3), the follower's dynamics (6.1) can be expressed as

$$\begin{aligned} \dot{x}_{i,1} &= f_{i,1}\left(x_{i,1}\right) + x_{i,2}, \\ \dot{x}_{i,j} &= f_{i,j}\left(\hat{\bar{x}}_{i,j}\right) + \tilde{f}_{i,j} + x_{i,j+1}, \quad j = 2, \ldots, n-1 \\ \dot{x}_{i,n} &= f_{i,n}\left(\hat{\bar{x}}_{i,n}\right) + \tilde{f}_{i,n} + \hat{\Gamma}_i(u_i) + h(u_i), \end{aligned} \tag{6.8}$$

where $\tilde{f}_{i,j} = f_{i,j}\left(\bar{x}_{i,j}\right) - f_{i,j}\left(\hat{\bar{x}}_{i,j}\right)$, and $\hat{x}_{i,j}$ is the estimation of $x_{i,j}$ with $\hat{\bar{x}}_{i,j} = \left[\hat{x}_{i,1} \cdots \hat{x}_{i,j}\right]^{\mathrm{T}}$. In the following, the fuzzy logic system is used for function approximation in the observer design.

The nonlinear follower dynamics can be written in terms of FLS,

$$f_{i,j}(\cdot) = \left(W_{i,j}^*\right)^{\mathrm{T}} \varphi_{i,j}(\cdot) + \varepsilon_{i,j}(\cdot) \tag{6.9}$$

where $W_{i,j}^* \in \mathbb{R}^{N_{i,j}}$ is the unknown weight and $\varphi_{i,j}(\cdot) : \mathbb{R}^j \to \mathbb{R}^{N_{i,j}}$ is the basis function. Based on (6.9) and (6.4), the follower dynamics (6.8) can be further expressed as

$$\begin{aligned} \dot{x}_{i,1} &= \left(W_{i,1}^*\right)^{\mathrm{T}} \varphi_{i,1}\left(x_{i,1}\right) + x_{i,2} + \varepsilon_{i,1} \\ \dot{x}_{i,j} &= \left(W_{i,j}^*\right)^{\mathrm{T}} \varphi_{i,j}\left(\hat{\bar{x}}_{i,j}\right) + x_{i,j+1} + \tilde{f}_{i,j} + \varepsilon_{i,j}, \\ \dot{x}_{i,n} &= \left(W_{i,n}^*\right)^{\mathrm{T}} \varphi_{i,n}\left(\hat{\bar{x}}_{i,n}\right) + \hat{\Gamma}_i(u_i) + h_i(u_i) + \tilde{f}_{i,n} + \varepsilon_{i,n}, \end{aligned} \tag{6.10}$$

where $\tilde{f}_{i,j} = f_{i,j}\left(\bar{x}_{i,j}\right) - f_{i,j}\left(\hat{\bar{x}}_{i,j}\right)$ for $j = 2, ..., n-1$. The observer is designed as

$$\begin{aligned}
\dot{\hat{x}}_{i,1} &= \hat{W}_{i,1}^{\mathrm{T}} \varphi_{i,1}\left(x_{i,1}\right) + \hat{x}_{i,2} + k_{i,1} r_i(t) \tilde{\xi}_i, \\
\dot{\hat{x}}_{i,j} &= \hat{W}_{i,j}^{\mathrm{T}} \varphi_{i,j}\left(\hat{\bar{x}}_{i,j}\right) + \hat{x}_{i,j+1} + k_{i,j}\left[r_i(t)\right]^j \tilde{\xi}_i, \\
\dot{\hat{x}}_{i,n} &= \hat{W}_{i,n}^{\mathrm{T}} \varphi_{i,n}\left(\hat{\bar{x}}_{i,n}\right) + \hat{\Gamma}_i\left(u_i\right) + k_{i,n}\left[r_i(t)\right]^n \tilde{\xi}_i, \\
\hat{y}_i &= \hat{x}_{i,1},
\end{aligned} \tag{6.11}$$

for $j = 2, \ldots, n-1$, where $\hat{W}_{i,j}$ is the estimation of $W_{i,j}^*$, $k_{i,j}$ and $r_i(t)$ are the observer feedback gain designed later, $\tilde{\xi}_i = \xi_i - \hat{\xi}_i$ with the local output synchronization error ξ_i and its estimation $\hat{\xi}_i$ defined as

$$\begin{aligned}
\xi_i &= \sum_{k \in N_i} a_{ik}\left(y_i - y_k\right) + b_i\left(y_i - y_0\right) \\
\hat{\xi}_i &= \sum_{k \in N_i} a_{ik}\left(\hat{y}_i - y_k\right) + b_i\left(\hat{y}_i - y_0\right).
\end{aligned}$$

In the observer design (6.11), $k_{i,j}$ is the fixed feedback gain and $r_i(t)$ is the adaptive feedback gain satisfying $r_i(t) \geq 1$ for $\forall t > 0$. Therefore, the observer design in (6.11) is referred to as the high-gain observer. The Luenberger observer design can be viewed as a special case of the high-gain observer when $r_i\,(t)$ is designed as a constant. In this chapter, the dynamics of the adaptive feedback gain $r_i\,(t)$ is designed as

$$\begin{aligned}
\dot{r}_i &= \phi_i\left(r_i, y_i\right) \\
&= -\frac{r_i}{a_i}\left[\frac{q_i}{3}\left(r_i - 1\right) - \frac{2(n-1)\,\iota_i\left(y_i\right)}{\sqrt{p_i}}\right],
\end{aligned}$$

where $r_i\,(0) = 1$. We will show that $r_i(t) \geq 1$ for $\forall t > 0$. Suppose that $r_i\,(t) < 1$, it follows that $\dot{r}_i > 0$. Furthermore, one can obtain $r_i\,(t) \geq r_i\,(0) = 1$, which contradicts with the fact $r_i\,(t) < 1$. Therefore, one can conclude that $r_i\,(t) \geq 1$ for $t \in [0, +\infty)$. In addition, $r_i(t)$ is bounded due to the following facts. Consider the Lyapunov function $V_{i,r} = \frac{1}{2} r_i(t)^2$, then

$$\begin{aligned}
\dot{V}_{i,r} &= r_i \dot{r}_i \\
&= -\frac{r_i^2}{a_i}\left[\frac{q_i}{3}\left(r_i - 1\right) - \frac{2(n-1)\,\iota_i\left(y_i\right)}{\sqrt{p_i}}\right] \\
&= -\frac{q_i}{3a_i} r_i^3 + \left(\frac{q_i}{3a_i} + \frac{2(n-1)\,\iota_i\left(y_i\right)}{a_i\sqrt{p_i}}\right) r_i^2.
\end{aligned}$$

Since $\dot{V}_{i,r}$ is a cubic function with $-\frac{q_i}{3a_i} < 0$. Combined with the fact that $r_i(t) \geq 1$, then $\dot{V}_{i,r}$ is negative when r_i is sufficiently large, which implies that $r_i\,(t)$ is bounded.

Define the estimation error for the unknown weight $W^*_{i,j}$ and state $x_{i,j}$ as

$$\begin{aligned}
\tilde{W}_{i,j} &= W^*_{i,j} - \hat{W}_{i,j}, \\
\tilde{\bar{x}}_{i,j} &= \left[\tilde{x}^{\mathrm{T}}_{i,1} \cdots \tilde{x}^{\mathrm{T}}_{i,j}\right]^{\mathrm{T}}, \\
\tilde{x}_{i,j} &= x_{i,j} - \hat{x}_{i,j}, \quad j = 1, ..., n \\
\tilde{\bar{x}}_i &= \tilde{\bar{x}}_{i,n} = \bar{x}_{i,n} - \hat{\bar{x}}_{i,n}.
\end{aligned} \tag{6.12}$$

Note that $\tilde{\xi}_i = \delta_{i,i}\tilde{x}_{i,1}$. Then, from (6.10) and (6.8), the dynamics of state estimation error can be denoted as

$$\begin{aligned}
\dot{\tilde{x}}_{i,1} &= \tilde{W}^{\mathrm{T}}_{i,1}\varphi_{i,1}\left(x_{i,1}\right) + \tilde{x}_{i,2} - r_i\bar{k}_{i,1}\tilde{x}_{i,1} + \varepsilon_{i,j} \\
\dot{\tilde{x}}_{i,j} &= \tilde{W}^{\mathrm{T}}_{i,j}\varphi_{i,j}\left(\hat{\bar{x}}_{i,j}\right) + \tilde{x}_{i,j+1} - r_i^j\bar{k}_{i,j}\tilde{x}_{i,1} + \tilde{f}_{i,j} + \varepsilon_{i,j}, \\
\dot{\tilde{x}}_{i,n} &= \tilde{W}^{\mathrm{T}}_{i,n}\varphi_{i,n}\left(\hat{\bar{x}}_{i,n}\right) + h_i\left(u_i\right) - r_i^n\bar{k}_{i,n}\tilde{x}_{i,1} + \tilde{f}_{i,n} + \varepsilon_{i,n}, \\
\bar{k}_{i,j} &= \delta_{i,i}k_{i,j}.
\end{aligned} \tag{6.13}$$

With linear matrix inequality technique, the fixed feedback gain $\bar{k}_{i,1}$ for observer design can be selected as follows.

Lemma 6.1 ([9]) *Consider the observer feedback gains $\bar{k}_{i,j}$ for $j = 1, ..., n$. There exist strictly positive constants a_i, p_i, and q_i and positive definite matrix $P_i = P_i^{\mathrm{T}}$ such that*

$$\begin{aligned}
&A_i^{\mathrm{T}} P_i + P_i A_i \le -q_i P_i, \\
&p_i I_{n\times n} \le P_i \le I_{n\times n}, \\
&0 < p_i \le 1, \\
&-a_i P_i \le S_i P_i + P_i S_i \le a_i P_i
\end{aligned} \tag{6.14}$$

where

$$A_i = \begin{bmatrix} -\bar{k}_{i,1} & 1\ 0 \cdots 0 \\ -\bar{k}_{i,2} & 0\ 1 \cdots 0 \\ \vdots & \vdots\ \vdots \ddots \vdots \\ -\bar{k}_{i,n-1} & 0\ 0 \cdots 1 \\ -\bar{k}_{i,n} & 0\ 0 \cdots 0 \end{bmatrix}_{n\times n}, \quad S = \begin{bmatrix} 0 \cdots 0 & 0 \\ \vdots \ddots \vdots & \vdots \\ 0 \cdots 0 & 0 \\ 0 \cdots 0 & n-1 \end{bmatrix}_{n\times n}$$

with $n \ge 2$.

For subsequent analysis, the following error transformation is introduced:

$$\begin{aligned}
e_i &= \left[e^{\mathrm{T}}_{i,1} \cdots e^{\mathrm{T}}_{i,n}\right]^{\mathrm{T}}, \\
e_{i,j} &= \frac{\tilde{x}_{i,j}}{\left[r_i(t)\right]^{j-1+a_i}},
\end{aligned} \tag{6.15}$$

where $a_i > 1, i \in \mathcal{F}$. Using (6.13) and (6.15), the dynamics of e_i is

$$\dot{e}_i = r_i(t) A_i e_i - \frac{\dot{r}_i(t)}{r_i(t)} (a_i I_{n\times n} + S) e_i + \Lambda_i \left[\Phi_i^{\mathrm{T}} \tilde{W}_i + \varepsilon_i + \tilde{f}_i + H_i \right] \quad (6.16)$$

where

$$\Lambda_i = \begin{bmatrix} \frac{1}{[r_i(t)]^{a_i}} & \cdots & 0 \\ \vdots & \ddots & \vdots \\ 0 & \cdots & \frac{1}{[r_i(t)]^{a_i+n-1}} \end{bmatrix}_{n\times n}, \quad \Phi_i = \begin{bmatrix} \varphi_{i,1} & \cdots & 0_{N_{i,j}\times 1} \\ \vdots & \ddots & \vdots \\ 0_{N_{i,n}\times 1} & \cdots & \varphi_{i,n} \end{bmatrix}_{N_i\times n},$$

$$H_i = \begin{bmatrix} 0 \\ \vdots \\ 0 \\ h(u_i) \end{bmatrix}_{n\times 1}, \varepsilon_i = \begin{bmatrix} \varepsilon_{i,1} \\ \varepsilon_{i,2} \\ \vdots \\ \varepsilon_{i,n} \end{bmatrix}_{n\times 1}, \tilde{f}_i = \begin{bmatrix} 0 \\ \tilde{f}_{i,2} \\ \vdots \\ \tilde{f}_{i,n} \end{bmatrix}_{n\times 1}, \tilde{W}_i = \begin{bmatrix} \tilde{W}_{i,1} \\ \vdots \\ \tilde{W}_{i,n} \end{bmatrix}_{N_i\times 1},$$

with $N_i = \sum_{j=1}^{n} N_{i,j}$.

To analyze the stability of the observer estimation error e_i, we consider the following Lyapunov function $V_{i,0} = e_i^{\mathrm{T}} P_i e_i$. From Lemma 6.1 and the fact that $r_i(t) \geq 1$ for $\forall t > 0$, one has

$$-\frac{\dot{r}_i(t)}{r_i(t)} e_i^{\mathrm{T}} (S_i P_i + P_i S_i) e_i \leq a_i \frac{|\dot{r}_i(t)|}{r_i(t)} e_i^{\mathrm{T}} P_i e_i.$$

Using (6.14), (6.16) and above fact, differentiating $V_{i,0}$ yields

$$\begin{aligned} \dot{V}_{i,0} &= \dot{e}_i^{\mathrm{T}} P_i e_i + e_i^{\mathrm{T}} P_i \dot{e}_i \\ &\leq 2e_i^{\mathrm{T}} P_i \Lambda_i \left(\Phi_i^{\mathrm{T}} \tilde{W}_i + \varepsilon_i + \tilde{f}_i + H_i \right) - q_i r_i(t) e_i^{\mathrm{T}} P_i e_i \\ &\quad -2a_i \frac{\dot{r}_i(t)}{r_i(t)} e_i^{\mathrm{T}} P_i e_i - \frac{\dot{r}_i(t)}{r_i(t)} e_i^{\mathrm{T}} (S_i P_i + P_i S_i) e_i \\ &\leq -\left(q_i r_i(t) + 2a_i \frac{\dot{r}_i(t)}{r_i(t)} - a_i \frac{|\dot{r}_i(t)|}{r_i(t)} \right) e_i^{\mathrm{T}} P_i e_i \\ &\quad +2e_i^{\mathrm{T}} P_i \Lambda_i \left(\Phi_i^{\mathrm{T}} \tilde{W}_i + \varepsilon_i + \tilde{f}_i + H_i \right). \end{aligned} \quad (6.17)$$

From Assumption 6.3 and Young's inequality, we have

$$\begin{aligned} \Lambda_{i,j} \tilde{f}_{i,j} &\leq \iota(y_i) \left(|e_{i,2}| + \cdots |e_{i,j}| \right), \\ 2e_i^{\mathrm{T}} P_i \Lambda_i \tilde{f}_i &\leq 2 \sum_{j=2}^{n} |e_i^{\mathrm{T}} P_i| \iota(y_i) \left(|e_{i,2}| + \cdots |e_{i,j}| \right) \\ &\leq \frac{2(n-1)\iota_i(y_i)}{\sqrt{p}_i} e_i^{\mathrm{T}} P_i e_i \end{aligned}$$

where p_i is defined in Lemma 6.1. Based on above facts, $\dot{V}_{i,0}$ further satisfies

$$\begin{aligned}\dot{V}_{i,0} &\leq 2e_i^{\mathrm{T}} P_i \Lambda_i \left(\Phi_i^{\mathrm{T}} W_i + \varepsilon_i + H_i\right) + Q_i \\ Q_i &= -\left[q_i r_i(t) + 2a_i \frac{\dot{r}_i(t)}{r_i(t)} - a_i \frac{|\dot{r}_i(t)|}{r_i(t)} - \frac{2(n-1)\iota_i(y_i)}{\sqrt{p_i}}\right] e_i^{\mathrm{T}} P_i e_i. \end{aligned} \tag{6.18}$$

Since $\iota_i(y_i)$ is strictly positive, this implies that, for $\dot{r}_i \geq 0$, one has

$$\begin{aligned} Q_i &= -\left(q_i r_i(t) + a_i \frac{\dot{r}_i(t)}{r_i(t)} - \frac{2(n-1)\iota_i(y_i)}{\sqrt{p_i}}\right) e_i^{\mathrm{T}} P_i e_i \\ &= -\frac{q_i}{3}(2r_i + 1) e_i^{\mathrm{T}} P_i e_i \\ &\leq -q_i e_i^{\mathrm{T}} P_i e_i \end{aligned}$$

where q_i is defined in Lemma 6.1. For $\dot{r}_i < 0$, we can obtain

$$\begin{aligned} Q_i &= -\left(q_i r_i(t) + 3a_i \frac{\dot{r}_i(t)}{r_i(t)} - \frac{2(n-1)\iota_i(y_i)}{\sqrt{p_i}}\right) e_i^{\mathrm{T}} P_i e_i \\ &= -\left(q_i r_i(t) + \frac{4(n-1)\iota_i(y_i)}{\sqrt{p_i}}\right) e_i^{\mathrm{T}} P_i e_i \\ &\leq -q_i e_i^{\mathrm{T}} P_i e_i. \end{aligned}$$

To summarize, $Q_i(t) \leq -q_i e_i(t)^{\mathrm{T}} P_i e_i(t)$ for $\forall t > 0$.

From the fact that $r_i(t) \geq 1$ for $\forall t > 0$, one has $\|\Lambda_i\| \leq 1$. Also, based on Lemma 1.5, one has $\|\varphi_{i,j}(\cdot)\|^2 \leq Y_{i,j}$. Therefore, applying Young's inequality one has

$$\begin{aligned} 2e_i^{\mathrm{T}} P_i \Lambda_i \Phi_i^{\mathrm{T}} \tilde{W}_i &\leq \|e_i\|^2 + \|P_i\|^2 \sum_{j=1}^{n} \left\|\varphi_{i,j}\left(x_{i,j}\right)\right\|^2 \left\|\tilde{W}_{i,j}\right\|^2 \\ &\leq \|e_i\|^2 + \|P_i\|^2 \sum_{j=1}^{n} Y_{i,j} \left\|\tilde{W}_{i,j}\right\|^2 \\ 2e_i^{\mathrm{T}} P_i \Lambda_i \varepsilon_i &\leq \|e_i\|^2 + \|P_i\|^2 \|\varepsilon_i\|^2 \\ 2e_i^{\mathrm{T}} P_i \Lambda_i H_i &\leq \|e_i\|^2 + \|P_i\|^2 \left\|H_i^*\right\|^2 \end{aligned}$$

where $H_i^* = \left[0 \cdots 0 \; h_i^*\right]^{\mathrm{T}}$. Then, based on above facts, one can obtain

$$\begin{aligned} & 2e_i^{\mathrm{T}} P_i \Lambda_i \left(\Phi_i^{\mathrm{T}} W_i + \varepsilon_i + H_i\right) \\ & \leq 3\|e_i\|^2 + \|P_i\|^2 \left(\sum_{j=1}^{n} Y_{i,j} \left\|\tilde{W}_{i,j}\right\|^2 + \|\varepsilon_i\|^2 + \left\|H_i^*\right\|^2\right). \end{aligned} \tag{6.19}$$

Combine (6.18) with (6.19) and the fact that $Q_i \leq -q_i e_i^{\mathrm{T}} P_i e_i$, one has

$$\begin{aligned}\dot{V}_{i,0} &\leq -q_i e_i^{\mathrm{T}} P_i e_i + 3\|e_i\|^2 + \|P_i\|^2 \left(\sum_{i=1}^{n} Y_{i,j} \|\tilde{W}_{i,j}\|^2 + \|\varepsilon_i\|^2 + \left\| H_i^* \right\|^2 \right) \\ &= -q_i e_i^{\mathrm{T}} P_i e_i + 3\|e_i\|^2 + M_{i,0} \end{aligned} \tag{6.20}$$

where $M_{i,0} = \|P_i\|^2 \left(\sum_{i=1}^{n} Y_{i,j} \|\tilde{W}_{i,j}\|^2 + \|\varepsilon_i\|^2 + \left\| H_i^* \right\|^2 \right)$.

To guarantee the prescribed performance of the tracking error $z_{i,1}$, a function ρ_i is chosen as

$$\rho_i(t) = \left(\rho_{i,0} - \rho_{i,\infty}\right) e_i^{-\kappa_i t} + \rho_{i,\infty}$$

where $\kappa > 0$ and $\rho_{i,0} > \rho_{i,\infty} > 0$. According to the above prescribed performance function (PPF) definition, $\rho_i(t)$ satisfies $\rho_i(0) = \rho_{i,0}$ and lim $\lim_{t\to\infty} \rho(t) = \rho_\infty$. In this chapter, the following prescribed transient performance is imposed on the output tracking error $z_{i,1}(t)$ as

$$-\beta_{i,1}\rho_i(t) \leq z_{i,1}(t) \leq \beta_{i,2}\rho_i(t) \tag{6.21}$$

where $\beta_{i,1} > 0$ and $\beta_{i,2} > 0$ are constants, and $-\beta_{i,1}\rho_{i,0}$ and $\beta_{i,2}\rho_{i,0}$ represent the constraints of the initial synchronization error $z_{i,1}(0)$. In addition, based on PPF definition, $-\beta_{i,1}\rho_{i,\infty}$ and $\beta_{i,2}\rho_{i,\infty}$ in condition (6.21) stand for the steady performance of the synchronization error $z_{i,1}(t)$ as $t \to \infty$.

To consider the predefined constraints (6.21), the asymmetric BLF is employed based on the properties as discussed above.

$$\bar{V}_i\left(z_{i,1}\right) = \frac{q\left(z_{i,1}\right)}{2} \log \frac{\beta_{i,b}^2(t)}{\beta_{i,b}^2(t) - z_{i,1}^2} + \frac{1 - q\left(z_{i,1}\right)}{2} \log \frac{\beta_{i,a}^2(t)}{\beta_{i,a}^2(t) - z_{i,1}^2} \tag{6.22}$$

where $\beta_{i,b}(t) = \beta_{i,1}\rho_i(t)$ and $\beta_{i,a}(t) = \beta_{i,2}\rho_i(t)$ represent the prescribed transient performance, $q\left(z_{i,1}\right)$ is defined as

$$q\left(z_{i,1}\right) = \begin{cases} 1, & \text{if } z_{i,1} > 0 \\ 0, & \text{if } z_{i,1} \leq 0 \end{cases}.$$

Then, the BLF (6.22) can be rewritten as $\bar{V}_i\left(z_{i,1}\right)=\frac{1}{2}\log\frac{1}{1-\eta_i^2}$ with

$$\begin{aligned}
\eta_i &= q\left(z_{i,1}\right)\eta_{i,b}+\left[1-q\left(z_{i,1}\right)\right]\eta_{i,a},\\
\eta_{i,a} &= \frac{z_{i,1}\left(t\right)}{\beta_{i,a}\left(t\right)},\\
\eta_{i,b} &= \frac{z_{i,1}\left(t\right)}{\beta_{i,b}\left(t\right)},\\
\dot{\eta}_{i,a} &= \frac{\dot{z}_{i,1}\beta_{i,a}(t)-z_{i,1}\dot{\beta}_{i,a}(t)}{\beta_{i,a}^2(t)},\\
\dot{\eta}_{i,b} &= \frac{\dot{z}_{i,1}\beta_{i,b}(t)-z_{i,1}\dot{\beta}_{i,b}(t)}{\beta_{i,b}^2(t)}.
\end{aligned}\tag{6.23}$$

The asymmetric BLF has time-varying upper and lower boundaries, which is in contrast to the asymmetric BLF with fixed boundary. The PPF $\rho_i(t)$ is satisfied if the asymmetric BLF $\bar{V}_i\left(z_{i,1}\right)$ remains bounded [21]. According to (6.22), one can obtain

$$\begin{aligned}
\lim_{z_{i,1}\to 0^-}\bar{V}_i\left(z_{i,1}\right) &= \lim_{z_{i,1}\to 0^-}\frac{1-q\left(z_{i,1}\right)}{2}\log\frac{\beta_{i,a}^2\left(t\right)}{\beta_{i,a}^2\left(t\right)-z_{i,1}^2}=0\\
\lim_{z_{i,1}\to 0^+}\bar{V}_i\left(z_{i,1}\right) &= \lim_{z_{i,1}\to 0^+}\frac{q\left(z_{i,1}\right)}{2}\log\frac{\beta_{i,b}^2\left(t\right)}{\beta_{i,b}^2\left(t\right)-z_{i,1}^2}=0\\
\bar{V}_i\left(z_{i,1}\right)\Big|_{z_{i,1}=0} &= \frac{1}{2}\log\frac{\beta_{i,a}^2\left(t\right)}{\beta_{i,a}^2\left(t\right)}=0.
\end{aligned}\tag{6.24}$$

Then $\lim\limits_{z_{i,1}\to 0^-}\bar{V}_i\left(z_{i,1}\right)=\lim\limits_{z_{i,1}\to 0^+}\bar{V}_i\left(z_{i,1}\right)=\bar{V}_i\left(0\right)$. Therefore, the Lyapunov function $\bar{V}_i\left(z_{i,1}\right)$ is a continuous function.

6.3 Adaptive Nussbaum Design

For each follower, we define the following coordinates change:

$$\begin{aligned}
z_{i,1} &= \xi_i=\sum_{k\in\mathcal{N}_i}a_{ik}\left(y_i-y_k\right)+b_i\left(y_i-y_0\right)\\
z_{i,j} &= \hat{x}_{i,j}-\hat{\alpha}_{i,j-1},\quad j=2,\dots,n\\
z_{i,n+1} &= \hat{\Gamma}_i(u_i)-\hat{\alpha}_{i,n}
\end{aligned}\tag{6.25}$$

where $\hat{\Gamma}_i(\cdot)$ and the signal u_i is to be designed in Sect. 6.3, $\hat{\alpha}_{i,j}$ is the DSC filter output to avoid the repeat differentiation of the virtual control input $\alpha_{i,j}$. For the DSC filter of jth subsystem for ith follower, we define the filter error as

$$\tilde{\alpha}_{i,j} = \hat{\alpha}_{i,j} - \alpha_{i,j}, \quad j = 1, ..., n \tag{6.26}$$

where $\alpha_{i,j}$ and $\hat{\alpha}_{i,j}$ are input and output of the DSC filter.

The adaptive output feedback controller for each follower is designed within the backstepping framework. The commonly used Nussbaum function selections are $x^2 \sin(x)$ and $x^2 \cos(x)$. In this work, we select

$$N(x) = x^2 \cos(x).$$

6.3.1 Step 1

Consider the following Lyapunov function:

$$V_{i,1} = V_{i,0} + \frac{1}{2}\log\frac{1}{1-\eta_i^2} + \frac{\underline{\tilde{W}}_{i,1}^{\mathrm{T}}\underline{\tilde{W}}_{i,1}}{2} + \frac{\tilde{W}_{i,1}^{\mathrm{T}}\tilde{W}_{i,1}}{2} + \frac{\tilde{\alpha}_{i,1}^2}{2}. \tag{6.27}$$

From (6.20), the time derivative of $V_{i,1}$ satisfies

$$\begin{aligned}
\dot{V}_{i,1} &= \dot{V}_{i,0} + \frac{\eta_i\dot{\eta}_i}{1-\eta_i^2} + \frac{\underline{\tilde{W}}_{i,1}^{\mathrm{T}}\underline{\tilde{W}}_{i,1}}{2} + \frac{\tilde{W}_{i,1}^{\mathrm{T}}\tilde{W}_{i,1}}{2} + \tilde{\alpha}_{i,1}\dot{\tilde{\alpha}}_{i,1} \\
&\le -q_i e_i^{\mathrm{T}} P_i e_i + 3\|e_i\|^2 + M_{i,0} \\
&\quad + \frac{\eta_i\dot{\eta}_i}{1-\eta_i^2} - \underline{\tilde{W}}_{i,1}^{\mathrm{T}}\dot{\hat{\underline{W}}}_{i,1} - \tilde{W}_{i,1}^{\mathrm{T}}\dot{\hat{W}}_{i,1} + \tilde{\alpha}_{i,1}\dot{\tilde{\alpha}}_{i,1}
\end{aligned}$$

with $\frac{\eta_i\dot{\eta}_i}{1-\eta_i^2} = \frac{q(z_{i,1})\eta_{i,b}}{1-\eta_{i,b}^2}\dot{\eta}_{i,b} + \frac{[1-q(z_{i,1})]\eta_{i,a}}{1-\eta_{i,a}^2}\dot{\eta}_{i,a}$. Based on (6.23), $\dot{V}_{i,1}$ satisfies

$$\begin{aligned}
\dot{V}_{i,1} &\le -q_i e_i^{\mathrm{T}} P_i e_i + 3\|e_i\|^2 + M_{i,0} - \underline{\tilde{W}}_{i,1}^{\mathrm{T}}\dot{\hat{\underline{W}}}_{i,1} - \tilde{W}_{i,1}^{\mathrm{T}}\dot{\hat{W}}_{i,1} + \tilde{\alpha}_{i,1}\dot{\tilde{\alpha}}_{i,1} \\
&\quad + \frac{q(z_{i,1})\eta_{i,b}}{\beta_{i,b}(t)\left(1-\eta_{i,b}^2\right)}\left[\dot{z}_{i,1} - z_{i,1}\frac{\dot{\beta}_{i,b}(t)}{\beta_{i,b}(t)}\right] \\
&\quad + \frac{[1-q(z_{i,1})]\eta_{i,a}}{\beta_{i,a}(t)\left(1-\eta_{i,a}^2\right)}\left[\dot{z}_{i,1} - z_{i,1}\frac{\dot{\beta}_{i,a}(t)}{\beta_{i,a}(t)}\right] \\
&= -q_i e_i^{\mathrm{T}} P_i e_i + 3\|e_i\|^2 + M_{i,0} - \underline{\tilde{W}}_{i,1}^{\mathrm{T}}\dot{\hat{\underline{W}}}_{i,1} - \tilde{W}_{i,1}^{\mathrm{T}}\dot{\hat{W}}_{i,1} + \tilde{\alpha}_{i,1}\dot{\tilde{\alpha}}_{i,1} \\
&\quad + \mu_i z_{i,1}\left[\dot{z}_{i,1} - q z_{i,1}\frac{\dot{\beta}_{i,b}(t)}{\beta_{i,b}(t)} - (1-q)z_{i,1}\frac{\dot{\beta}_{i,a}(t)}{\beta_{i,a}(t)}\right]
\end{aligned} \tag{6.28}$$

with

$$\mu_i z_{i,1} = \frac{q\left(z_{i,1}\right)\eta_{i,b}}{\beta_{i,b}(t)\left(1-\eta_{i,b}^2\right)} + \frac{\left[1-q\left(z_{i,1}\right)\right]\eta_{i,a}}{\beta_{i,a}(t)\left(1-\eta_{i,a}^2\right)},$$

$$\mu_i = \frac{q\left(z_{i,1}\right)}{\beta_{i,b}^2(t)-z_{i,1}^2} + \frac{1-q\left(z_{i,1}\right)}{\beta_{i,a}^2(t)-z_{i,1}^2}.$$

Based on (6.12), we have $x_{i,2} = \tilde{x}_{i,2} + \hat{x}_{i,2}$. Then,

$$\begin{aligned}\dot{z}_{i,1} &= \sum_{k\in\mathcal{N}_i} a_{ik}\left(\dot{y}_i - \dot{y}_k\right) + b_i\left(\dot{y}_i - \dot{y}_0\right)\\ &= \delta_{i,i}\left[x_{i,2} + f_{i,1}\left(x_{i,1}\right) + \underline{f}_{i,1}\left(\underline{x}_{i,1}\right)\right] - \sum_{k\in\mathcal{N}_i} a_{i,k}x_{k,2} - b_i\dot{y}_0\end{aligned} \qquad (6.29)$$

where $\underline{f}_{i,1}\left(\underline{x}_{i,1}\right) = -\frac{1}{\delta_{i,i}}\sum_{k\in\mathcal{N}_i} a_{i,k}f_{k,1}\left(x_{k,1}\right)$, and $\underline{x}_{i,1}$ is the stack column vector of $x_{k,1}$ with $\forall k \in \mathcal{N}_i$.

According to Lemma 1.5, given $\bar{\epsilon}_{i,1} > 0$, there exists a constant vector $\underline{W}_{i,1}^*$ and a function $\psi_{i,1}(\cdot)$ with the same dimension such that

$$\underline{f}_{i,1}\left(\underline{x}_{i,1}\right) = \left(\underline{W}_{i,1}^*\right)^{\mathrm{T}}\psi_{i,1}\left(\underline{x}_{i,1}\right) + \epsilon_{i,1},\ \left\|\epsilon_{i,1}\right\| \le \bar{\epsilon}_{i,1},$$

where $\bar{\epsilon}_{i,1}$ is a finite approximation residual.

Denote the estimation of $\underline{W}_{i,1}^*$ as $\hat{\underline{W}}_{i,1}$ with estimation error $\tilde{\underline{W}}_{i,1} = \underline{W}_{i,1}^* - \hat{\underline{W}}_{i,1}$. Combining (6.29) and the FLS representation of the function $\underline{f}_{i,1}(\cdot)$ yields

$$\begin{aligned}\dot{z}_{i,1} &= \delta_{i,i}\left[\tilde{\underline{W}}_{i,1}^{\mathrm{T}}\psi_{i,1}\left(\underline{x}_{i,1}\right) + \underline{W}_{i,1}^{\mathrm{T}}\psi_{i,1}\left(\underline{x}_{i,1}\right) + \epsilon_{i,1}\right]\\ &\quad + \delta_{i,i}\left[\tilde{W}_{i,1}^{\mathrm{T}}\varphi_{i,1}\left(x_{i,1}\right) + W_{i,1}^{\mathrm{T}}\varphi_{i,1}\left(x_{i,1}\right) + \varepsilon_{i,1} + x_{i,2}\right]\\ &\quad - \sum_{k\in\mathcal{N}_i} a_{i,k}x_{k,2} - b_i\dot{y}_0\\ &= \delta_{i,i}\left[\tilde{\underline{W}}_{i,1}^{\mathrm{T}}\psi_{i,1}\left(\underline{x}_{i,1}\right) + \hat{\underline{W}}_{i,1}^{\mathrm{T}}\psi_{i,1}\left(\underline{x}_{i,1}\right) + \epsilon_{i,1} + \tilde{W}_{i,1}^{\mathrm{T}}\varphi_{i,1}\left(x_{i,1}\right)\right]\\ &\quad + \delta_{i,i}\left[\hat{W}_{i,1}^{\mathrm{T}}\varphi_{i,1}\left(x_{i,1}\right) + \varepsilon_{i,1} + x_{i,2}\right]\\ &\quad - \sum_{k\in\mathcal{N}_i} a_{i,k}\left(x_{k,2} - \hat{x}_{k,2} + \hat{x}_{k,2}\right) - b_i\dot{y}_0.\end{aligned}$$

From the variables definition in (6.12), (6.25), and (6.26), the dynamics of $z_{i,1}$ can be further expressed as

$$\begin{aligned}\dot{z}_{i,1} =& \delta_{i,i}\underline{\tilde{W}}_{i,1}^{\mathrm{T}}\psi_{i,1}\left(\underline{x}_{i,1}\right) + \delta_{i,i}\underline{\hat{W}}_{i,1}^{\mathrm{T}}\psi_{i,1}\left(\underline{x}_{i,1}\right) + \delta_{i,i}\epsilon_{i,1} \\ &+\delta_{i,i}\tilde{W}_{i,1}^{\mathrm{T}}\varphi_{i,1}\left(x_{i,1}\right) + \delta_{i,i}\hat{W}_{i,1}^{\mathrm{T}}\varphi_{i,1}\left(x_{i,1}\right) + \delta_{i,i}\varepsilon_{i,1} \\ &+\delta_{i,i}\tilde{x}_{i,2} + \delta_{i,i}z_{i,2} + \delta_{i,i}\tilde{\alpha}_{i,1} + \delta_{i,i}\alpha_{i,1} \\ &-\sum_{k\in\mathcal{N}_i} a_{i,k}\hat{x}_{k,2} - \sum_{k\in\mathcal{N}_i} a_{i,k}\tilde{x}_{k,2} - b_i\dot{y}_0. \end{aligned} \tag{6.30}$$

Inserting (6.30) into $\dot{V}_{i,1}$ yields

$$\begin{aligned}\dot{V}_{i,1} \le& \mu_i z_{i,1}\Big[\delta_{i,i}\underline{\tilde{W}}_{i,1}^{\mathrm{T}}\psi_{i,1}\left(\underline{x}_{i,1}\right) + \delta_{i,i}\underline{\hat{W}}_{i,1}^{\mathrm{T}}\psi_{i,1}\left(\underline{x}_{i,1}\right) + \delta_{i,i}\tilde{W}_{i,1}^{\mathrm{T}}\varphi_{i,1}\left(x_{i,1}\right) \\ &+\delta_{i,i}\hat{W}_{i,1}^{\mathrm{T}}\varphi_{i,1}\left(x_{i,1}\right) + \delta_{i,i}\epsilon_{i,1} + \delta_{i,i}\varepsilon_{i,1} + \delta_{i,i}\alpha_{i,1} \\ &-\sum_{k\in\mathcal{N}_i} a_{i,k}\hat{x}_{k,2} - \sum_{k\in\mathcal{N}_i} a_{i,k}\tilde{x}_{k,2} + \delta_{i,i}\tilde{x}_{i,2} + \delta_{i,i}z_{i,2} \\ &+\delta_{i,i}\tilde{\alpha}_{i,1} - b_i\dot{y}_0 - qz_{i,1}\frac{\dot{\beta}_{i,b}(t)}{\beta_{i,b}(t)} - (1-q)z_{i,1}\frac{\dot{\beta}_{i,a}(t)}{\beta_{i,a}(t)}\Big] \\ &+M_{i,0} - q_i e_i^{\mathrm{T}}P_i e_i + 3\|e_i\|^2 - \underline{\tilde{W}}_{i,1}^{\mathrm{T}}\dot{\underline{\hat{W}}}_{i,1} - \tilde{W}_{i,1}^{\mathrm{T}}\dot{\hat{W}}_{i,1} + \tilde{\alpha}_{i,1}\dot{\tilde{\alpha}}_{i,1}. \end{aligned}$$

Applying Young's inequality, we have

$$\begin{aligned}\mu_i z_{i,1}z_{i,2} &\le \frac{1}{4}\left(\mu_i z_{i,1}\right)^2 + z_{i,2}^2, \\ \mu_i z_{i,1}\tilde{\alpha}_{i,1} &\le \frac{1}{4}\left(\mu_i z_{i,1}\right)^2 + \tilde{\alpha}_{i,1}^2, \\ \mu_i z_{i,1}\left(\epsilon_{i,1} + \varepsilon_{i,1}\right) &\le \frac{1}{4}\left(\mu_i z_{i,1}\right)^2 + 2\left(\left\|\epsilon_{i,1}\right\|^2 + \left\|\varepsilon_{i,1}\right\|^2\right), \\ \mu_i z_{i,1}\left[\delta_{i,i}\tilde{x}_{i,2} - \sum_{k\in\mathcal{N}_i} a_{i,k}\tilde{x}_{k,2}\right] &\le \frac{1}{4}\left(\mu_i z_{i,1}\right)^2 + \left(m^2 + m - 1\right)\|\tilde{x}_i\|^2. \end{aligned}$$

Then, $\dot{V}_{i,1}$ further satisfies

$$\begin{aligned}\dot{V}_{i,1} \le& -q_i e_i^{\mathrm{T}}P_i e_i + 3\|e_i\|^2 + M_{i,0} + \underline{\tilde{W}}_{i,1}^{\mathrm{T}}\left[\delta_{i,i}\mu_i z_{i,1}\psi_{i,1}\left(\underline{x}_{i,1}\right) - \dot{\underline{\hat{W}}}_{i,1}\right] \\ &+\tilde{W}_{i,1}^{\mathrm{T}}\left[\delta_{i,i}\mu_i z_{i,1}\varphi_{i,1}\left(x_{i,1}\right) - \dot{\hat{W}}_{i,1}\right] + \tilde{\alpha}_{i,1}\dot{\tilde{\alpha}}_{i,1} \\ &+\delta_{i,i}\left[z_{i,2}^2 + \tilde{\alpha}_{i,1}^2 + 2\left(\left\|\epsilon_{i,1}\right\|^2 + \left\|\varepsilon_{i,1}\right\|^2\right) + \left(m^2 + m - 1\right)\|\tilde{x}_i\|^2\right] \\ &+\mu_i z_{i,1}\Big[\delta_{i,i}\alpha_{i,1} + \delta_{i,i}\mu_i z_{i,1} + \delta_{i,i}\underline{\hat{W}}_{i,1}^{\mathrm{T}}\psi_{i,1}\left(\underline{x}_{i,1}\right) \\ &+\delta_{i,i}\hat{W}_{i,1}^{\mathrm{T}}\varphi_{i,1}\left(x_{i,1}\right) - b_i\dot{y}_0 - \sum_{k\in\mathcal{N}_i} a_{i,k}\hat{x}_{k,2} - \ell_i\left(z_{i,1}\right)\Big]. \end{aligned} \tag{6.31}$$

The virtual control law $\alpha_{i,1}$ is designed as

$$\alpha_{i,1} = \frac{1}{\delta_{i,i}} \sum_{k \in \mathcal{N}_i} a_{i,k} \hat{x}_{k,2} + \frac{b_i \dot{y}_0}{\delta_{i,i}} + \frac{\ell_i\left(z_{i,1}\right)}{\delta_{i,i}} - \mu_{i,1} z_{i,1}$$
$$- \frac{c_{i,1}}{\delta_{i,i}} z_{i,1} - \hat{W}_{i,1}^{\mathrm{T}} \varphi_{i,1}\left(x_{i,1}\right) - \underline{\hat{W}}_{i,1}^{\mathrm{T}} \psi_{i,1}\left(\underline{x}_{i,1}\right),$$
$$\ell_i\left(z_{i,1}\right) = q z_{i,1} \frac{\dot{\beta}_{i,b}(t)}{\beta_{i,b}(t)} + (1-q) z_{i,1} \frac{\dot{\beta}_{i,a}(t)}{\beta_{i,a}(t)}. \tag{6.32}$$

In addition, according to (6.32),

$$\lim_{z_{i,1} \to 0^-} \ell\left(z_{i,1}\right) = \lim_{z_{i,1} \to 0^-} z_{i,1} \frac{\dot{k}_{i,a}(t)}{\beta_{i,a}(t)} = 0$$
$$\lim_{z_{i,1} \to 0^+} \ell\left(z_{i,1}\right) = \lim_{z_{i,1} \to 0^-} z_{i,1} \frac{\dot{k}_{i,b}(t)}{\beta_{i,b}(t)} = 0$$
$$\ell\left(z_{i,1}\right)\Big|_{z_{i,1}=0} = z_{i,1} \frac{\dot{k}_{i,a}(t)}{\beta_{i,a}(t)} = 0.$$

Then $\lim_{z_{i,1} \to 0^-} \ell\left(z_{i,1}\right) = \lim_{z_{i,1} \to 0^+} \ell\left(z_{i,1}\right) = \ell\left(z_{10}\right)$. Because $z_{i,1}$, $\theta_{i,1}^{\mathrm{T}} \varphi_{i,1}\left(x_{i,1}\right)$, $\dot{y}_d$, and $\mu_i z_{i,1}$ are continuous functions. Then $\alpha_{i,1}$ is continuous. Therefore, the virtual control input $\alpha_{i,1}$ is continuous in its argument.

Based on the virtual control design (6.32), (6.31) is equivalent to

$$\dot{V}_{i,1} \leq -q_i e_i^{\mathrm{T}} P_i e_i + 3\|e_i\|^2 + M_{i,0} + \underline{\tilde{W}}_{i,1}^{\mathrm{T}} \left[\delta_{i,i} \mu_i z_{i,1} \psi_{i,1}\left(\underline{x}_{i,1}\right) - \dot{\underline{\hat{W}}}_{i,1}\right]$$
$$+ \tilde{W}_{i,1}^{\mathrm{T}} \left[\delta_{i,i} \mu_i z_{i,1} \varphi_{i,1}\left(x_{i,1}\right) - \dot{\hat{W}}_{i,1}\right] + \tilde{\alpha}_{i,1} \dot{\tilde{\alpha}}_{i,1} - c_{i,1} \mu_i z_{i,1}^2$$
$$+ \delta_{i,i} \left[z_{i,2}^2 + \tilde{\alpha}_{i,1}^2 + 2\left(\left\|\epsilon_{i,1}\right\|^2 + \left\|\varepsilon_{i,1}\right\|^2\right) + \left(m^2 + m - 1\right) \|\tilde{x}_i\|^2\right] \tag{6.33}$$

The parameter adaption law is designed as

$$\dot{\underline{\hat{W}}}_{i,1} = \delta_{i,i} \psi_{i,1}\left(\underline{x}_{i,1}\right) \mu_i z_{i,1} - \sigma_{i,1} \underline{\hat{W}}_{i,1},$$
$$\dot{\hat{W}}_{i,1} = \delta_{i,i} \varphi_{i,1}\left(x_{i,1}\right) \mu_i z_{i,1} - \sigma_{i,1} \hat{W}_{i,1} \tag{6.34}$$

where $\sigma > 0$ are constant parameters. Then,

$$\dot{V}_{i,1} \leq -q_i e_i^{\mathrm{T}} P_i e_i + 3\|e_i\|^2 + M_{i,0} + \sigma_{i,1} \left(\underline{\tilde{W}}_{i,1}^{\mathrm{T}} \underline{\hat{W}}_{i,1} + \tilde{W}_{i,1}^{\mathrm{T}} \hat{W}_{i,1}\right)$$
$$+ \delta_{i,i} \left[z_{i,2}^2 + 2\left(\left\|\epsilon_{i,1}\right\|^2 + \left\|\varepsilon_{i,1}\right\|^2\right) + \left(m^2 + m - 1\right) \|\tilde{x}_i\|^2\right]$$
$$+ \tilde{\alpha}_{i,1} \dot{\tilde{\alpha}}_{i,1} + \delta_{i,i} \tilde{\alpha}_{i,1}^2 - c_{i,1} \mu_i z_{i,1}^2.$$

Design the first-order DCS filter as

$$\tau_{i,1}\dot{\hat{\alpha}}_{i,1}+\hat{\alpha}_{i,1}=\alpha_{i,1},\quad \hat{\alpha}_{i,1}(0)=\alpha_{i,1}(0) \tag{6.35}$$

where $\tau_{i,1}>0$ is the time constant. From (6.26), we have $\dot{\hat{\alpha}}_{i,1}=-\frac{\tilde{\alpha}_{i,1}}{\tau_{i,1}}$. Then, $\dot{\tilde{\alpha}}_{i,1}=-\frac{\tilde{\alpha}_{i,1}}{\tau_{i,1}}+K_{i,1}(\cdot)$, where

$$\begin{aligned}
K_{i,1}(\cdot)=&\frac{c_{i,1}}{\delta_{i,i}}\dot{z}_{i,1}+\mu_{i,1}\dot{z}_{i,1}-\frac{1}{\delta_{i,i}}\sum_{k\in N_i}a_{i,k}\dot{\hat{x}}_{k,2}-\frac{b_i\ddot{y}_0}{\delta_{i,i}}\\
&+\frac{\partial\mu_{i,1}}{\partial z_{i,1}}\dot{z}_{i,1}z_{i,1}-\frac{\dot{z}_{i,1}}{\delta_{i,i}}\left[q\frac{\dot{\beta}_{i,b}(t)}{\beta_{i,b}(t)}+(1-q)\frac{\dot{\beta}_{i,a}(t)}{\beta_{i,a}(t)}\right]\\
&-\frac{z_{i,1}}{\delta_{i,i}}\left[q\frac{\ddot{\beta}_{i,b}\beta_{i,b}-\dot{\beta}_{i,b}^2}{\beta_{i,b}}+(1-q)\frac{\ddot{\beta}_{i,a}\beta_{i,a}-\dot{\beta}_{i,a}^2}{\beta_{i,a}}\right]\\
&+\dot{\hat{W}}_{i,1}^{\mathrm{T}}\varphi_{i,1}(x_{i,1})+\hat{W}_{i,1}^{\mathrm{T}}\frac{\partial\varphi_{i,1}(x_{i,1})}{\partial x_{i,1}}\dot{x}_{i,1}\\
&+\dot{\underline{\hat{W}}}_{i,1}^{\mathrm{T}}\psi_{i,1}(_{i,1})+\hat{\underline{W}}_{i,1}^{\mathrm{T}}\frac{\partial\psi_{i,1}(x_{i,1})}{\partial x_{i,1}}\dot{x}_{i,1}
\end{aligned}$$

which is a continuous function with arguments $z_{i,1}$, $\dot{z}_{i,1}$, $\theta_{i,1}$, y_d, $\dot{y}_d$, $\ddot{y}_d$, ρ, and $\dot{\rho}$.

Applying Young's inequality to the term $\tilde{\alpha}_{i,1}\dot{\tilde{\alpha}}_{i,1}$ yields

$$\begin{aligned}
\tilde{\alpha}_{i,1}\dot{\tilde{\alpha}}_{i,1}&=-\frac{\tilde{\alpha}_{i,1}^2}{\tau_{i,1}}+K_{i,1}\tilde{\alpha}_{i,1}\\
&\le-\frac{\tilde{\alpha}_{i,1}^2}{\tau_{i,1}}+\left|K_{i,1}\tilde{\alpha}_{i,1}\right|\\
&\le-\frac{\tilde{\alpha}_{i,1}^2}{\tau_{i,1}}+\frac{K_{i,1}^2\tilde{\alpha}_{i,1}^2}{2\pi_{i,1}}+\frac{\pi_{i,1}}{2}.
\end{aligned} \tag{6.36}$$

Then,

$$\begin{aligned}
\dot{V}_{i,1}\le&-q_ie_i^{\mathrm{T}}P_ie_i+3\|e_i\|^2+M_{i,0}+\sigma_{i,1}\left(\tilde{\underline{W}}_{i,1}^{\mathrm{T}}\hat{\underline{W}}_{i,1}+\tilde{W}_{i,1}^{\mathrm{T}}\hat{W}_{i,1}\right)\\
&-\left[\frac{1}{\tau_{i,1}}-\frac{K_{i,1}^2}{2\pi_{i,1}}-\delta_{i,i}\right]\tilde{\alpha}_{i,1}^2-c_{i,1}\mu_iz_{i,1}^2+\delta_{i,i}z_{i,2}^2\\
&+\delta_{i,i}\left[\left(m^2+m-1\right)\left\|\sum_{i\in\mathcal{F}}\tilde{x}_{i,2}^2\right\|+\frac{\pi_{i,1}}{2\delta_{i,i}}+2\left(\|\epsilon_{i,1}\|^2+\|\varepsilon_{i,1}\|^2\right)\right]\\
=&-q_ie_i^{\mathrm{T}}P_ie_i+3\|e_i\|^2+\sigma_{i,1}\left(\tilde{\underline{W}}_{i,1}^{\mathrm{T}}\hat{\underline{W}}_{i,1}+\tilde{W}_{i,1}^{\mathrm{T}}\hat{W}_{i,1}\right)\\
&-\left[\frac{1}{\tau_{i,1}}-\frac{K_{i,1}^2}{2\pi_{i,1}}-\delta_{i,i}\right]\tilde{\alpha}_{i,1}^2-c_{i,1}\mu_iz_{i,1}^2+\delta_{i,i}z_{i,2}^2+M_{i,1}
\end{aligned} \tag{6.37}$$

with $M_{i,1}=\left(m^2+m-1\right)\|\tilde{x}_i\|^2+2\left(\left\|\epsilon_{i,1}\right\|^2+\left\|\varepsilon_{i,1}\right\|^2\right)+M_{i,0}+\frac{\pi_{i,1}}{2\delta_{i,i}}$.

6.3.2 Step j $(2 \leq j \leq n)$

Consider the following Lyapunov function candidate,

$$V_{i,j} = V_{i,j-1} + \frac{1}{2}z_{i,j}^2 + \frac{\tilde{W}_{i,j}^{\mathrm{T}}\tilde{W}_{i,j}}{2} + \frac{\tilde{\alpha}_{i,j}\tilde{\alpha}_{i,j}}{2}.$$

According to (6.25) and (6.11), the dynamics of $z_{i,j}$ for $j = 2, ..., n-1$ and $j = n$ can be written, respectively, as

$$\begin{aligned}
\dot{z}_{i,j} &= \hat{W}_{i,j}^{\mathrm{T}}\varphi_{i,j}\left(\hat{\bar{x}}_{i,j}\right) + \hat{x}_{i,j+1} + k_{i,j}\left[r_i(t)\right]^j\tilde{\xi}_i - \dot{\hat{\alpha}}_{i,j-1} \\
&= \hat{W}_{i,j}^{\mathrm{T}}\varphi_{i,j}\left(\hat{\bar{x}}_{i,j}\right) + z_{i,j+1} + \hat{\alpha}_{i,j} + k_{i,j}\left[r_i(t)\right]^j\tilde{\xi}_i - \dot{\hat{\alpha}}_{i,j-1}, \quad j = 2, ..., n-1 \\
\dot{z}_{i,n} &= \hat{W}_{i,n}^{\mathrm{T}}\varphi_{i,n}\left(\hat{\bar{x}}_{i,n}\right) + g\left(u_i\right) + k_{i,n}\left[r_i(t)\right]^n\tilde{\xi}_i - \dot{\hat{\alpha}}_{i,n-1} \\
&= \hat{W}_{i,n}^{\mathrm{T}}\varphi_{i,n}\left(\hat{\bar{x}}_{i,n}\right) + z_{i,n+1} + \hat{\alpha}_{i,n} + k_{i,n}\left[r_i(t)\right]^n\tilde{\xi}_i - \dot{\hat{\alpha}}_{i,n-1}.
\end{aligned}$$

To summarize, one has

$$\dot{z}_{i,j} = \hat{W}_{i,j}^{\mathrm{T}}\varphi_{i,j}\left(\hat{\bar{x}}_{i,j}\right) + z_{i,j+1} + \alpha_{i,j} + \tilde{\alpha}_{i,j} + k_{i,j}\left[r_i(t)\right]^j\tilde{\xi}_i - \dot{\hat{\alpha}}_{i,j-1}, \quad j = 2, ..., n.$$

According to the dynamics of $z_{i,j}$, differentiating $V_{i,j}$ yields

$$\begin{aligned}
\dot{V}_{i,j} =& \dot{V}_{i,j-1} + z_{i,j}\dot{z}_{i,j} - \tilde{W}_{i,j}^{\mathrm{T}}\dot{\hat{W}}_{i,j} + \tilde{\alpha}_{i,j}\dot{\tilde{\alpha}}_{i,j} \\
=& \dot{V}_{i,j-1} + z_{i,j}\left[\hat{W}_{i,j}^{\mathrm{T}}\varphi_{i,j}\left(\hat{\bar{x}}_{i,j}\right) + z_{i,j+1} + \alpha_{i,j} + \tilde{\alpha}_{i,j} + k_{i,j}\left[r_i(t)\right]^j\tilde{\xi}_i - \dot{\hat{\alpha}}_{i,j-1}\right] \\
&- \tilde{W}_{i,j}^{\mathrm{T}}\dot{\hat{W}}_{i,j} + \tilde{\alpha}_{i,j}\dot{\tilde{\alpha}}_{i,j} - \tilde{W}_{i,j}^{\mathrm{T}}\varphi_{i,j}\left(\hat{\bar{x}}_{i,j}\right)z_{i,j} + \tilde{W}_{i,j}^{\mathrm{T}}\varphi_{i,j}\left(\hat{\bar{x}}_{i,j}\right)z_{i,j}.
\end{aligned}$$

Design the virtual control input $\alpha_{i,j}$ and the parameter adaptation law for $\hat{W}_{i,j}$ as

$$\begin{aligned}
\alpha_{i,j} =& -\frac{1}{2}z_{i,j} - c_{i,j}z_{i,j} - \hat{W}_{i,j}^{\mathrm{T}}\varphi_{i,j}\left(\hat{\bar{x}}_{i,j}\right) \\
&- z_{i,j+1} - k_{i,j}\left[r_i(t)\right]^j\tilde{\xi}_i + \dot{\hat{\alpha}}_{i,j-1} \\
\dot{\hat{W}}_{i,j} =& z_{i,j}\varphi_{i,j}\left(\hat{\bar{x}}_{i,j}\right) - \sigma_i\hat{W}_{i,j}, \sigma_i > 0.
\end{aligned} \tag{6.38}$$

Using the virtual control design (6.38) and the adaptive parameter update law (6.38), one can obtain

$$\begin{aligned}\dot{V}_{i,j} = &\dot{V}_{i,j-1} - \frac{1}{2}z_{i,j}^2 - c_{i,j}z_{i,j}^2 + z_{i,j}\tilde{\alpha}_{i,j} + \tilde{\alpha}_{i,j}\dot{\tilde{\alpha}}_{i,j} \\ &-\tilde{W}_{i,j}^{\mathrm{T}}\left[\dot{\hat{W}}_{i,j} - \varphi_{i,j}\left(\hat{\bar{x}}_{i,j}\right)z_{i,j}\right] - \tilde{W}_{i,j}^{\mathrm{T}}\varphi_{i,j}\left(\hat{\bar{x}}_{i,j}\right)z_{i,j} \\ \le &\dot{V}_{i,j-1} - c_{i,j}z_{i,j}^2 + \sigma_i\tilde{W}_{i,j}^{\mathrm{T}}\hat{W}_{i,j} + Y_{i,j}\left\|\tilde{W}_{i,j}\right\|^2 + \tilde{\alpha}_{i,j}^2 + \tilde{\alpha}_{i,j}\dot{\tilde{\alpha}}_{i,j}\end{aligned}$$

due to the following facts based on Young's inequality:

$$\begin{aligned}z_{i,j}\tilde{\alpha}_{i,j} &\le \frac{1}{4}z_{i,j}^2 + \tilde{\alpha}_{i,j}^2 \\ -z_{i,j}\tilde{W}_{i,j}^{\mathrm{T}}\varphi_{i,j}\left(\hat{\bar{x}}_{i,j}\right) &\le \frac{1}{4}z_{i,j}^2 + Y_{i,j}\left\|\tilde{W}_{i,j}\right\|^2.\end{aligned}$$

To avoid the repeated differentiation of the virtual control input $\alpha_{i,j}$, we defined the first-order DSC filter as

$$\tau_i\dot{\hat{\alpha}}_i + \hat{\alpha}_i = \alpha_i, \quad \hat{\alpha}_i(0) = \alpha_i(0), \tau_i > 0. \tag{6.39}$$

Similar to Step 1, we have

$$\dot{\hat{\alpha}}_{i,j} = -\frac{\tilde{\alpha}_{i,j}}{\tau_{i,j}}. \tag{6.40}$$

Then, from (6.26), one has

$$\dot{\tilde{\alpha}}_{i,j} = \dot{\hat{\alpha}}_{i,j} - \dot{\alpha}_{i,j} = -\frac{\tilde{\alpha}_{i,j}}{\tau_{i,j}} + K_{i,j}(\cdot)$$

with

$$\begin{aligned}K_{i,j}(\cdot) = &\left(\frac{1}{2} + c_{i,j}\right)\dot{z}_{i,j} + \dot{z}_{i,j+1} + \ddot{\hat{\alpha}}_{i,j-1} + \dot{\hat{W}}_{i,j}^{\mathrm{T}}\varphi_{i,j}\left(\hat{\bar{x}}_{i,j}\right) \\ &+\hat{W}_{i,j}^{\mathrm{T}}\frac{\partial\varphi_{i,j}\left(\hat{\bar{x}}_{i,j}\right)}{\partial\hat{\bar{x}}_{i,j}}\dot{\hat{\bar{x}}}_{i,j} + k_{i,j}jr_i^{j-1}\dot{r}_i(t)\tilde{\xi}_i + k_{i,j}r_i^j\dot{\tilde{\xi}}_i.\end{aligned}$$

Applying Young's inequality to the term $\tilde{\alpha}_{i,j}\dot{\tilde{\alpha}}_{i,j}$ yields

$$\tilde{\alpha}_{i,j}\dot{\tilde{\alpha}}_{i,j} \le -\frac{\tilde{\alpha}_{i,j}^2}{\tau_{i,j}} + \frac{K_{i,j}^2(\cdot)}{2\pi_{i,j}}\tilde{\alpha}_{i,j}^2 + \frac{\pi_{i,j}}{2}.$$

Therefore, $\dot{V}_{i,j}$ further satisfies

$$\begin{aligned}\dot{V}_i &\leq \dot{V}_{i,j-1} - c_{i,j} z_{i,j}^2 + \sigma_i \tilde{W}_{i,j}^{\mathrm{T}} \hat{W}_{i,j} + Y_{i,j} \left\| \tilde{W}_{i,j} \right\|^2 + \tilde{\alpha}_{i,j}^2 + \tilde{\alpha}_{i,j} \dot{\hat{\alpha}}_{i,j} \\ &\leq -q_i e_i^{\mathrm{T}} P_i e_i + 3\|e_i\|^2 + \sigma_i \left(\underline{\tilde{W}}_{i,1}^{\mathrm{T}} \underline{\hat{W}}_{i,1} + \sum_{l=1}^{j} \tilde{W}_{i,j}^{\mathrm{T}} \hat{W}_{i,j} \right) \\ &\quad - \left[\frac{1}{\tau_{i,1}} - \frac{K_{i,1}^2(\cdot)}{2\pi_{i,1}} - \delta_{i,i} \right] \tilde{\alpha}_{i,1}^2 - \sum_{l=2}^{j} \left[\frac{1}{\tau_{i,l}} - \frac{K_{i,l}^2(\cdot)}{2\pi_{i,l}} - 1 \right] \tilde{\alpha}_{i,l}^2 \\ &\quad - c_{i,1} \mu_i z_{i,1}^2 - \left(c_{i,2} - \delta_{i,i} \right) z_{i,2}^2 + \sum_{l=3}^{j} c_{i,l} z_{i,l}^2 + M_{i,j},\end{aligned}$$

where $M_{i,j} = M_{i,j-1} + Y_{i,j} \left\| \tilde{W}_{i,j} \right\|^2 + \frac{\pi_{i,j}}{2}$.

6.3.3 Step n + 1

Consider the Lyapunov candidate as

$$V_{i,n+1} = V_{i,n} + \frac{1}{2} z_{i,n+1}^2. \tag{6.41}$$

In this step, we introduce an additional dynamical system for the control input signal u_i as

$$\dot{u}_i = -c_i u_i + g_i \tag{6.42}$$

with

$$\begin{aligned}\bar{g}_i &= -c_{i,n+1} z_{i,n+1} + \dot{\hat{\alpha}}_{i,n} + c_i u_i \frac{\partial \hat{\Gamma}(u_i)}{\partial u_i}, \\ \dot{\omega}_i &= \gamma_i z_{i,n+1} \bar{g}_i, \\ g_i &= N(\omega_i) \bar{g}_i,\end{aligned} \tag{6.43}$$

where $N(\cdot)$ is the Nussbaum function defined in (6.27).

Then, differentiating $z_{i,n+1}$ yields

$$\dot{z}_{i,n+1} = \frac{\partial \hat{\Gamma}(u_i)}{\partial u_i} \dot{u}_i - \dot{\hat{\alpha}}_{i,n} = \frac{\partial \hat{\Gamma}(u_i)}{\partial u_i} (-c_i u_i + g_i) - \dot{\hat{\alpha}}_{i,n}.$$

According to (6.43) and above fact, we have

$$\begin{aligned}\dot{z}_{i,n+1}+\bar{g}_i &= \frac{\partial\hat{\Gamma}(u_i)}{\partial u_i}(-c_i u_i+g_i)-\dot{\hat{\alpha}}_{i,n}-c_{i,n+1}z_{i,n+1}+\dot{\hat{\alpha}}_n+c_i u_i\frac{\partial\hat{\Gamma}(u_i)}{\partial u_i}\\ &= \frac{\partial\hat{\Gamma}(u_i)}{\partial u_i}g_i-c_{i,n+1}z_{i,n+1}.\end{aligned}$$

Differentiating $V_{i,n+1}$ yields (6.44) with $\varpi_i=\frac{\partial\hat{\Gamma}_i(u_i)}{\partial u_i}$.

$$\begin{aligned}\dot{V}_{i,n+1} \le& -q_i e_i^{\mathrm{T}}P_i e_i+3\|e_i\|^2+\sigma_i\left(\underline{\tilde{W}}_{i,1}^{\mathrm{T}}\underline{\hat{W}}_{i,1}+\sum_{j=1}^{n}\tilde{W}_{1,j}^{\mathrm{T}}\hat{W}_{1,j}\right)\\ &-\left[\frac{1}{\tau_{i,1}}-\frac{K_{i,1}^2(\cdot)}{2\pi_{i,1}}-\delta_{i,i}\right]\tilde{\alpha}_{i,1}^2-\sum_{l=2}^{n}\left[\frac{1}{\tau_{i,l}}-\frac{K_{i,l}^2(\cdot)}{2\pi_{i,l}}-1\right]\tilde{\alpha}_{i,l}^2\\ &-c_{i,1}\mu_i z_{i,1}^2-\left(c_{i,2}-\delta_{i,i}\right)z_{i,2}^2+\sum_{l=3}^{n+1}c_{i,l}^2 z_{i,l}^2\\ &+M_{i,n}+\frac{1}{\gamma_i}\left[\varpi_i N(\omega_i)-1\right]\dot{\omega}_i. \end{aligned} \tag{6.44}$$

For $j=1,\ldots,n$, applying Young's inequality yields

$$\begin{aligned}\tilde{W}_{i,j}^{\mathrm{T}}\hat{W}_{i,j} &\le -\frac{1}{2}\tilde{W}_{i,j}^{\mathrm{T}}\tilde{W}_{i,j}+\frac{1}{2}W_{i,j}^{*\mathrm{T}}W_{i,j}^{*},\\ \underline{\tilde{W}}_{i,1}^{\mathrm{T}}\underline{\hat{W}}_{i,1} &\le -\frac{1}{2}\underline{\tilde{W}}_{i,1}^{\mathrm{T}}\underline{\tilde{W}}_{i,1}+\frac{1}{2}\underline{W}_{i,1}^{*\mathrm{T}}\underline{W}_{i,1}^{*}.\end{aligned} \tag{6.45}$$

Substituting (6.45) into (6.44) yields

$$\begin{aligned}\dot{V}_{i,n+1} \le& -q_i e_i^{\mathrm{T}}P_i e_i+3\|e_i\|^2-\left[\frac{1}{\tau_{i,1}}-\frac{K_{i,1}^2(\cdot)}{2\pi_{i,1}}-\delta_{i,i}\right]\tilde{\alpha}_{i,1}^2\\ &-\sum_{l=2}^{n}\left[\frac{1}{\tau_{i,l}}-\frac{K_{i,l}^2(\cdot)}{2\pi_{i,l}}-1\right]\tilde{\alpha}_{i,l}^2+M_{i,n+1}-c_{i,1}\mu_i z_{i,1}^2-\sum_{l=3}^{n+1}c_{i,l}z_{i,l}^2\\ &-\frac{1}{2}\sigma_i\underline{\tilde{W}}_{i,1}^{\mathrm{T}}\underline{\tilde{W}}_{i,1}-\frac{1}{2}\sigma_i\sum_{j=1}^{n}\tilde{W}_{i,j}^{\mathrm{T}}\tilde{W}_{i,j}-\left(c_{i,2}-\delta_{i,i}\right)z_{i,2}^2\\ &+\frac{1}{\gamma_i}\left[\varpi_i N(\omega_i)-1\right]\dot{\omega}_i \end{aligned} \tag{6.46}$$

where $M_{i,n+1} = M_{i,n} + \frac{\sigma_i}{2}\left(\sum_{j=1}^{n} W_{i,j}^{*\mathrm{T}} W_{i,j}^{*} + \underline{W}_{i,1}^{*\mathrm{T}} \underline{W}_{i,1}^{*}\right)$ with detailed expression as

$$M_{i,n+1} = \frac{\sigma_i}{2}\left(\sum_{j=1}^{n} \|W_{i,j}^{*}\|^2 + \|\underline{W}_{i,1}^{*}\|^2\right) + \sum_{j=2}^{n} Y_{i,j}\left\|\tilde{W}_{i,j}\right\|^{+} \frac{\pi_{i,j}}{2\delta_{i,i}} + \sum_{j=2}^{n} \frac{\pi_{i,j}}{2}$$
$$+2\left(\left\|\epsilon_{i,1}\right\|^2 + \left\|\varepsilon_{i,1}\right\|^2\right) + \left(m^2 + m - 1\right)\|\tilde{x}_i\|^2$$
$$+\|P_i\|^2\|\varepsilon_i\|^2 + \|P_i\|^2 \sum_{i=1}^{n} Y_{i,j}\|\tilde{W}_{i,j}\|^2 + \|P_i\|^2\|H_i^{*}\|^2. \tag{6.47}$$

Define

$$\mathcal{S}_{i,j} = \left\{\chi_{i,j}\,\middle|\,\mathcal{K}_{i,j}\left(\chi_{i,j}\right) \le 2p\right\}$$
$$\chi_{i,j} = \left[e_i^{\mathrm{T}}\; z_{i,1} \cdots z_{i,n}\; \tilde{W}_{i,1}^{\mathrm{T}} \cdots \tilde{W}_{i,j}^{\mathrm{T}}\; \underline{\tilde{W}}_{i,1}^{\mathrm{T}} \cdots \underline{\tilde{W}}_{i,j}^{\mathrm{T}}\; \tilde{\alpha}_{i,1} \cdots \tilde{\alpha}_{i,j}\right]^{\mathrm{T}}$$
$$\mathcal{K}_{i,j}\left(\chi_{i,j}\right) = \sum_{l=1}^{j} z_{i,l}^2 + \sum_{l=1}^{j} \tilde{W}_{i,l}^{\mathrm{T}}\tilde{W}_{i,l} + \underline{\tilde{W}}_{i,1}^{\mathrm{T}}\underline{\tilde{W}}_{i,1} + \sum_{l=1}^{j} \tilde{\alpha}_{i,l}^2 + e_i^{\mathrm{T}} P_i e_i.$$

Then $\mathcal{S}_{i,j}$ is a compact set. Because $K_{i,j}(\cdot)$ is a continuous function, there exists a positive constant $\bar{K}_{i,j}$ such that $\left|K_{i,j}(\cdot)\right| \le \bar{K}_{i,j}$ on $\mathcal{S}_{i,j}$. Therefore, we have

$$\dot{V}_{i,n+1} \le -q_i e_i^{\mathrm{T}} P_i e_i + 3\|e_i\|^2 - \frac{1}{2}\sigma_i \underline{\tilde{W}}_{i,1}^{\mathrm{T}}\underline{\tilde{W}}_{i,1} - \frac{1}{2}\sigma_i \sum_{j=1}^{n} \tilde{W}_{i,j}^{\mathrm{T}} \tilde{W}_{i,j}$$
$$-\left[\frac{1}{\tau_{i,1}} - \frac{\bar{K}_{i,1}^2}{2\pi_{i,1}} - \delta_{i,i}\right]\tilde{\alpha}_{i,1}^2 - \sum_{l=2}^{n}\left[\frac{1}{\tau_{i,l}} - \frac{\bar{K}_{i,l}^2}{2\pi_{i,l}} - 1\right]\tilde{\alpha}_{i,l}^2$$
$$-c_{i,1}\mu_i z_{i,1}^2 - \left(c_{i,2} - \delta_{i,i}\right) z_{i,2}^2 - \sum_{l=3}^{n+1} c_{i,l} z_{i,l}^2$$
$$+\frac{1}{\gamma_i}\left[\varpi_i N(\omega_i) - 1\right]\dot{\omega}_i + M_{i,n+1}. \tag{6.48}$$

6.4 Stability Analysis

Definition 6.1 If for $\forall x(t_0) = x_0 \in \sum$ where $\sum$ is a compact subset in R^n, there exist a $\lambda > 0$ and a number $T(\lambda, x_0)$ such that $\|x(t)\| < \lambda$ for all $t > t_0 + T$, then each agent is semi-globally uniformly ultimately bounded (SGUUB).

The closed-loop stability analysis for the adaptive Nussbaum design in Sect. 6.3 can be summarized as the following theorem.

Theorem 6.1 *Under Assumptions 6.1–6.3, consider the virtual control input design* (6.32) *and* (6.38) *with the dynamic surface filter design* (6.35) *and* (6.39)*, the adaptive parameter estimation* (6.34) *and* (6.38) *for MAS with followers* (6.1)*, and the control input design* (6.42)*. Then, the output synchronization error of MAS* (6.1) *and the closed-loop system is semi-global uniformly ultimately bounded provided that the following conditions are satisfied for each follower:*

$$\begin{aligned} &\lambda_{\min}(P_i) - 3 - r_i^{2(a_i+n-1)} > 0, \quad c_{i,1} > 0, \\ &c_{i,2} > \delta_{i,i}, \quad c_i > 0, \quad \frac{1}{\tau_j} - \frac{\bar{K}_{i,j}^2}{2\pi_{i,j}} > \delta_{i,i}. \end{aligned} \tag{6.49}$$

Also, the output tracking error satisfies the prescribed performance (6.21).

Proof The Lyapunov candidate $V_{i,n+1}$ for each follower can be written as

$$V_{i,n+1} = V_{i,e} + V_{i,\eta} + V_{i,z} + V_{i,W} + V_{i,\alpha} \tag{6.50}$$

with

$$\begin{aligned} V_{i,e} &= e_i^{\mathrm{T}} P_i e_i, \quad V_{i,\eta} = \frac{1}{2}\log\frac{1}{1-\eta_i^2}, \\ V_{i,z} &= \frac{1}{2}\sum_{j=2}^{n+1} z_{i,j}^2, \quad V_{i,\alpha} = \frac{1}{2}\sum_{j=1}^{n} \tilde{\alpha}_j^2 \\ V_{i,W} &= \frac{1}{2}\left(\sum_{j=1}^{n} \tilde{W}_{i,j}^{\mathrm{T}}\tilde{W}_{i,j} + \underline{\tilde{W}}_{i,1}^{\mathrm{T}}\underline{\tilde{W}}_{i,1}\right). \end{aligned}$$

It is obvious that the Lyapunov candidate $V_{i,n+1}$ is positive definite. According to (6.47) and (6.48), $\dot{V}_{i,n+1}$ satisfies (6.51)

$$\dot{V}_{i,n+1} \le \dot{V}_{i,e} + \dot{V}_{i,\eta} + \dot{V}_{i,z} + \dot{V}_{i,W} + \dot{V}_{i,\alpha} + \frac{1}{\gamma_i}\left[\varpi_i N(\omega_i) - 1\right]\dot{\omega}_i + D_i \tag{6.51}$$

with

$$
\begin{aligned}
\dot{V}_{i,e} &= -q_i e_i^{\mathrm{T}} P_i e_i + 3\|e_i\|^2 + \left(m^2 + m - 1\right) \|\tilde{x}_i\|^2 \\
\dot{V}_{i,\eta} &= -c_{i,1}\mu_i z_{i,1}^2 \\
\dot{V}_{i,z} &= -\left(c_{i,2} - \delta_{i,i}\right) z_{i,2}^2 - \sum_{j=3}^{n+1} c_{i,j} z_{i,j}^2 \\
\dot{V}_{i,W} &= -\frac{1}{2}\sigma_i \sum_{j=1}^{n} \underline{\tilde{W}}_{i,1}^{\mathrm{T}} \underline{\tilde{W}}_{i,1} + \left(\|P_i\|^2 Y_{i,j} - \frac{1}{2}\sigma_i\right) \tilde{W}_{i,j}^{\mathrm{T}} \tilde{W}_{i,j} \\
&\quad + \sum_{j=2}^{n} \left(\|P_i\|^2 Y_{i,j} - \frac{1}{2}\sigma_i + Y_{i,j}\right) \tilde{W}_{i,j}^{\mathrm{T}} \tilde{W}_{i,j} \\
\dot{V}_{i,\alpha} &= -\left(\frac{1}{\tau_{i,1}} - \frac{\bar{K}_{i,1}^2}{2\pi_{i,1}} - \delta_{i,i}\right) \tilde{\alpha}_{i,1}^2 - \sum_{j=2}^{n} \left(\frac{1}{\tau_{i,j}} - \frac{\bar{K}_{i,j}^2}{2\pi_{i,j}} - 1\right) \tilde{\alpha}_{i,j}^2
\end{aligned}
$$

$$
\begin{aligned}
D_i &= \frac{\sigma_i}{2} \left(\sum_{j=1}^{n} \|W_{i,j}^*\|^2 + \frac{\sigma_i}{2} \|\underline{W}_{i,1}^*\|^2 \right) \\
&\quad + \frac{\pi_{i,j}}{2\delta_{i,i}} + \sum_{j=2}^{n} \frac{\pi_{i,j}}{2} + 2\left(\left\|\epsilon_{i,1}\right\|^2 + \left\|\varepsilon_{i,1}\right\|^2\right) \\
&\quad + \|P_i\|^2 \sum_{i=1}^{n} Y_{i,j} \|\tilde{W}_{i,j}\|^2 + \|P_i\|^2 \|\varepsilon_i\|^2 + \|P_i\|^2 \|H_i^*\|^2.
\end{aligned}
$$

Note the fact that

$$
\begin{aligned}
\lambda_{\min}(P_i) \|e_i\|^2 &\le e_i^{\mathrm{T}} P_i e_i \le \lambda_{\max}(P_i) \|e_i\|^2, \\
\|e_i\|^2 &\le \frac{e_i^{\mathrm{T}} P_i e_i}{\lambda_{\min}(P_i)}, \\
q_i &> 0, \\
\Lambda_i^{-1} &= \mathrm{d}iag\left(r_i^{a_i}, \ldots, r_i^{a_i+n-1}\right), \\
\left\|\Lambda_i^{-1}\right\| &= r_i^{a_i+n-1}, \\
\tilde{x}_i &= \Lambda_i^{-1} e_i, \\
\left(m^2 + m - 1\right) \|\tilde{x}_i\|^2 &\le \left(m^2 + m - 1\right) \|\Lambda_i^{-1}\|^2 \|e_i\|^2 \\
&= \left(m^2 + m - 1\right) r_i^{2(a_i+n-1)} \|e_i\|^2.
\end{aligned}
$$

Then,

$$\dot{V}_{i,e} \leq -q_i e_i^{\mathrm{T}} P_i e_i + \left(3 + r_i^{2(a_i+n-1)}\right) \frac{e_i^{\mathrm{T}} P_i e_i}{\lambda_{\min}(P_i)}$$
$$= -K_{i,e} V_{i,e}$$

with

$$K_{i,e} = \frac{q_i \lambda_{\min}(P_i) - 3 - r^{2(a_i+n-1)}}{\lambda_{\min}(P_i)}. \tag{6.52}$$

According to Lemma 1.11 and (6.51), V_η can be rewritten as

$$\dot{V}_{i,\eta} \leq -c_{i,1} \log \frac{1}{1-\eta_i^2} = -K_{i,\eta} V_{i,\eta} \tag{6.53}$$

with $K_{i,\eta} = 2c_{i,1}$. Similarly, one can obtain (6.54)

$$\begin{aligned} \dot{V}_{i,W} &\leq -K_{i,W} V_{i,W} \\ K_{i,W} &= 2\min\left\{\frac{1}{2}\sigma_i - \|P_i\|^2 Y_{i,j}, \frac{1}{2}\sigma_i - Y_{i,j} - \|P_i\|^2 Y_{i,j}, \frac{1}{2}\sigma_i\right\} \end{aligned} \tag{6.54}$$

and

$$\dot{V}_{i,z} \leq -K_{i,z} V_{i,z} \quad \dot{V}_{i,\alpha} \leq -K_{i,\alpha} V_{i,\alpha} \tag{6.55}$$

with

$$\begin{aligned} K_{i,z} &= 2\min\left\{c_{i,2} - \delta_{i,i}, c_{i,3}, \dots, c_{i,n+1}\right\} \\ K_{i,\alpha} &= 2\min_{j=2,\dots,n}\left\{\frac{1}{\tau_{i,j}} - \frac{\bar{K}_{i,j}^2}{2\pi_{i,j}} - \delta_{i,i}\right\}. \end{aligned} \tag{6.56}$$

Then, one has

$$\begin{aligned} \dot{V}_{i,n+1} &\leq \dot{V}_{i,e} + \dot{V}_{i,\eta} + \dot{V}_{i,z} + \dot{V}_{i,\alpha} + \dot{V}_{i,W} + D_i + \frac{1}{\gamma_i}\left[\varpi_i N(\omega_i) - 1\right]\dot{\omega}_i \\ &\leq -C_i\left(V_{i,\alpha} + V_{i,\eta} + V_{i,z} + V_{i,\alpha} + V_{i,W}\right) + D_i + \frac{1}{\gamma_i}\left[\varpi_i N(\omega_i) - 1\right]\dot{\omega}_i \end{aligned}$$

where $C_i = \min\left\{K_{i,e}, K_{i,\eta}, K_{i,z}, K_{i,W}, K_{i,\alpha}\right\}$.

Based on the conditions in (6.49), $\dot{V}_{i,n+1}$ satisfies

$$\dot{V}_{i,n+1} \leq -C_i V_{i,n+1} + \frac{1}{\gamma_i}\left[\varpi_i N(\omega_i) - 1\right]\dot{\omega}_i + D_i.$$

Integrating the above differential inequality yields

$$
\begin{aligned}
V_{i,n+1} &\le V_{i,n+1}(0)\, e^{-C_i t} + \frac{D_i}{C_i}\left(1 - e^{-C_i t}\right) \\
&\quad + \frac{e^{-C_i t}}{\gamma_i} \int_0^t [\varpi_i N(\omega_i) - 1]\dot{\omega}_i e^{C_i s} ds.
\end{aligned}
\tag{6.57}
$$

Based on [17], we can obtain that $\int_0^t [\varpi_i N(\omega_i) - 1]\dot{\omega}_i ds$ and $\omega_i(t)$ are bounded on $[0, t_f]$ using proof by contradiction. According to (6.57), $V_{i,n+1}(t)$ is bounded. This implies that $\left\{z_{i,j}, \tilde{W}_{i,j}, \alpha_{i,j}\right\}_{j=2}^{n+1}$ and $\left\{\underline{\tilde{W}}_{i,1}, e_i, \eta_i\right\}$ are bounded for each follower. From the fact that $r_i(t) > 0$ remains bounded and (6.15), e is also bounded. From Assumption 6.2, there exists a class-$\mathcal{K}$ function [22] $\Psi_i(\cdot)$ such that

$$
\sup_{0\le s\le t} \|x_i(s)\| \le K_i + \Psi_i\left(\sup_{0\le s\le t} \|u_i(s)\|\right) \le K_i + \Psi_i(\bar{u}_i)
$$

where K_i is a constant. Thus the global boundedness of $x_{i,1}, x_{i,2}, \ldots, x_n$ is established. Based on the facts in (6.12), (6.25), (6.26), and (6.40), we can obtain that $\{\hat{x}_{i,2}\}_{i=2}^{n}$, $\{\hat{\alpha}_{i,j}, \alpha_{i,j}\}_{j=1}^{n}$ are bounded due to the boundedness of $\{e_{i,j}, \alpha_{i,j}, z_{i,j}\}_{j=2}^{n+1}$. From (6.6)–(6.7) and (6.43), $\bar{g}_i$ in (6.43) is bounded due to the boundedness of $z_{i,n+1}$, $\hat{\alpha}_{i,n}$. This further implies that g_i in (6.43) is bounded because ω_i is bounded. Then u_i in (6.42) is bounded. Thus all closed-loop signals are semi-global uniformly ultimately bounded. This completes the proof. □

Remark 6.3 From (6.57), one can observe that larger parameters $c_{i,j}, \sigma_i$, and smaller $\tau_{i,j}$ would increase the decaying rate C_i of the closed-loop signals. On the other hand, larger parameter π_i would increase the parameter D_i, which might result in larger convergence residual set for closed-loop signals.

6.5 Simulation Example

To illustrate the proposed design method described above, we consider the MAS with the directed communication topology as shown in Fig. 6.1. One can find that only followers 1 and 2 are informed by the reference signal $y_d(t) = sin(t)$ and this directed communication graph satisfies Assumption 6.1.

The MAS is considered with the following nonlinear strict-feedback dynamics for each follower:

$$
\begin{aligned}
\dot{x}_{i,1} &= x_{i,2} + f_{i,1}(x_{i,1}) \\
\dot{x}_{i,2} &= \Gamma_i(u_i) + f_{i,2}(x_{i,1}, x_{i,2}) \\
y_i &= x_{i,1}
\end{aligned}
\tag{6.58}
$$

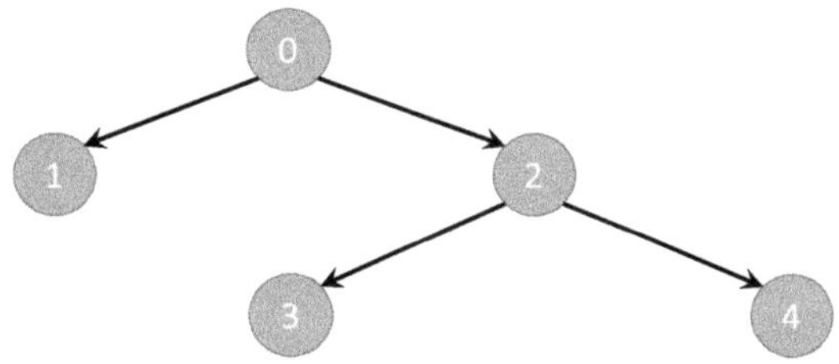

Fig. 6.1 The communication topology

where $f_{i,1}(\cdot)$ and $f_{i,2}(\cdot)$ are unknown smooth functions. The controller design aims at finding a signal u_i such that the output of the system can approximately track $y_d = \sin(t)$ asymptotically. The heterogeneous MAS with agent dynamics in (6.58) for each follower takes the following form:

$$\begin{aligned}
f_{1,1} &= \left(x_1 - x_1^3\right) / \left(1 + x_1^4\right), \\
f_{1,2} &= -e^{-x_1^2} \sin(4x_1) \\
f_{2,1} &= \left(x_1 - x_1^3\right) / \left(3 + 0.9x_1^4\right), \\
f_{2,2} &= -e^{-x_1^2} \sin(5x_1) \\
f_{3,1} &= \left(x_1 - x_1^3\right) / \left(2 + 0.95x_1^4\right), \\
f_{3,2} &= -e^{-x_1^2} \sin(3x_1) \\
f_{4,1} &= 0, \\
f_{4,2} &= -5\sin(x_1)
\end{aligned}$$

with the saturation function

$$\Gamma_i(u_i) = \begin{cases} sign(u_i)\,\bar{u}_i, & |u| \geq \bar{u}_i \\ u_i, & |u| \geq \bar{u}_i \end{cases}$$

where $\bar{u}_1 = 16.5$, $\bar{u}_2 = 17$, $\bar{u}_3 = 12$ and $\bar{u}_4 = 20$.

The prescribed transient tracking performance for each follower is designed as $\kappa_i = 0.5$, $\rho_{i,0} = 3.2$, $\rho_{i,\infty} = 0.3$, $\beta_{i,1} = 1.5$, $\beta_{i,2} = 1.2$.

The high-gain observer is designed as

$$\begin{aligned}
\dot{\hat{x}}_{i,1} &= \hat{W}_{i,1}^{\mathrm{T}} \varphi_{i,1}\left(x_{i,1}\right) + \hat{x}_{i,2} + r_i \bar{k}_{i,1} \tilde{x}_{i,1} \\
\dot{\hat{x}}_{i,2} &= \hat{W}_{i,2}^{\mathrm{T}} \varphi_{i,2}\left(\hat{\bar{x}}_{i,2}\right) + \bar{\Gamma}_i(u_i) + r_i^2 \bar{k}_{i,2} \tilde{x}_{i,1} \\
\dot{r}_i &= -\frac{r_i}{a_i}\left[\frac{\delta}{3}(r_i - 1) - \frac{2L_i(y_i)}{\sqrt{P_i}}\right],
\end{aligned}$$

with the heterogeneous parameters $a_1 = 10, a_2 = 12, a_3 = 11, a_4 = 14, \bar{k}1, 1 = 20$, $\bar{k}2, 1 = 25, \bar{k}3, 1 = 23, \bar{k}4, 1 = 22, \bar{k}1, 2 = 22, \bar{k}2, 2 = 24, \bar{k}3, 2 = 20, \bar{k}4, 2 = 21$,

$L_1(y_1) = e^{(0.1y_1)}$, $L_2(y_2) = e^{(0.11y_2)}$, $L_3(y_3) = e^{(0.09y_3)}$, $L_4(y_4) = 5$, $\delta_1 = 21$, $\delta_2 = 23$, $\delta_3 = 21.5$, $\delta_4 = 22$, $p_1 = 0.1$, $p_2 = 0.11$, $p_3 = 0.13$, $p_4 = 0.12$.

Next, the FLS is used to compensate for the nonlinear functions $f_{i,1}(\cdot)$ and $f_{i,2}(\cdot)$ for each follower. Here, we use two sets of fuzzy membership functions with accordingly adaptive parameter estimation for the followers dynamics estimation, i.e., the fuzzy membership functions

$$\mu_{F_1^l}(x_1) = \exp\left[-\frac{(x_1 - 3 + l)^2}{16}\right], \quad l = 1.\ldots.5$$

$$\mu_{F_2^l}(x_1) = \exp\left[-\frac{(x_1 - 3 + l)^2}{4}\right], \quad l = 1.\ldots.5$$

with the adaptive parameter update laws for followers 1, 2, and 3 as

$$\begin{aligned}
\dot{\hat{W}}_{1,1} &= \varphi_{1,1}(x_1)\mu_1 z_{1,1} - \sigma_1 \hat{W}_{1,1}, \\
\dot{\hat{W}}_{1,2} &= z_{1,2}\varphi_{1,2}\left(\hat{\bar{x}}_2\right) - \sigma_1 \hat{W}_{1,2}, \quad \sigma_2 = 1.5 \\
\dot{\hat{W}}_{2,1} &= \varphi_{2,1}(x_1)\mu_2 z_{2,1} - \sigma_2 \hat{W}_{2,1}, \\
\dot{\hat{W}}_{2,2} &= z_{2,2}\varphi_{2,2}\left(\hat{\bar{x}}_2\right) - \sigma_2 \hat{W}_{2,2}, \quad \sigma_2 = 1.2 \\
\dot{\hat{W}}_{3,1} &= \varphi_{3,1}(x_1)\mu_3 z_{3,1} - \sigma_3 \hat{W}_{3,1}, \\
\dot{\hat{W}}_{3,2} &= z_{3,2}\varphi_{3,2}\left(\hat{\bar{x}}_2\right) - \sigma_3 \hat{W}_{3,2}, \quad \sigma_3 = 1
\end{aligned}$$

and the fuzzy membership functions

$$\begin{aligned}
\mu_{F_1^1}(x_1) &= \frac{1}{1 + \exp[5(x_1 + 0.6)]}, \\
\mu_{F_1^2}(x_1) &= \exp\left[-(x_1 + 0.4)^2\right], \\
\mu_{F_1^3}(x_1) &= \exp\left[-(x_1 + 0.2)^2\right], \\
\mu_{F_1^4}(x_1) &= \exp\left(-x_1^2\right), \\
\mu_{F_1^5}(x_1) &= \exp\left[-(x_1 - 0.4)^2\right], \\
\mu_{F_1^6}(x_1) &= \exp\left[-(x_1 - 0.2)^2\right], \\
\mu_{F_1^7}(x_1) &= \frac{1}{1 + \exp[5(x_1 + 0.6)]}
\end{aligned}$$

with the adaptive parameter update laws for follower 4 as

$$\dot{\hat{W}}_{4,2} = z_{4,2}\varphi_{4,2}\left(\hat{\bar{x}}_2\right) - \sigma_4 \hat{W}_{4,2}, \quad \sigma_4 = 3.$$

The virtual control for each follower is designed as

$$\begin{aligned}
\alpha_{i,1} &= \frac{1}{\delta_{i,i}} \sum_{k \in \mathcal{N}_i} a_{i,k}\hat{x}_{k,2} + \frac{b_i \dot{y}_0}{\delta_{i,i}} + \frac{\ell_i\left(z_{i,1}\right)}{\delta_{i,i}} - \mu_{i,1} z_{i,1} \\
&\quad - \frac{c_{i,1}}{\delta_{i,i}} z_{i,1} - \hat{W}_{i,1}^{\mathrm{T}} \varphi_{i,1}\left(x_{i,1}\right) - \underline{\hat{W}}_{i,1}^{\mathrm{T}} \psi_{i,1}\left(\underline{x}_{i,1}\right), \\
\ell_i\left(z_{i,1}\right) &= q z_{i,1} \frac{\dot{\beta}_{i,b}(t)}{\beta_{i,b}(t)} + (1-q) z_{i,1} \frac{\dot{\beta}_{i,a}(t)}{\beta_{i,a}(t)}, \\
\alpha_{i,2} &= -\frac{1}{2} z_{i,2} - c_{i,2} z_{i,2} - \hat{W}_{i,2}^{\mathrm{T}} \varphi_{i,2}\left(\hat{\bar{x}}_{i,2}\right) - z_{i,3} \\
&\quad - k_{i,2}\left[r_i(t)\right]^2 \tilde{\xi}_i + \dot{\hat{\alpha}}_{i,1},
\end{aligned}$$

where b_i and $\delta_{i,i}$ are the topology parameters and $c_{1,1} = 0.1, c_{2,1} = 0.12, c_{3,1} = 0.13$, $c_{4,1} = 0.5$, $c_{1,2} = 1.1$, $c_{2,2} = 1.15$, $c_{3,2} = 1.13$, $c_{4,2} = 1.2$.

The dynamic surface filter is designed as

$$\begin{aligned}
\tau_{i,1}\dot{\hat{\alpha}}_{i,1} + \hat{\alpha}_{i,1} &= \alpha_{i,1}, \\
\tau_{i,2}\dot{\hat{\alpha}}_{i,2} + \hat{\alpha}_{i,2} &= \alpha_{i,2},
\end{aligned}$$

with filter parameters for each follower as $\tau_{1,1} = 0.04$, $\tau_{2,1} = 0.03$, $\tau_{3,1} = 0.06$, $\tau_{4,1} = 0.05$, $\tau_{1,2} = 0.04$, $\tau_{2,2} = 0.06$, $\tau_{3,2} = 0.03$, $\tau_{4,2} = 0.05$.

To deal with the input saturation, the Nussbaum function is selected as $N(\omega_i) = \omega_i^2 \cos(\omega_i)$ with auxiliary systems

$$\begin{aligned}
\dot{\omega}_i &= \gamma_i z_{i,3} \bar{g}_i, \\
\bar{g}_i &= -c_{i,3} z_{i,3} + \dot{\hat{\alpha}}_{i,2} + c_i u_i \frac{\partial \hat{\Gamma}_i(u_i)}{\partial u_i}.
\end{aligned}$$

By applying the developed adaptive output feedback controller design to MAS (6.58), the simulation results are shown in Figs. 6.2, 6.3, 6.4, 6.5, and 6.6. As shown in Fig. 6.2, all the followers' outputs are synchronized to the reference signal $y_d(t)$. From Fig. 6.3, the tracking errors z_1 are bounded by the asymmetric time-varying boundaries $k_a(t)$ and $-k_b(t)$, which are denoted as the dashed lines at the bottom of Fig. 6.5a. Also, one can observe from Fig. 6.4 that, with the novel high-gain observer, the state $x_{i,1}$ and $x_{i,2}$ can be well estimated by $\hat{x}_{i,1}$ and $\hat{x}_{i,2}$. The adaptive parameter

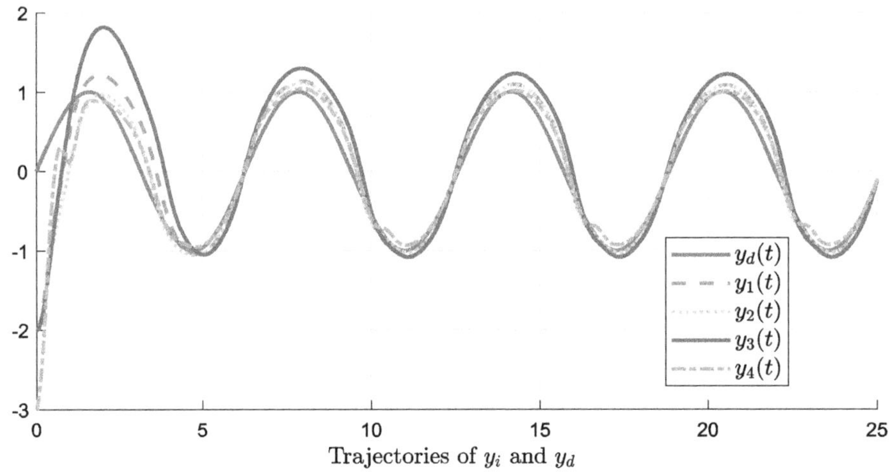

Fig. 6.2 System output and reference signal

updates $\hat{W}_{i,1}(t)$ and $\hat{W}_{i,2}(t)$ in Fig. 6.5, which are guaranteed to be bounded, are designed to compensate the unknown dynamics. The control input signal $\Gamma_i(u_i(t))$ and the additional signal $u_i(t)$ are shown in Fig. 6.6, where the input saturation constraints are guaranteed.

6.6 Conclusion

In this chapter, the output synchronization problem with prescribed performance for uncertain nonlinear strict-feedback MAS is solved by a high-gain observer-based adaptive output feedback design. The unmeasured states of each follower are estimated by the novel distributed observer with adaptive feedback gain, which can efficiently avoid the dependence of controller design on the Lipschitz constant. The input saturation is solved by the auxiliary dynamical system approach. The output synchronization performance is guaranteed by a novel BLF with asymmetric and time-varying tracking performance. The complexity of the virtual control design is reduced by employing the DSC approach. It is shown that the presented scheme can guarantee the boundedness of all the closed-loop signals.

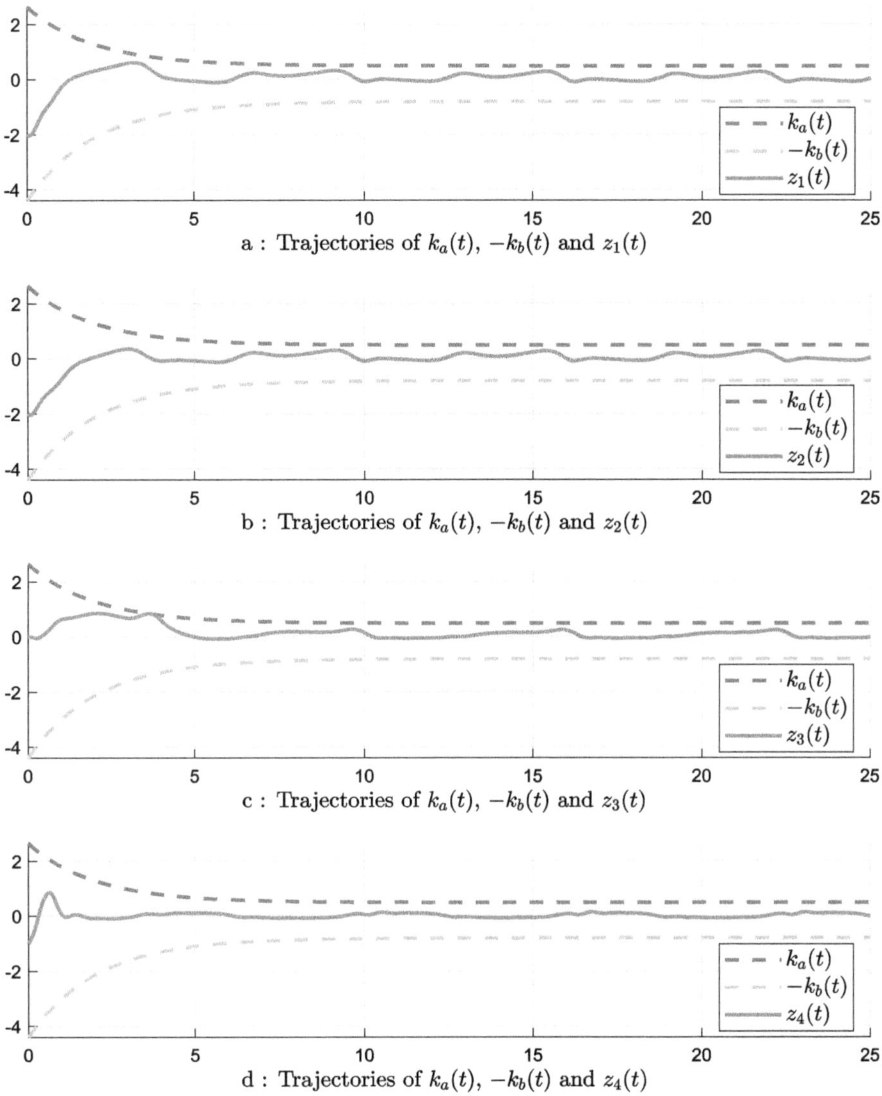

Fig. 6.3 Tracking error and transient performance

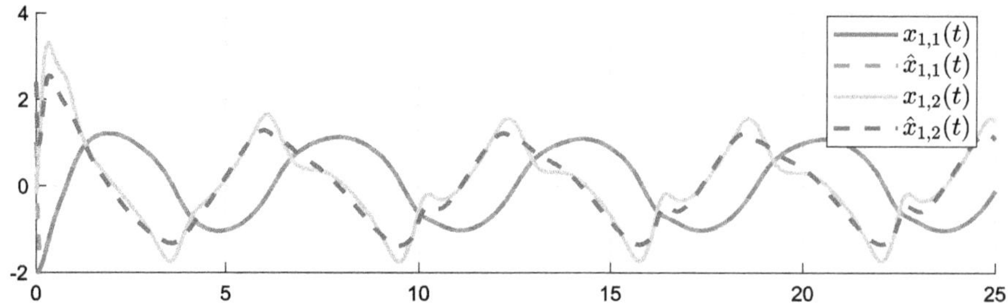

a : Trajectories of $x_{1,1}(t)$, $\hat{x}_{1,1}(t)$, $x_{1,2}(t)$ and $\hat{x}_{1,2}(t)$

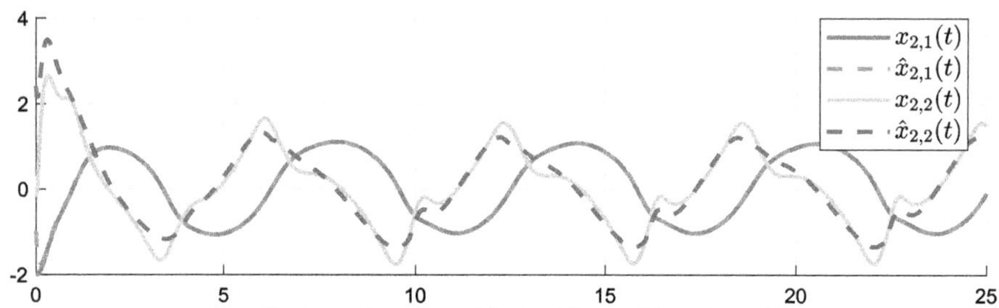

b : Trajectories of $x_{2,1}(t)$, $\hat{x}_{2,1}(t)$, $x_{2,2}(t)$ and $\hat{x}_{2,2}(t)$

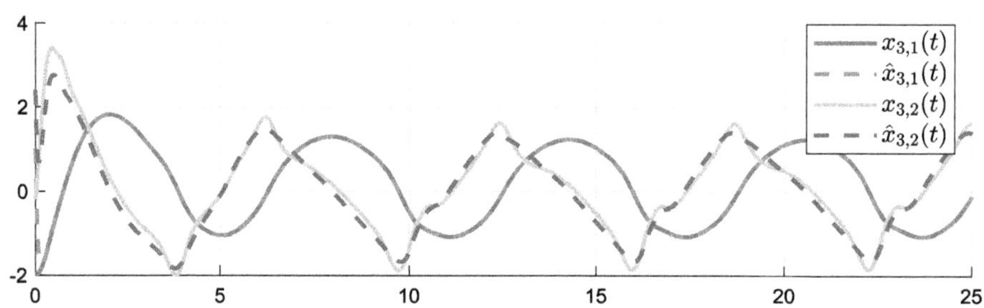

c : Trajectories of $x_{3,1}(t)$, $\hat{x}_{3,1}(t)$, $x_{3,2}(t)$ and $\hat{x}_{3,2}(t)$

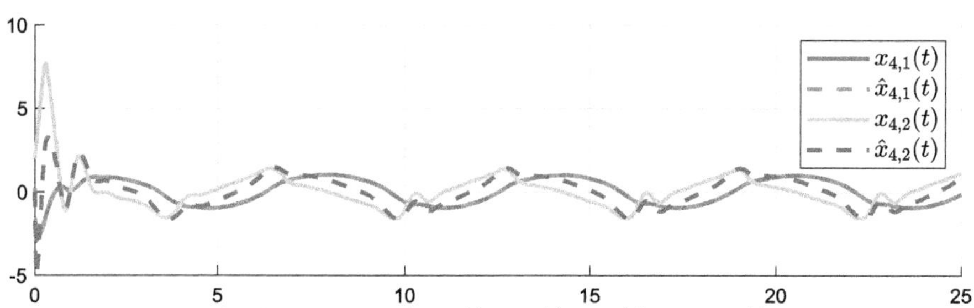

d : Trajectories of $x_{4,1}(t)$, $\hat{x}_{4,1}(t)$, $x_{4,2}(t)$ and $\hat{x}_{4,2}(t)$

Fig. 6.4 System state and state observation

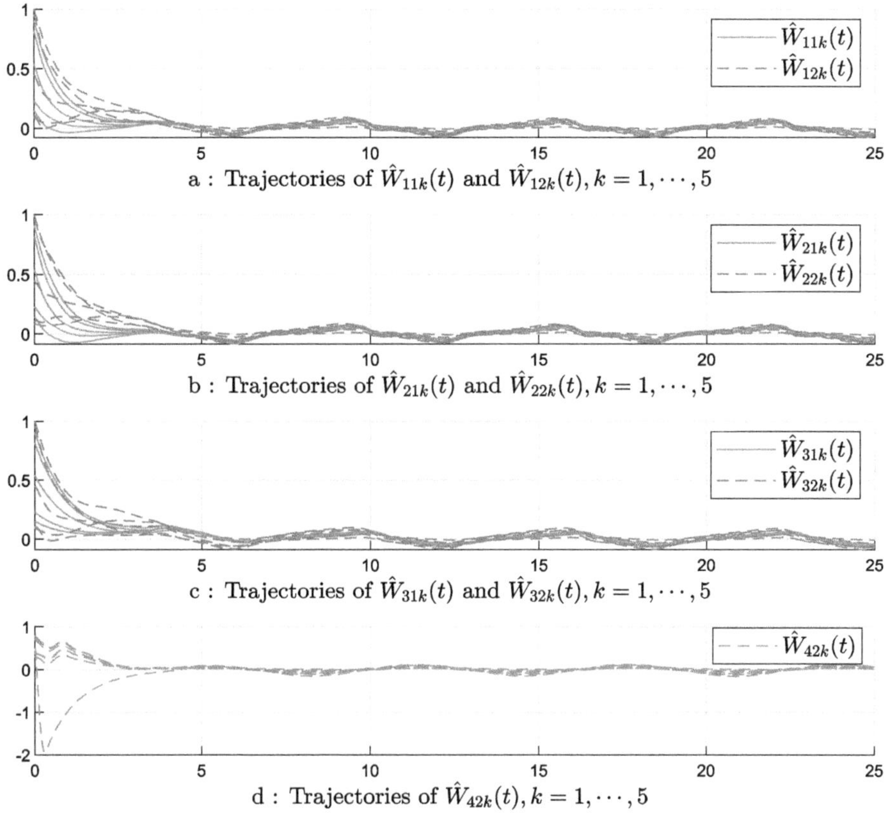

Fig. 6.5 The FLS weight

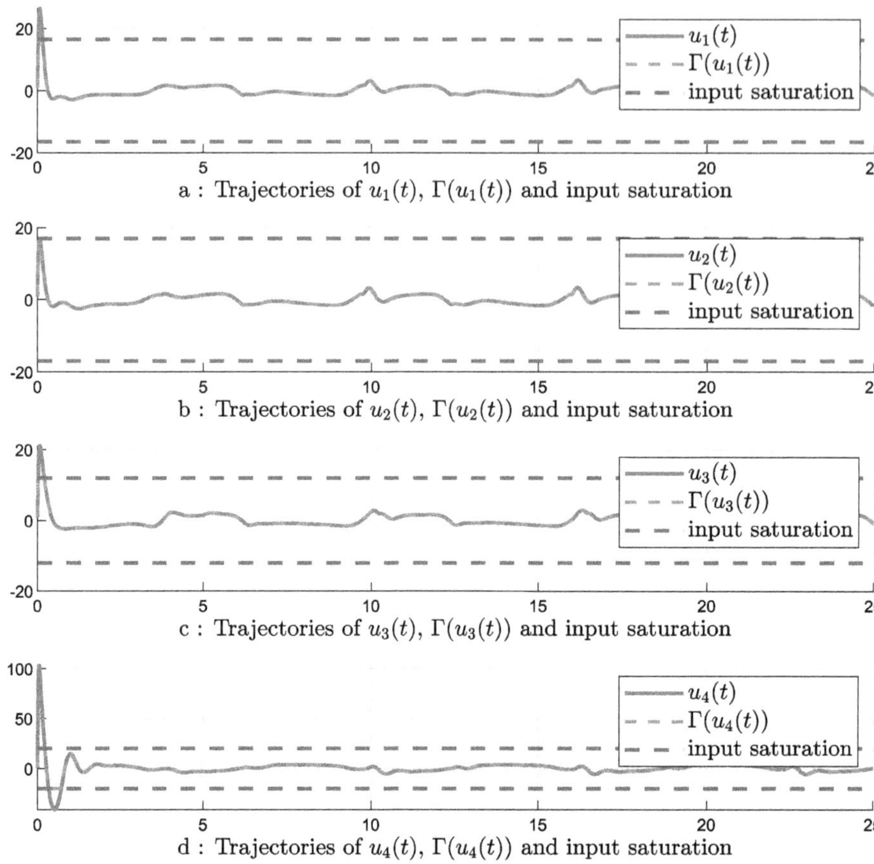

Fig. 6.6 Control input signals

References

1. Liu, G., Yang, Y., Gao, W., Wunsch, D.C.: Adaptive Nussbaum design for nonholonomic systems with asymptotic stabilization against false data injection. IEEE Trans. Cybern. **55**(11), 5189–5201 (2025)
2. Tong, S.C., Li, Y.M., Feng, G., Li, T.S.: Observer-based adaptive fuzzy backstepping dynamic surface control for a class of MIMO nonlinear systems. IEEE Trans. Syst. Man Cybern. Part B (Cybern.). **41**(4), 1124–1135 (2011)
3. Tong, S., Li, Y., Shi, P.: Observer-based adaptive fuzzy backstepping output feedback control of uncertain MIMO pure-feedback nonlinear systems. IEEE Trans. Fuzzy Syst. **20**(4), 771–785 (2012)
4. Wang, W., Tong, S.: Observer-based adaptive fuzzy containment control for multiple uncertain nonlinear systems. IEEE Trans. Fuzzy Syst. **27**(11), 2079–2089 (2019)
5. Wang, W., Wang, D., Peng, Z.H.: Cooperative learning neural network output feedback control of uncertain nonlinear multi-agent systems under directed topologies. Int. J. Syst. Sci. **48**(12), 2590–2598 (2017)

6. Sui, S., Tong, S., Chen, C.L.P.: Finite-time filter decentralized control for nonstrict-feedback nonlinear large-scale systems. IEEE Trans. Fuzzy Syst. **26**(6), 3289–3300 (2018)
7. Sui, S., Chen, C.L.P., Tong, S.: Neural network filtering control design for nontriangular structure switched nonlinear systems in finite time. IEEE Trans. Neural Netw. Learn. Syst. **30**(7), 2153–2162 (2019)
8. Li, Y., Tong, S.: Adaptive fuzzy backstepping output feedback control of nonlinear uncertain systems with unknown virtual control coefficients using MT-filters. Neurocomputing **74**(10), 1557–1563 (2011)
9. Praly, L.: Asymptotic stabilization via output feedback for lower triangular systems with output dependent incremental rate. IEEE Trans. Autom. Control **48**(6), 1103–1108 (2003)
10. Li, C., Tong, S., Wang, W.: Fuzzy adaptive high-gain-based observer backstepping control for SISO nonlinear systems. Inf. Sci. **181**(11), 2405–2421 (2011)
11. Wang, P., Zhang, X., Zhu, J.: Online performance-based adaptive fuzzy dynamic surface control for nonlinear uncertain systems under input saturation. IEEE Trans Fuzzy Syst. **27**(2), 209–220 (2019)
12. Abu-Khalaf, M., Lewis, F.L.: Nearly optimal control laws for nonlinear systems with saturating actuators using a neural network HJB approach. Automatica **41**(5), 779–791 (2005)
13. Modares, H., Lewis, F.L.: Optimal tracking control of nonlinear partially-unknown constrained-input systems using integral reinforcement learning. Automatica **50**(7), 1780–1792 (2014)
14. Yang, Y., Ding, D.W., Xiong, H., Yin, Y., Wunsch, D.C.: Online barrier-actor-critic learning for H_∞ control with full-state constraints and input saturation. J. Frankl. Inst. **357**(6), 3316–3344 (2020)
15. Yang, Y., Vamvoudakis, K.G., Modares, H., Yin, Y., Wunsch, D.C.: Safe intermittent reinforcement learning with static and dynamic event generators. IEEE Trans. Neural Netw. Learn. Syst. **31**(12), 5441–5455 (2020)
16. Wang, W., Tong, S.: Adaptive fuzzy bounded control for consensus of multiple strict-feedback nonlinear systems. IEEE Trans. Cybern. **48**(2), 522–531 (2018)
17. Wen, C., Zhou, J., Liu, Z., Su, H.: Robust adaptive control of uncertain nonlinear systems in the presence of input saturation and external disturbance. IEEE Trans. Autom. Control **56**(7), 1672–1678 (2011)
18. Yang, Y., Modares, H., Vamvoudakis, K., Yin, Y., Wunsch, D.C.: Dynamic intermittent feedback design for H_∞ containment control on a directed graph. IEEE Trans. Cybern. **50**(8), 3752–3765 (2020)
19. Wang, W., Li, Y.: Observer-based event-triggered adaptive fuzzy control for leader-following consensus of nonlinear strict-feedback systems. IEEE Trans. Cybern. **51**(4), 2131–2141 (2021)
20. Qu, F., Tong, S., Li, Y.: Observer-based adaptive fuzzy output constrained control for uncertain nonlinear multi-agent systems. Inf. Sci. **467**, 446–463 (2018)
21. Li, D.P., Liu, Y.J., Tong, S., Chen, C., Li, D.J.: Neural networks-based adaptive control for nonlinear state constrained systems with input delay. IEEE Trans. Cybern. **49**(4), 1249–1258 (2019)
22. Khalil, H.K.: Nonlinear Systems. Prentice Hall, New Jersey (2002)

Chapter 7
Adaptive Nussbaum Design for Bipartite Synchronization of Multiagent Systems

7.1 Introduction

The control problem of MASs was introduced to address the challenges presented when coordinating the actions of multiple agents [1]. The primary objective is to design control strategies that coordinate the behaviors of multiple agents, enabling them to achieve common goals or objectives [2]. Considering the collaborative relationships within MASs, the system efficiency and robustness can be enhanced [3]. With cooperative control targets, the interactions between agents are captured by the non-negative communication link, which cannot describe a competitive configuration, such as social networks [4–6]. In contrast to the representation of friend, ally, and trust relationship by the positive connection, the negative link on topology can stand for the concept of the enemy, adversary, distrust, etc. [7]. Such a signed graph is widely used in many antagonistic systems, such as the polarization of the community, and the teams opposed in a sports match, to name a few [8]. In this chapter, we focus on the bipartite consensus for a group of MASs with heterogeneous nonlinear dynamics. To investigate competitive distributed control, Altafini studied the bipartite synchronization problem [9], which entails synchronizing two independent groups of agents, each with their own objectives. The key purpose of this problem is to ensure that the two sets of agents in the MASs can reach consensus within each subgroup, while simultaneously allowing them to maintain a degree of independence [10]. This approach offers broader applicability because each group can independently make decisions and exercise controls, ultimately enhancing the flexibility and adaptability of the MASs. Based on this advantage, bipartite synchronization control is applied across diverse domains, such as the bipartite synchronization problem with adversarial inputs [11] and under deception attacks [12]. However, challenging factors remain to be addressed, including the communication efficiency between multiple agents and the uncertainty of control within dynamic systems.

Recently, event-triggered scheduling was successfully applied to networked control to tackle the increasing pressure on communication networks with massive

Y. Yang et al., *Adaptive Nussbaum Design for Nonlinear Dynamical Systems*,
https://doi.org/10.1007/978-981-95-4997-9_7

signal transmission. The event-triggered mechanism represents a flexible control strategy, wherein control actions are initiated in response to specific events that transpire within the system [13]. Compared with the periodic sampling technique, event-triggered mechanisms can reduce the communication bandwidth and avoid any unnecessary signal updates [14, 15]. It is particularly noteworthy that the importance of event-triggered mechanisms is further highlighted in MASs, where information is exchanged between the agents to achieve synchronization [16, 17]. In this scenario, employing the event-triggered mechanism can alleviate communication workload. In addition to the cooperative protocol design, event-triggered mechanisms are also widely used in competitive control [18]. Nevertheless, considering the previously outlined situation, event-triggered bipartite synchronization control protocols fail to consider the presence of multiple unknown control directions. The contributions of this chapter can be summarized as follows. In previous studies [19–21], the researchers studied the Nussbaum-type function as a means of addressing the challenge posed by unknown control directions. However, the existing functions lack sufficient processing capability when confronted with problems where the interactions cancel each other out. This chapter introduces a novel Nussbaum-type function to research into the resolution requirements of multi-directional control problems. Furthermore, a novel Lyapunov stability theorem is established based on this function. Existing research [22, 23] introduced various complex variations of event-triggered mechanisms. However, the complexity of such mechanisms renders them inapplicable in the presence of unknown control directions. The aim of this study is to streamline event-triggered mechanisms to enable their effective utilization within systems featuring unknown control directions.

The remaining sections are organized as follows. Section 7.2 contains the preliminaries. Section 7.2 provides the problem definition and related assumptions. In Sect. 7.3, a novel Nussbaum-type function is established for multiple unknown control directions in the group of MASs. Section 7.4 investigates theoretical discussions about the presented design scheme. The effectiveness of the controller is then verified in Sect. 7.5. Finally, the conclusions are made in Sect. 7.6.

7.2 Problem Formulation

In the group of MASs, the dynamics of each follower is considered as

$$\begin{aligned} \dot{x}_{i,j} &= f_{i,j}\left(\bar{x}_{i,j}\right) + g_{i,j}\left(\bar{x}_{i,j}\right) x_{i,j+1},\ j = 1, ..., n_i - 1, \\ \dot{x}_{i,n_i} &= f_{i,n_i}\left(\bar{x}_{i,n_i}\right) + g_{i,n_i}\left(\bar{x}_{i,n_i}\right) u_i, \\ y_i &= x_{i,1}, \end{aligned} \tag{7.1}$$

where $i \in \mathcal{F}$, $x_{i,j} \in \left[x_{i,1} \ldots x_{i,n_i}\right]^{\mathrm{T}} \in \mathbb{R}^{n_i}$ is the follower state, and $u_i \in \mathbb{R}$ and $y_i \in \mathbb{R}$ are the input and output of the follower, respectively. The functions $f_{i,j}\left(\bar{x}_{i,j}\right)$ and $g_{i,j}\left(\bar{x}_{i,j}\right)$ are supposed to be smooth but unknown.

To address the issue in MAS with unknown control directions, the following Assumptions, Definitions, and Lemma are required.

Assumption 7.1 The signs of the functions $g_{i,j}(\cdot)$ are unknown and there exist two positive constants $\underline{g}_{i,j}$ and $\bar{g}_{i,j}$ satisfying $0 < \underline{g}_{i,j} \leq |g_{i,j}(x_{i,j})| \leq \bar{g}_{i,j}$.

Assumption 7.2 It is assumed that the information about neighbor agents is available for each agent.

Assumption 7.3 There exists a directed spanning tree in the graph that consists of a single leader and N followers, where the sub-graph among N followers is connected and structurally balanced.

Definition 7.1 Given a smooth reference signal $y_d(t)$, the MASs are said to achieve bipartite consensus if

$$
\begin{aligned}
\lim_{t\to\infty} \|y_i - y_d\|_2 &= 0, i \in \Upsilon_p, \\
\lim_{t\to\infty} \|y_i + y_d\|_2 &= 0, i \in \Upsilon_q,
\end{aligned}
$$

where Υ_1 and Υ_2 is defined in Definition 1.10 with $p, q \in \{1, 2\}$ and $p \neq q$.

In the context of a connected signed graph ($\mathcal{G}$), the MASs attain a bipartite consensus solution if and only if the signed graph $\mathcal{G}$ demonstrates structural balance. In the bipartite consensus problem considered in this chapter, the reference signal for the followers is represented by the leader with output $y_0 = y_d$, which is supposed to be a smooth signal.

Lemma 7.1 *Consider the local output interaction between agents*

$$
e_{i,1} = \sum_{k\in N_i} |a_{i,k}| \left[y_i - \text{sign}(a_{i,k})\, y_k\right] + |b_i| \left[y_i - \text{sign}(b_i)\, y_0\right]. \tag{7.2}
$$

The MASs can achieve bipartite consensus if $e_{i,1} = 0$.

Proof From Lemma 1 in [9], if there exists a linear combination of the variables y_i such that $e_{i,1} = 0$, structural balance can be obtained. Therefore, it must be ensured that the MASs can achieve a bipartite consensus in Definition 7.1. This completes the proof. □

From Lemma 7.1, the convergence of the signal $e_{i,1}$ is equivalent to bipartite consensus. Based on this fact, the term $e_{i,1}$ should be regulated to the origin and referred to as the bipartite consensus error. To relax the bipartite consensus condition in Definition 7.1 for general applications, we considered the following novel bipartite consensus concept.

Definition 7.2 Given a smooth reference signal $y_d(t)$, the MASs are said to achieve the practical bipartite consensus if $\|e_{i,1}\| \leq \varsigma$ for sufficiently small $\varsigma > 0$.

To overcome the problem of unknown control directions, we develop a novel Nussbaum-type function as

$$N_j(\xi) = 2\xi e^{\xi^2}\cos\left(2^{j-1}\xi\right) - 2^{j-1}\left(e^{\xi^2}-1\right)\sin\left(2^{j-1}\xi\right) \tag{7.3}$$

where $N_j(\cdot)$ with different subscript j denotes different functions. Such Nussbaum design aims to avoid the counteract of multiple Nussbaum-type functions. It can be observed that the novel Nussbaum-type function satisfies $\lim_{s\to\infty}\sup \frac{1}{s}\int_0^s N(\xi)\mathrm{d}\xi = \infty$ and $\lim_{s\to\infty}\inf \frac{1}{s}\int_0^s N(\xi)\mathrm{d}\xi = -\infty$. Therefore, one can conclude that the novel Nussbaum-type function in (7.3) satisfies the conditions in Definition 1.1.

Accordingly, a novel Lyapunov-based stability theorem can be developed as follows.

Lemma 7.2 *Suppose that there exists a positive definite, radially unbounded function $V(t)$ for all $t \in [0, t_f)$ with the Nussbaum-type functions $\{N_j(\cdot)\}_{j=1}^M$. Then, $V(t)$, ξ_j, and $\int_0^t e^{-a(t-\sigma)}\left[1 + g_j(\sigma) N_j(\xi_j(\sigma))\dot{\xi}_j(\sigma)\right]\mathrm{d}\sigma$ are bounded if*

$$\begin{aligned} V(t) \le & \sum_{j=1}^{M}\int_0^t g_j(\sigma) N_j(\xi_j(\sigma))\dot{\xi}_j(\sigma) e^{-a(t-\sigma)}\mathrm{d}\sigma \\ & + \sum_{j=1}^{M}\int_0^t \dot{\xi}_j(\sigma) e^{-a(t-\sigma)}\mathrm{d}\sigma + c_0, \end{aligned} \tag{7.4}$$

where $a > 0$, $c_0 > 0$, and $\{g_j(\cdot)\}_{j=1}^M$ are continuous functions satisfying $|g_j(\cdot)| \in [g_{j,\min}, g_{j,\max}]$ with constants $g_{j,\min} > 0$ and $g_{j,\max} > 0$.

Proof Without loss of generality, consider the general case when $\xi_1, \cdots, \xi_\vartheta$ are unbounded and $\xi_{\vartheta+1}, \cdots, \xi_M$ are bounded. Therefore, the fact (7.4) indicates

$$\begin{aligned} V(t) \le & \sum_{j=1}^{\vartheta}\int_0^t |g_j| \operatorname{sign}(g_j) N_j(\xi_j(\sigma))\dot{\xi}_j(\sigma) \mathrm{e}^{-a(t-\sigma)}\mathrm{d}\sigma \\ & + \sum_{j=1}^{\vartheta}\int_0^t \mathrm{e}^{-a(t-\sigma)}\dot{\xi}_j(\sigma)\,\mathrm{d}\sigma + c_1, \end{aligned} \tag{7.5}$$

with the bounded constant $c_1 = \sum_{j=\vartheta+1}^{M}\int_0^t g_j(\sigma) N_j(\xi_j(\sigma))\dot{\xi}_j(\sigma)\mathrm{e}^{-a(t-\sigma)}\mathrm{d}\sigma + \sum_{j=\vartheta+1}^{M}\int_0^t \mathrm{e}^{-a(t-\sigma)}\dot{\xi}_j(\sigma)\,\mathrm{d}\sigma + c_0$. In the following, we investigate two cases to prove the boundedness of $V(t)$ and $\xi_j(t)$ by contradiction.

First Case. Consider the case that $\xi_1, \cdots, \xi_\vartheta$ do not have an upper bound on $[0, t_f)$. Define three monotone-increasing sequences $\Xi_{j,r} = \frac{(4r+1)\pi}{2^j} + v_j\pi - \frac{\theta_j}{2^{j-1}}$, $\Psi_{j,r} = \frac{(4r-1)\pi}{2^j} + v_j\pi + \frac{\theta_j}{2^{j-1}}$, and $\Phi_{j,r} = \frac{(4r-1)\pi}{2^j} + v_j\pi$, where $v_j = \sum_{k=1}^{j} \left(2^{-k} + 2^{-k}\text{sign}\left(g_k\right)\right) \geq 0$ and $0 \leq \theta_j \leq \frac{\pi}{2}$. Accordingly, define the increasing sequences satisfying $\left\{t_{1,j,r}\right\}_{r=1}^{\infty}$, $\left\{t_{2,j,r}\right\}_{r=1}^{\infty}$, and $\left\{t_{3,j,r}\right\}_{r=1}^{\infty}$ with $\xi_j\left(t_{1,j,r}\right) = \Phi_{j,r}$, $\xi_j\left(t_{2,j,r}\right) = \Psi_{j,r}$, and $\xi_j\left(t_{3,j,r}\right) = \Xi_{j,r}$, respectively. From this, we have $\lim_{r\to\infty} t_{1,j,r} = \lim_{r\to\infty} t_{2,j,r} = \lim_{r\to\infty} t_{3,j,r} = t_f$ and $\lim_{r\to\infty} \Phi_{j,r} = \lim_{r\to\infty} \Psi_{j,r} = \lim_{r\to\infty} \Xi_{j,r} = \infty$. Furthermore, define the increasing sequences $\left\{d_{j,w}\right\}_{w=0}^{\infty}$ satisfying $\xi_j\left(d_{j,w}\right) = \frac{(2w-1)\pi}{2^j} + v_j\pi$ for $w = 1, 2, \cdots, 2r-1, 2r$, $\xi_j\left(d_{j,0}\right) = v_j\pi$ for $w = 0$, and the following intervals

$$
\begin{aligned}
\Omega_{j,0} &= \left[v_j\pi, \frac{\pi}{2^j} + v_j\pi\right], \\
\Omega_{j,1} &= \left[\frac{\pi}{2^j} + v_j\pi, \frac{3\pi}{2^j} + v_j\pi\right], \\
\Omega_{j,w} &= \left[\frac{(2w-1)\pi}{2^j} + v_j\pi, \frac{(2w+1)\pi}{2^i} + v_j\pi\right].
\end{aligned} \tag{7.6}
$$

With above notations, the fact (7.5) implies

$$
\begin{aligned}
V(t) &\leq \Theta_1 + \Theta_2 + \Theta_3 + \Theta_4, \\
\Theta_1 &= \sum_{j=1}^{\vartheta} \int_{t_{2,j,r}}^{t_{3,j,r}} \left|g_j\right| \text{sign}\left(g_j\right) N_j\left(\xi_j\right) \dot{\xi}_j \mathrm{e}^{-a\left(t_{3,j,r}-\sigma\right)} \mathrm{d}\sigma, \\
\Theta_2 &= \sum_{j=1}^{\vartheta} \int_{t_{1,j,r}}^{t_{2,j,r}} \left|g_j\right| \text{sign}\left(g_j\right) N_j\left(\xi_j\right) \dot{\xi}_j \mathrm{e}^{-a\left(t_{3,j,r}-\sigma\right)} \mathrm{d}\sigma, \\
\Theta_3 &= \sum_{j=1}^{\vartheta} \int_{0}^{t_{1,j,r}} \left|g_j\right| \text{sign}\left(g_j\right) N_j\left(\xi_j\right) \dot{\xi}_j \mathrm{e}^{-a\left(t_{3,j,r}-\sigma\right)} \mathrm{d}\sigma, \\
\Theta_4 &= \sum_{j=1}^{\vartheta} \int_{0}^{t_{3,j,r}} \mathrm{e}^{-a\left(t_{3,j,r}-\sigma\right)} \dot{\xi}_j \mathrm{d}\sigma + c_1.
\end{aligned} \tag{7.7}
$$

By integrating ξ_j over the interval $\left[0, \Phi_{j,r}\right]$, one has

$$
\begin{aligned}
&\int_0^{\Phi_{j,r}} \text{sign}\left(g_j\right) N_j\left(\xi_j\right) \mathrm{d}\xi_j \\
&= \text{sign}\left(g_j\right)\left(\mathrm{e}^{\Phi_{j,r}^2} - 1\right) \cos\left(2r\pi - \frac{1}{2}\pi + 2^{j-1}v_j\pi\right).
\end{aligned} \tag{7.8}
$$

Note that $2^{j-1}v_j$ should be a positive integer. Then, one can obtain

$$\int_0^{\Phi_{j,r}} \operatorname{sign}\left(g_j\right) N_j\left(\xi_j\right) \mathrm{d}\xi_j = 0. \tag{7.9}$$

Similarly, with the fact $\int_0^{\frac{(4r+1)\pi}{2^j}+v_j\pi} \operatorname{sign}\left(g_j\right) N_j\left(\xi_j\right) \mathrm{d}\xi_j = 0$, one has

$$2^{j-1}v_j = 2^{j-1}\sum_{k=1}^{j-1}\left[2^{-k}+2^{-k}\operatorname{sign}\left(g_k\right)\right] + \frac{1}{2}\left[1+\operatorname{sign}\left(g_j\right)\right]$$

with $2^{j-1}\sum\limits_{k=1}^{j-1}\left(2^{-k}+2^{-k}\operatorname{sign}\left(g_k\right)\right)$ being an even integer. It should be noted that $\frac{1}{2}\left(1+\operatorname{sign}\left(g_j\right)\right)$ is an odd number if $\operatorname{sign}\left(g_j\right)=1$ and is an even number if $\operatorname{sign}\left(g_j\right)=-1$. Then, $2^{j-1}v_j$ is odd if $\operatorname{sign}\left(g_j\right)=1$ and is even if $\operatorname{sign}\left(g_j\right)=-1$. Based on these facts, on the interval $\left[0, \frac{r\pi}{2^{j-2}}+v_j\pi\right]$, we have

$$\begin{aligned}
&\int_0^{\frac{r\pi}{2^{j-2}}+v_j\pi} \operatorname{sign}\left(g_j\right) N_j\left(\xi_j\right) \mathrm{d}\xi_j \\
&= \operatorname{sign}\left(g_j\right)\left(\mathrm{e}^{\left(\frac{r\pi}{2^{j-2}}+v_j\pi\right)^2}-1\right)\cos\left(2^{j-1}v_j\pi\right) < 0.
\end{aligned} \tag{7.10}$$

Furthermore, we have $\int_0^{\xi_j} \operatorname{sign}\left(g_j\right) N_j\left(\sigma\right) \mathrm{d}\sigma \leq 0$ for $\xi_j \in \Omega_{j,2r}$.

Therefore, the term Θ_1 in (7.7) satisfies

$$\begin{aligned}
\Theta_1 &\leq \sum_{j=1}^{\vartheta} \mathrm{e}^{-a\left(t_{3,j,r}-t_{2,j,r}\right)} \int_{\Psi_{j,r}}^{\Xi_{j,r}} \left|g_j\right| \operatorname{sign}\left(g_j\right) N_j\left(\xi_j\right) \mathrm{d}\xi_j \\
&\leq \sum_{j=1}^{\vartheta} G_{\min}\mathrm{e}^{-a\left(t_{3,j,r}-t_{2,j,r}\right)} \int_{\Psi_{j,r}}^{\Xi_{j,r}} \operatorname{sign}\left(g_j\right) N_j\left(\xi_j\right) \mathrm{d}\xi_j \\
&= -\sum_{j=1}^{\vartheta} G_{\min}\mathrm{e}^{-a\left(t_{3,j,r}-t_{2,j,r}\right)} \sin\left(\theta_j\right)\left(\mathrm{e}^{\Xi_{j,r}^2}+\mathrm{e}^{\Psi_{j,r}^2}\right),
\end{aligned} \tag{7.11}$$

with $G_{\min} = \min\limits_j g_{j,\min}$.

Moreover, the term Θ_2 in (7.7) satisfies

$$\begin{aligned}
\Theta_2 &\leq G_{\min}\sum_{j=1}^{\vartheta} \mathrm{e}^{-a\left(t_{2,j,r}-t_{1,j,r}\right)} \int_{\Phi_{j,r}}^{\Psi_{j,r}} \operatorname{sign}\left(g_j\right) N_j\left(\xi_j\right) \mathrm{d}\xi_j \\
&\leq 0.
\end{aligned} \tag{7.12}$$

Similar to the discussion in (7.8)–(7.10), one has $\int_0^{\xi_j} \text{sign}\left(g_j\right) N_j\left(\sigma\right) \mathrm{d}\sigma \geq 0$ for $\xi_j \in \Omega_{j,2r-1}$. Then, one can obtain

$$\begin{aligned}&\int_{d_{j,r-1}}^{d_{j,r}} \text{sign}\left(g_j\right) N_j\left(\xi_j\right) \dot{\xi}_j \mathrm{e}^{-a\left(d_{j,r}-\sigma\right)} \mathrm{d}\sigma \\ &\leq \int_{\frac{(4r-3)\pi}{2^j}+v_j\pi}^{\frac{(4r-1)\pi}{2^j}+v_j\pi} \text{sign}\left(g_j\right) N_j\left(\xi_j\right) \mathrm{d}\xi_j = 0.\end{aligned} \tag{7.13}$$

Similar to the analysis on $\left[d_{j,r}, d_{j,r-1}\right]$, one has

$$\int_{d_{j,w}}^{d_{j,w+1}} \text{sign}\left(g_j\right) N_j\left(\xi_j\right) \dot{\xi}_j \mathrm{e}^{-a\left(d_{j,w+1}-\sigma\right)} \mathrm{d}\sigma \leq 0, \tag{7.14}$$

on $\left[d_{j,w}, d_{j,w+1}\right]$ for $w = 0, 1, \cdots, 2r-1$. Additionally, we know that

$$\int_0^{\frac{\pi}{2^j}} \text{sign}\left(g_j\right) N_j\left(\xi_j\right) \mathrm{d}\xi_j = 0. \tag{7.15}$$

Then, when $g_j > 0$, the following holds

$$\begin{aligned}&\int_0^{d_{j,c}} \text{sign}\left(g_j\right) N_j\left(\xi_j\right) \dot{\xi}_j \mathrm{e}^{-a\left(d_{j,0}-\sigma\right)} \mathrm{d}\sigma \\ &\leq \int_0^{d_{j,c}} \text{sign}\left(g_j\right) N_j\left(\xi_j\right) \dot{\xi}_j \mathrm{d}\sigma = 0,\end{aligned} \tag{7.16}$$

with $\xi_j\left(\underline{d}_j\right) = \frac{\pi}{2}$. In contrast, when $g_j < 0$, we have

$$\begin{aligned}&\int_0^{\underline{d}_j} \text{sign}\left(g_j\right) N_j\left(\xi_j\right) \dot{\xi}_j \mathrm{e}^{-a\left(d_{j,0}-\sigma\right)} \mathrm{d}\sigma \\ &\leq \mathrm{e}^{-a\left(d_{j,0}-\underline{d}_j\right)} \int_0^{d_{j,c}} \text{sign}\left(g_j\right) N_j\left(\xi_j\right) \dot{\xi}_j \mathrm{d}\sigma = 0.\end{aligned} \tag{7.17}$$

Accordingly, we divide the interval $\left[0, d_{j,0}\right]$ into multiple sub-intervals with the same size as $\left[0, \underline{d}_j\right]$, which yields

$$\int_0^{d_{j,0}} \text{sign}\left(g_j\right) N_j\left(\xi_j\right) \dot{\xi}_j \mathrm{e}^{-a\left(d_{j,0}-\sigma\right)} \mathrm{d}\sigma \leq 0. \tag{7.18}$$

Therefore, the term Θ_3 in (7.7) satisfies

$$\Theta_3 \leq \sum_{w=0}^{2r-1} \sum_{j=1}^{\vartheta} \int_{d_{j,w}}^{d_{j,w+1}} \operatorname{sign}\left(g_j\right) N_j\left(\xi_j\right) \mathrm{e}^{-a\left(d_{j,w+1}-\sigma\right)} \mathrm{d}\sigma$$
$$+\sum_{j=1}^{\vartheta} \int_0^{d_{j,0}} \operatorname{sign}\left(g_j\right) N_j\left(\xi_j\right) \mathrm{e}^{-a\left(d_{j,0}-\sigma\right)} \mathrm{d}\sigma \leq 0. \tag{7.19}$$

Consider (7.11), (7.12), (7.19), and the fact that

$$\sum_{j=1}^{\vartheta} \int_0^{t_{3,j,r}} \mathrm{e}^{-a\left(t_{3,j,r}-\sigma\right)} \dot{\xi}_j \mathrm{d}\sigma \leq \sum_{j=1}^{\vartheta} \Xi_{j,r} - \sum_{j=1}^{\vartheta} \xi_j\,(0), \tag{7.20}$$

the function $V\,(t)$ satisfies

$$V\,(t) \leq c_1 - \sum_{i=1}^{\vartheta} \xi_i\,(0) + \sum_{i=1}^{\vartheta} \Xi_{j,r} - G_{\min} \sum_{j=1}^{\vartheta} \sin\left(\theta_j\right) \left(\mathrm{e}^{\Xi_{j,r}^2} + \mathrm{e}^{\Psi_{j,r}^2}\right) \tag{7.21}$$

which is equivalent to

$$V\,(t) \leq -\sum_{j=1}^{\vartheta} c_3 \mathrm{e}^{\Xi_{j,r}^2} \left(1 + \mathrm{e}^{\Psi_{j,r}^2 - \Xi_{j,r}^2} - \frac{\Xi_{j,r}}{c_3 \mathrm{e}^{\Xi_{j,r}^2}}\right) + c_2,$$

with $c_2 = -\sum_{i=1}^{\vartheta} \xi_i\,(0) + c_1$ and $c_3 = G_{\min} \sin\left(\theta_j\right)$. It is evident that $\lim_{t\to t_f} V\,(t) = -\infty$, which contradicts the condition that $V\,(t) \geq 0$. Therefore, the terms $\xi_1, \cdots, \xi_\vartheta$ are upper-bounded on $\left[0, t_f\right)$.

Second Case. Consider the case that $\xi_1, \cdots, \xi_\vartheta$ do not have a lower bound on $\left[0, t_f\right)$, which is equivalent to $-\xi_1, \cdots, -\xi_\vartheta$ not having an upper bound on $\left[0, t_f\right)$. Note that the term $\int_0^{\xi_j} \operatorname{sign}\left(g_j\right) N_j\,(\tau)\, d\tau$ is an even function. Then, the Lyapunov candidate can be expressed as

$$V\,(t) \leq \sum_{j=1}^{\vartheta} \int_0^t \left|g_j\right| \operatorname{sign}\left(g_j\right) N_j\left(\xi_j\,(\sigma)\right) \dot{\xi}_j\,(\sigma)\, \mathrm{e}^{-a(t-\sigma)} \mathrm{d}\sigma$$
$$-\sum_{j=1}^{\vartheta} \int_0^t \mathrm{e}^{-a(t-\sigma)} \dot{\xi}_j\,(\sigma)\, \mathrm{d}\sigma + l_1,$$

with

$$l_1 = \sum_{j=\vartheta+1}^{M} \int_0^t g_j(\sigma) N_j(\xi_j(\sigma)) \dot{\xi}_j(\sigma) \mathrm{e}^{-a(t-\sigma)} \mathrm{d}\sigma$$
$$- \sum_{j=\vartheta+1}^{M} \int_0^t \mathrm{e}^{-a(t-\sigma)} \dot{\xi}_j(\sigma) \mathrm{d}\sigma + c_0.$$

Following the discussions in the first case, one has

$$V(t) \leq -\sum_{j=1}^{\vartheta} l_2 \mathrm{e}^{\Xi_{j,r}^2} \left(1 + \mathrm{e}^{\Psi_{j,r}^2 - \Xi_{j,r}^2} + \frac{\Xi_{j,r}}{d_3 \mathrm{e}^{\Xi_{j,r}^2}}\right) + l_3$$

with $l_2 = G_{\min} \mathrm{e}^{-a(t_{3,j,r} - t_{2,j,r})} \sin(\theta_j)$ and $l_3 = \sum_{j=1}^{\vartheta} \xi_j(0) + d_1$. Then, terms $\xi_1, \cdots, \xi_\vartheta$ are lower-bounded on $[0, t_f)$. Finally, we can conclude the boundedness of $V(t)$ and $\xi_j(t)$. This completes the proof. □

7.3 Adaptive Nussbaum Design

To achieve bipartite consensus for a group of MASs, a fuzzy Nussbaum design is developed based on the novel Nussbaum-type function (7.3) with intermittent feedback. It should be noted that each follower dynamics is in a strict-feedback nonlinear form, and the Nussbaum design is established in the fuzzy backstepping framework.

Define the tracking error $e_{i,j}$ as $e_{i,j} = x_{i,j} - \alpha_{i,j-1}$ with $\alpha_{i,j-1}$ being the dynamic surface control (DSC) filter output for $j = 1, 2, \cdots, n_i$ with $\alpha_{i,0} = y_0$. For an arbitrary continuous function $F_{i,j}(X_{i,j})$, given $\bar{\varepsilon}_{i,j} > 0$ and based on Lemma 1.5, there exists an FLS such that

$$F_{i,j}(X_{i,j}) = W_{i,j}^{\mathrm{T}} \varphi_{i,j}(X_{i,j}) + \varepsilon_{i,j}(X_{i,j}), \tag{7.22}$$

where $W_{i,j}$ is the ideal network weight vector and $\varepsilon_{i,j} \leq \bar{\varepsilon}_{i,j}$. To estimate the unknown weight $W_{i,j}$, $\hat{W}_{i,j}$ is designed with the estimation error $\tilde{W}_{i,j} = W_{i,j} - \hat{W}_{i,j}$.

7.3.1 Step 1

For the first subsystem of the ith follower, define

$$F_{i,1}(X_{i,1}) = (d_i + |b_i|)\, f_{i,1}\left(x_{i,1}\right) - \sum_{k\in\mathcal{N}_i} a_{i,k}\big(f_{k,1}\left(x_{k,1}\right) + g_{k,1}\left(x_{k,1}\right) x_{k,2}\big), \tag{7.23}$$

with $X_{i,1} = \left[x_{i,1}\ x_{k,1}\ x_{k,2}\right]^{\mathrm{T}}$. To compensate for the unknown function in (7.23), the adaptive fuzzy learning is designed as

$$\dot{\hat{W}}_{i,1} = e_{i,1}\varphi_{i,1}\left(X_{i,1}\right) - \gamma_{i,1}\hat{W}_{i,1}. \tag{7.24}$$

To address the challenge posed by an unknown control direction, the adaptive Nussbaum-type gain is developed as

$$\dot{\xi}_{i,1} = h_{i,1}e_{i,1}r_{i,1} \tag{7.25}$$

with $r_{i,1} = -k_{i,1}e_{i,1} - \hat{W}_{i,1}^{\mathrm{T}}(t)\,\varphi_{i,1}\left(X_{i,1}\right) + b_i\dot{y}_0$ and constant parameter $h_{i,1} > 0$. Then, the virtual control is designed as

$$s_{i,1} = \frac{1}{h_{i,1}}N_1\left(\xi_{i,1}\right)r_{i,1}. \tag{7.26}$$

To avoid recursive differentiation of $s_{i,1}$, the DSC filter is introduced as

$$\dot{\alpha}_{i,1} = \frac{1}{\tau_{i,1}}s_{i,1} - \frac{1}{\tau_{i,1}}\alpha_{i,1} \tag{7.27}$$

with the filtered error $\tilde{\alpha}_{i,1} = \alpha_{i,1} - s_{i,1}$ and constant $\tau_{i,1} > 0$.

From (7.2), differentiating $e_{i,1}$ yields

$$\dot{e}_{i,1} = (d_i + |b_i|)\, f_{i,1}\left(x_{i,1}\right) - \sum_{k\in\mathcal{N}_i} a_{i,k}\left(f_{k,1}\left(x_{k,1}\right) + g_{k,1}\left(x_{k,1}\right)x_{k,2}\right) + (d_i + |b_i|)\, g_{i,1}\left(x_{i,1}\right)\left(e_{i,2} + \tilde{\alpha}_{i,1} + s_{i,1}\right) - b_i\dot{y}_0. \tag{7.28}$$

From $\tilde{\alpha}_{i,1} = \alpha_{i,1} - s_{i,1}$, differentiating $\tilde{\alpha}_{i,1}$ yields

$$\dot{\tilde{\alpha}}_{i,1} = \dot{\alpha}_{i,1} - \dot{s}_{i,1} = -\frac{1}{\tau_{i,1}}\tilde{\alpha}_{i,1} - \dot{s}_{i,1}. \tag{7.29}$$

Define $V_{i,1} = \frac{1}{2}e_{i,1}^2 + \frac{1}{2}\tilde{\alpha}_{i,1}^2 + \frac{1}{2}\tilde{W}_{i,1}^{\mathrm{T}}\tilde{W}_{i,1}$. Differentiating $V_{i,1}$ yields

$$\dot{V}_{i,1} = e_{i,1}\dot{e}_{i,1} + \tilde{\alpha}_{i,1}\dot{\tilde{\alpha}}_{i,1} - \tilde{W}_{i,1}^{\mathrm{T}}\dot{\hat{W}}_{i,1}. \tag{7.30}$$

Then, inserting (7.22), (7.23), (7.28), and (7.29) into (7.30) yields

$$\begin{aligned}\dot{V}_{i,1} = & e_{i,1}\hat{W}_{i,1}^{\mathrm{T}}(t)\varphi_{i,1}\left(X_{i,1}\right) + e_{i,1}\tilde{W}_{i,1}^{\mathrm{T}}(t)\varphi_{i,1}\left(X_{i,1}\right)\\ &+e_{i,1}\left[(d_i + |b_i|)\,g_{i,1}\left(x_{i,1}\right)s_{i,1} - b_i\dot{y}_0\right]\\ &+e_{i,1}(d_i + |b_i|)\,g_{i,1}\left(x_{i,1}\right)\left(e_{i,2} + \tilde{\alpha}_{i,1}\right)\\ &-\frac{1}{\tau_{i,1}}\tilde{\alpha}_{i,1}^2 - \tilde{\alpha}_{i,1}\dot{s}_{i,1} - \tilde{W}_{i,1}^{\mathrm{T}}\dot{\hat{W}}_{i,1} + e_{i,1}\varepsilon_{i,1}.\end{aligned} \tag{7.31}$$

With the design of $\xi_{i,1}$ in (7.25), $s_{i,1}$ in (7.26), and the adaptive fuzzy law $\hat{W}_{i,j}$ in (7.24), one has

$$\begin{aligned}\dot{V}_{i,1} = & \frac{1}{h_{i,1}}\left[\frac{(d_i + |b_i|)}{h_{i,1}}g_{i,1}\left(x_{i,1}\right)N\left(\xi_{i,1}\right) + 1\right]\dot{\xi}_{i,1}\\ &-k_{i,1}e_{i,1}^2 + (d_i + |b_i|)\,g_{i,1}\left(x_{i,1}\right)e_{i,1}e_{i,2}\\ &+(d_i + |b_i|)\,g_{i,1}\left(x_{i,1}\right)e_{i,1}\tilde{\alpha}_{i,1} - \frac{1}{\tau_{i,1}}\tilde{\alpha}_{i,1}^2\\ &-\tilde{\alpha}_{i,1}\dot{s}_{i,1} + \gamma_{i,1}\tilde{W}_{i,1}^{\mathrm{T}}\hat{W}_{i,1} + e_{i,1}\varepsilon_{i,1}.\end{aligned} \tag{7.32}$$

Taking the derivative of virtual control $s_{i,1}$ yields

$$\begin{aligned}\dot{s}_{i,1} &= \frac{1}{h_{i,1}}\left[N_1\left(\xi_{i,1}\right)\dot{r}_{i,1} + \frac{\partial N_1\left(\xi_{i,1}\right)}{\partial \xi_{i,1}}\dot{\xi}_{i,1}r_{i,1}\right],\\ \dot{r}_{i,1} &= -k_{i,1}\dot{e}_{i,1} - \dot{\hat{W}}_{i,1}^{\mathrm{T}}(t)\varphi_{i,1}\left(X_{i,1}\right)\\ &\quad -\hat{W}_{i,1}^{\mathrm{T}}(t)\dot{\varphi}_{i,1}\left(X_{i,1}\right) + b_i\ddot{y}_0.\end{aligned} \tag{7.33}$$

Consider the sets

$$\begin{aligned}\Lambda_{i,1} &= \left\{\left(e_{i,1}, \tilde{W}_{i,1}, \tilde{\alpha}_1\right) \middle| \frac{1}{2}\left(e_{i,1}^2 + \tilde{W}_{i,1}^{\mathrm{T}}\tilde{W}_{i,1} + \tilde{\alpha}_{i,1}^2\right) \le R_{i,1}\right\},\\ \Lambda_{y_0} &= \left\{(y_0, \dot{y}_0, \ddot{y}_0) \middle| \, y_0^2 + \dot{y}_0^2 + \ddot{y}_0^2 \le R_y\right\},\end{aligned} \tag{7.34}$$

with positive constants R_y and $R_{i,1}$. Note that $\dot{s}_{i,1}$ is continuous on the compact sets $\Lambda_{i,1} \times \Lambda_{y_0}$. Then, $\left|\dot{s}_{i,1}\right| \le \bar{s}_{i,1}$ holds with $\bar{s}_{i,1} > 0$ on the compact sets $\Lambda_{i,1} \times \Lambda_{y_0}$. Based on Young's inequality, one has

$$
\begin{aligned}
-\tilde{\alpha}_{i,1}\dot{s}_{i,1} &\le \left|\tilde{\alpha}_{i,1}\dot{s}_{i,1}\right| \le \frac{\rho_{i,1}}{2}\tilde{\alpha}_{i,1}^{2} + \frac{1}{2\rho_{i,1}}\bar{s}_{i,1}^{2}, \\
\tilde{W}_{i,1}^{\mathrm{T}}\hat{W}_{i,1} &\le \frac{1}{2\Gamma_{i,1}}W_{i,1}^{\mathrm{T}}W_{i,1} - \frac{\Gamma_{i,1}}{2}\tilde{W}_{i,1}^{\mathrm{T}}\tilde{W}_{i,1}, \\
e_{i,1}\varepsilon_{i,1}\left(X_{i,1}\right) &\le \frac{\iota_{i,1}}{2}e_{i,1}^{2} + \frac{1}{2\iota_{i,1}}\bar{\varepsilon}_{i,1}^{2}.
\end{aligned}
\tag{7.35}
$$

Moreover, by the boundedness of $g_{i,1}$ and Young's inequality, one can also obtain

$$
\begin{aligned}
g_{i,1}\left(x_{i,1}\right)e_{i,1}e_{i,2} &\le \frac{\upsilon_{i,1}}{2}e_{i,1}^{2} + \frac{1}{2\upsilon_{i,1}}\bar{g}_{i,1}^{2}e_{i,2}^{2}, \\
g_{i,1}\left(x_{i,1}\right)e_{i,1}\tilde{\alpha}_{i,1} &\le \frac{\chi_{i,1}}{2}e_{i,1}^{2} + \frac{1}{2\chi_{i,1}}\bar{g}_{i,1}^{2}\tilde{\alpha}_{i,1}^{2}.
\end{aligned}
\tag{7.36}
$$

Substituting (7.35) and (7.36) into (7.32) gives

$$
\begin{aligned}
\dot{V}_{i,1} \le\; & \frac{1}{h_{i,1}}\left[\frac{\left(d_i + |b_i|\right)}{h_{i,1}}g_{i,1}\left(x_{i,1}\right)N_1\left(\xi_{i,1}\right) + 1\right]\dot{\xi}_{i,1} \\
& - \left[k_{i,1} - \frac{\upsilon_{i,1} + \chi_{i,1}}{2}\left(d_i + |b_i|\right) - \frac{\iota_{i,1}}{2}\right]e_{i,1}^{2} \\
& - \left[\frac{1}{\tau_{i,1}} - \frac{\rho_{i,1}}{2} - \frac{\left(d_i + |b_i|\right)\bar{g}_{i,1}^{2}}{2\chi_{i,1}}\right]\tilde{\alpha}_{i,1}^{2} \\
& - \frac{\gamma_{i,1}\Gamma_{i,1}}{2}\tilde{W}_{i,1}^{\mathrm{T}}\tilde{W}_{i,1} + \frac{\left(d_i + |b_i|\right)}{2\upsilon_{i,1}}\bar{g}_{i,1}^{2}e_{i,2}^{2} \\
& + \frac{1}{2\iota_{i,1}}\bar{\varepsilon}_{i,1}^{2} + \frac{\gamma_{i,1}}{2\Gamma_{i,1}}W_{i,1}^{\mathrm{T}}W_{i,1} + \frac{1}{2\rho_{i,1}}\bar{s}_{i,1}^{2}.
\end{aligned}
\tag{7.37}
$$

7.3.2 Step j ($2 \le j \le n_i - 1$)

For step j, define $F_{i,j}\left(X_{i,j}\right) = f_{i,j}\left(\bar{x}_{i,j}\right)$ with $X_{i,j} = \bar{x}_{i,j}$. To compensate for $F_{i,j}\left(\bar{x}_{i,j}\right)$, the adaptive fuzzy law is considered as

$$
\dot{\hat{W}}_{i,j} = e_{i,j}\varphi_{i,j}\left(\bar{x}_{i,j}\right) - \gamma_{i,j}\hat{W}_{i,j}. \tag{7.38}
$$

To address the problem of an unknown control direction in step j, the adaptive Nussbaum-type gain $\xi_{i,j}$ can be designed as

$$
\dot{\xi}_{i,j} = h_{i,j}e_{i,j}r_{i,j} \tag{7.39}
$$

with $r_{i,j} = -k_{i,j}e_{i,j} - \hat{W}_{i,j}^{\mathrm{T}}\left(\bar{x}_{i,j}\right)\varphi_{i,j}\left(\bar{x}_{i,j}\right) + \dot{\alpha}_{i,j-1}$. The virtual control can be developed as

$$s_{i,j} = \frac{1}{h_{i,j}} N_j\left(\xi_{i,j}\right) r_{i,j}. \tag{7.40}$$

Similarly, the DSC filter is designed as

$$\dot{\alpha}_{i,j} = \frac{1}{\tau_{i,j}} s_{i,j} - \frac{1}{\tau_{i,j}} \alpha_{i,j} \tag{7.41}$$

with the filtered error $\tilde{\alpha}_{i,j} = \alpha_{i,j} - s_{i,j}$.

From (7.2), differentiating $e_{i,j}$ yields

$$\begin{aligned}\dot{e}_{i,j} = & f_{i,j}\left(\bar{x}_{i,j}\right) + g_{i,j}\left(\bar{x}_{i,j}\right) e_{i,j+1} \\ & + g_{i,j}\left(\bar{x}_{i,j}\right)\tilde{\alpha}_{i,j} + g_{i,j}\left(\bar{x}_{i,j}\right) s_{i,j} - \dot{\alpha}_{i,j-1}.\end{aligned} \tag{7.42}$$

Differentiating $\tilde{\alpha}_{i,j}$ with $\tilde{\alpha}_{i,j} = \alpha_{i,j} - s_{i,j}$ yields

$$\dot{\tilde{\alpha}}_{i,j} = \dot{\alpha}_{i,j} - \dot{s}_{i,j} = -\frac{1}{\tau_{i,j}}\tilde{\alpha}_{i,j} - \dot{s}_{i,j}. \tag{7.43}$$

The Lyapunov function for step j can be defined as $V_{i,j} = \frac{1}{2}e_{i,j}^2 + \frac{1}{2}\tilde{\alpha}_{i,j}^2 + \frac{1}{2}\tilde{W}_{i,j}^{\mathrm{T}}\tilde{W}_{i,j}$. Differentiating $V_{i,j}$ gives

$$\dot{V}_{i,j} = e_{i,j}\dot{e}_{i,j} + \tilde{\alpha}_{i,j}\dot{\tilde{\alpha}}_{i,j} - \tilde{W}_{i,j}^{\mathrm{T}}\dot{\hat{W}}_{i,j}. \tag{7.44}$$

Inserting (7.42) and (7.43) into (7.44) yields

$$\begin{aligned}\dot{V}_{i,j} = & e_{i,j}\tilde{W}_{i,j}^{\mathrm{T}}(t)\varphi_{i,j}\left(\bar{x}_{i,j}\right) + e_{i,j}\hat{W}_{i,j}^{\mathrm{T}}(t)\varphi_{i,j}\left(\bar{x}_{i,j}\right) \\ & + e_{i,j}\varepsilon_{i,j}\left(\bar{x}_{i,j}\right) + g_{i,j}\left(\bar{x}_{i,j}\right) e_{i,j}e_{i,j+1} \\ & + g_{i,j}\left(\bar{x}_{i,j}\right) e_{i,j}s_{i,j} + g_{i,j}\left(\bar{x}_{i,j}\right) e_{i,j}\tilde{\alpha}_{i,j} \\ & - e_{i,j}\dot{\alpha}_{i,j-1} - \frac{1}{\tau_{i,j}}\tilde{\alpha}_{i,j}^2 - \tilde{\alpha}_{i,j}\dot{s}_{i,j} - \tilde{W}_{i,j}^{\mathrm{T}}\dot{\hat{W}}_{i,j}.\end{aligned} \tag{7.45}$$

Using the design of $\xi_{i,j}$, $s_{i,j}$, and $\hat{W}_{i,j}$, we can obtain

$$\begin{aligned}\dot{V}_{i,j} = & \frac{1}{h_{i,j}}\left[\frac{1}{h_{i,j}} g_{i,j}\left(\bar{x}_{i,j}\right) N_j\left(\xi_{i,j}\right) + 1\right]\dot{\xi}_{i,j} \\ & - k_{i,j}e_{i,j}^2 + g_{i,j}\left(\bar{x}_{i,j}\right) e_{i,j}e_{i,j+1}\end{aligned}$$

$$
\begin{aligned}
&+g_{i,j}\left(\bar{x}_{i,j}\right) e_{i,j}\tilde{\alpha}_{i,j} + e_{i,j}\varepsilon_{i,j}\left(\bar{x}_{i,j}\right) \\
&-\frac{1}{\tau_{i,j}}\tilde{\alpha}_{i,j}^2 - \tilde{\alpha}_{i,j}\dot{s}_{i,j} + \gamma_{i,j}\tilde{W}_{i,j}^{\mathrm{T}}\hat{W}_{i,j}.
\end{aligned} \tag{7.46}
$$

Differentiating virtual control $s_{i,j}$ yields

$$
\begin{aligned}
\dot{s}_{i,j} &= \frac{1}{h_{i,j}}\left[N_j\left(\xi_{i,j}\right)\dot{r}_{i,j} + \frac{\partial N_j\left(\xi_{i,j}\right)}{\partial \xi_{i,j}}\dot{\xi}_{i,j} r_{i,j}\right], \\
\dot{r}_{i,j} &= -k_{i,j}\dot{e}_{i,j} - \dot{\hat{W}}_{i,j}^{\mathrm{T}}\left(\bar{x}_{i,j}\right)\varphi_{i,j}\left(\bar{x}_{i,j}\right) \\
&\quad -\hat{W}_{i,j}^{\mathrm{T}}\left(\bar{x}_{i,j}\right)\dot{\varphi}_{i,j}\left(\bar{x}_{i,j}\right) + \ddot{\alpha}_{i,j-1}.
\end{aligned} \tag{7.47}
$$

Consider the sets

$$
\begin{aligned}
\Lambda_{i,j} &= \left\{\left(e_{i,j}, \tilde{W}_{i,1}, \tilde{\alpha}_{i,j}\right) \middle| \frac{1}{2}e_{i,j}^2 + \frac{1}{2}\tilde{W}_{i,j}^{\mathrm{T}}\tilde{W}_{i,j} + \frac{1}{2}\tilde{\alpha}_{i,j}^2 \le R_{i,j}\right\}, \\
\Lambda_{y_0} &= \left\{(y_0, \dot{y}_0, \ddot{y}_0) \middle| \, y_0^2 + \dot{y}_0^2 + \ddot{y}_0^2 \le R_y\right\},
\end{aligned} \tag{7.48}
$$

with positive constants R_y and $R_{i,j}$. Note that $\dot{s}_{i,j}$ is continuous on the compact sets $\Lambda_{i,j} \times \Lambda_{y_0}$. Then, $\left|\dot{s}_{i,j}\right| \le \bar{s}_{i,j}$ holds with $\bar{s}_{i,j} > 0$ on the compact sets $\Lambda_{i,j} \times \Lambda_{y_0}$. From Young's inequality, one can obtain

$$
\begin{aligned}
-\tilde{\alpha}_{i,j}\dot{s}_{i,j} &\le \left|\tilde{\alpha}_{i,j}\dot{s}_{i,j}\right| \le \frac{\rho_{i,j}}{2}\tilde{\alpha}_{i,j}^2 + \frac{1}{2\rho_{i,j}}\bar{s}_{i,j}^2, \\
\tilde{W}_{i,j}^{\mathrm{T}}\hat{W}_{i,j} &\le \frac{1}{2\Gamma_{i,j}}W_{i,j}^{\mathrm{T}}W_{i,j} - \frac{\Gamma_{i,j}}{2}\tilde{W}_{i,j}^{\mathrm{T}}\tilde{W}_{i,j}, \\
e_{i,j}\varepsilon_{i,j}\left(\bar{x}_{i,j}\right) &\le \frac{\iota_{i,j}}{2}e_{i,j}^2 + \frac{1}{2\iota_{i,j}}\bar{\varepsilon}_{i,j}^2.
\end{aligned} \tag{7.49}
$$

Further, using the boundedness of $g_{i,j}$ yields

$$
\begin{aligned}
g_{i,j}\left(\bar{x}_{i,j}\right) e_{i,j} e_{i,j+1} &\le \frac{\upsilon_{i,j}}{2}e_{i,j}^2 + \frac{1}{2\upsilon_{i,j}}\bar{g}_{i,j}^2 e_{i,j+1}^2, \\
g_{i,j}\left(\bar{x}_{i,j}\right) e_{i,j}\tilde{\alpha}_{i,j} &\le \frac{\chi_{i,j}}{2}e_{i,j}^2 + \frac{1}{2\chi_{i,j}}\bar{g}_{i,j}^2\tilde{\alpha}_{i,j}^2.
\end{aligned} \tag{7.50}
$$

Substituting (7.49) and (7.50) into (7.46) gives

$$
\begin{aligned}
\dot{V}_{i,j} \le\ & \frac{1}{h_{i,j}}\left[\frac{1}{h_{i,j}}g_{i,j}\left(\bar{x}_{i,j}\right)N_j\left(\xi_{i,j}\right) + 1\right]\dot{\xi}_{i,j} - \frac{\Gamma_{i,j}\gamma_{i,j}}{2}\tilde{W}_{i,j}^{\mathrm{T}}\tilde{W}_{i,j} \\
&-\left(k_{i,j} - \frac{\upsilon_{i,j} + \chi_{i,j} + \iota_{i,j}}{2}\right)e_{i,j}^2 - \left(\frac{1}{\tau_{i,j}} - \frac{\bar{g}_{i,j}^2}{2\chi_{i,j}} - \frac{\rho_{i,j}}{2}\right)\tilde{\alpha}_{i,j}^2 \\
&+\frac{1}{2\upsilon_{i,j}}\bar{g}_{i,j}^2 e_{i,j+1}^2 + \frac{1}{2\iota_{i,j}}\bar{\varepsilon}_{i,j}^2 + \frac{\gamma_{i,j}}{2\Gamma_{i,j}}W_{i,j}^{\mathrm{T}}W_{i,j} + \frac{1}{2\rho_{i,j}}\bar{s}_{i,j}^2.
\end{aligned} \tag{7.51}
$$

7.3.3 Step n_i

Define $F_{i,n_i}\left(X_{i,n_i}\right) = f_{i,n_i}\left(\bar{x}_{i,n_i}\right)$ with $X_{i,n_i} = \bar{x}_{i,n_i}$ in step n_i. The adaptive fuzzy law is then designed as

$$\dot{\hat{W}}_{i,n_i} = e_{i,n_i}\varphi_{i,n_i}\left(\bar{x}_{i,n_i}\right) - \gamma_{i,n_i}\hat{W}_{i,n_i}. \tag{7.52}$$

For step n_i, the Nussbaum-type gain ξ_{i,n_i} is designed as

$$\dot{\xi}_{i,n_i} = h_{i,n_i}e_{i,n_i}r_{i,n_i} \tag{7.53}$$

with $r_{i,n_i} = -k_{i,n_i}e_{i,n_i} - \hat{W}_{i,n_i}^{\mathrm{T}}\left(\bar{x}_{i,n_i}\right)\varphi_{i,n_i}\left(\bar{x}_{i,n_i}\right)$. To ensure the stability of the system, the signal v_i can be designed as

$$v_i = \frac{1}{h_{i,n_i}}N_{n_i}\left(\xi_{i,n_i}\right)r_{i,n_i}. \tag{7.54}$$

The fixed event-triggered mechanism can be designed as

$$t_{i,k+1} = \inf\left\{t > t_{i,k} \middle| \left|\tilde{v}_i(t)\right| \ge m_i\right\}, \tag{7.55}$$

with a positive constant m_i. To reduce the communication burden, let the event-triggered control be designed as $u_i(t) = v_i\left(t_{i,k}\right)$ with $\tilde{v}_i(t) = v_i(t) - u_i(t)$.

7.4 Stability Analysis

By employing the controller design detailed in Sect. 7.3, the stability analysis of the MASs is discussed with fixed and relative threshold strategies as follows.

Define $\eta_i(t) = \frac{v_i(t)-u_i(t)}{m_i}$. Then, we can obtain $|\eta_i(t)| \le 1$ from (7.55). Differentiating e_{i,n_i} yields

$$\begin{aligned}\dot{e}_{i,n_i} = {} & W_{i,n_i}^{\mathrm{T}}\varphi_{i,n_i}\left(\bar{x}_{i,n_i}\right) + \varepsilon_{i,n_i}\left(\bar{x}_{i,n_i}\right) + g_{i,n_i}\left(\bar{x}_{i,n_i}\right)v_i \\ & -g_{i,n_i}\left(\bar{x}_{i,n_i}\right)\eta_i(t)m_i - \dot{\alpha}_{i,n_i-1}.\end{aligned} \tag{7.56}$$

For the n_ith subsystem, define $V_{i,n_i} = \frac{1}{2}e_{i,n_i}^2 + \frac{1}{2}\tilde{W}_{i,n_i}^{\mathrm{T}}\tilde{W}_{i,n_i}$. Differentiation of the V_{i,n_i} gives

$$\begin{aligned}\dot{V}_n = {} & e_{i,n_i}W_{i,n_i}^{\mathrm{T}}\varphi_{i,n_i}\left(\bar{x}_{i,n_i}\right) + e_{i,n_i}\varepsilon_{i,n_i}\left(\bar{x}_{i,n_i}\right) + g_{i,n_i}\left(\bar{x}_{i,n_i}\right)e_{i,n_i}v_i \\ & -\eta_i(t)m_ig_{i,n_i}\left(\bar{x}_{i,n_i}\right)e_{i,n_i} - e_{i,n_i}\dot{\alpha}_{i,n_i-1} - \tilde{W}_{i,n_i}^{\mathrm{T}}\dot{\hat{W}}_{i,n_i}.\end{aligned} \tag{7.57}$$

Using the design of ξ_{i,n_i} in (7.53), v_i in (7.54), and the adaptive fuzzy law $\hat{W}_{i,n_i}$ in (7.52), we have

$$\begin{aligned}\dot{V}_{n_i} =& \frac{1}{h_{i,n_i}}\left[\frac{1}{h_{i,n_i}} g_{i,n_i} N_{n_i}\left(\xi_{i,n_i}\right)+1\right] \dot{\xi}_{i,n_i} - k_{i,n_i} e_{i,n_i}^2 + e_{i,n_i} \varepsilon_{i,n_i}\left(\bar{x}_{i,n_i}\right) \\ & -\eta_i(t) g_{i,n_i}\left(\bar{x}_{i,n_i}\right) m_i e_{i,n_i} - \gamma_{i,n_i} \tilde{W}_{i,n_i}^{\mathrm{T}} \hat{W}_{i,n_i}. \end{aligned} \tag{7.58}$$

According to Young's inequality and the boundedness of g_{i,n_i}, one has

$$\begin{aligned} \tilde{W}_{i,n_i}^{\mathrm{T}} \hat{W}_{i,n_i} &\le \frac{1}{2\Gamma_{i,n_i}} W_{i,n_i}^{\mathrm{T}} W_{i,n_i} - \frac{\Gamma_{i,n_i}}{2} \tilde{W}_{i,n_i}^{\mathrm{T}} \tilde{W}_{i,n_i}, \\ e_{i,n_i} \varepsilon_{i,n_i}\left(\bar{x}_{i,n_i}\right) &\le \frac{\iota_{i,n_i}}{2} e_{i,n_i}^2 + \frac{1}{2\iota_{i,n_i}} \bar{\varepsilon}_{i,n_i}^2, \\ -\eta_i(t) g_{i,n_i}\left(\bar{x}_{i,n_i}\right) m_i e_{i,n_i} &\le \frac{\nu_{i,n_i}}{2} m_i^2 + \frac{\bar{g}_{i,n_i}}{2\nu_{i,n_i}} e_{i,n_i}^2. \end{aligned} \tag{7.59}$$

Inserting (7.59) into (7.58) yields

$$\begin{aligned} \dot{V}_{i,n_i} \le & \frac{1}{h_{i,n_i}}\left[\frac{1}{h_{i,n_i}} g_{i,n_i} N_{n_i}\left(\xi_{i,n_i}\right)+1\right] \dot{\xi}_{i,n_i} \\ & -\left(k_{i,n_i} - \frac{\bar{g}_{i,n_i}}{2\nu_{i,n_i}} - \frac{\iota_{i,n_i}}{2}\right) e_{i,n_i}^2 + \frac{1}{2\iota_{i,n_i}} \bar{\varepsilon}_{i,n_i}^2 \\ & -\frac{\Gamma_{i,n_i}\gamma_{i,n_i}}{2} \tilde{W}_{i,n_i}^{\mathrm{T}} \tilde{W}_{i,n_i} + \frac{\gamma_{i,n_i}}{2\Gamma_{i,n_i}} W_{i,n_i}^{\mathrm{T}} W_{i,n_i} + \frac{\nu_{i,n_i}}{2} m_i^2. \end{aligned} \tag{7.60}$$

To this end, the following theoretical results can be obtained.

Theorem 7.1 *Under Assumptions 7.1–7.3, for system* (7.1) *with the bipartite synchronization design under fixed threshold strategy* (7.55)*, the MASs could achieve the practical bipartite consensus with the closed-loop signals being bounded.*

Proof Define the Lyapunov function for subsystem i as

$$V_i = \frac{1}{2}\left(\sum_{j=1}^{n_i} e_{i,j}^2 + \sum_{j=1}^{n_i-1} \tilde{\alpha}_{i,j}^2 + \sum_{j=1}^{n_i} \tilde{W}_{i,j}^{\mathrm{T}} \tilde{W}_{i,j}\right).$$

The derivative of V_i can be represented by (7.37), (7.51), and (7.60) as

$$\begin{aligned} \dot{V}_i \le & \sum_{j=1}^{n_i} \frac{1}{h_{i,j}}\left[\beta_{i,j} g_{i,j}\left(\bar{x}_{i,j}\right) N_j\left(\xi_{i,j}\right)+1\right] \dot{\xi}_{i,j} \\ & -\left[k_{i,1} - \frac{\upsilon_{i,1}+\chi_{i,1}}{2}\left(d_i+|b_i|\right) - \frac{\iota_{i,1}}{2}\right] e_{i,1}^2 \end{aligned}$$

$$
\begin{aligned}
&-\left[\frac{1}{\tau_{i,1}}-\frac{\rho_{i,1}}{2}-\frac{(d_i+|b_i|)\,\bar{g}_{i,1}^2}{2\chi_{i,1}}\right]\tilde{\alpha}_{i,1}^2\\
&-\frac{\gamma_{i,1}\Gamma_{i,1}}{2}\tilde{W}_{i,1}^{\mathrm{T}}\tilde{W}_{i,1}+\frac{(d_i+|b_i|)}{2\upsilon_{i,1}}\bar{g}_{i,1}^2e_{i,2}^2\\
&+\frac{1}{2\iota_{i,1}}\bar{\varepsilon}_{i,1}^2+\frac{\gamma_{i,1}}{2\Gamma_{i,1}}W_{i,1}^{\mathrm{T}}W_{i,1}+\frac{1}{2\rho_{i,1}}\bar{s}_{i,1}^2\\
&-\sum_{j=2}^{n_i-1}\left[\left(k_{i,j}-\frac{\upsilon_{i,j}+\chi_{i,j}+\iota_{i,j}}{2}\right)e_{i,j}^2\right]
\end{aligned}
$$

$$
\begin{aligned}
&-\sum_{j=2}^{n_i-1}\left[\left(\frac{1}{\tau_{i,j}}-\frac{\bar{g}_{i,j}^2}{2\chi_{i,j}}-\frac{\rho_{i,j}}{2}\right)\tilde{\alpha}_{i,j}^2\right]\\
&+\sum_{j=2}^{n_i-1}\left[\frac{1}{2\upsilon_{i,j}}\bar{g}_{i,j}^2e_{i,j+1}^2-\frac{\Gamma_{i,j}\gamma_{i,j}}{2}\tilde{W}_{i,j}^{\mathrm{T}}\tilde{W}_{i,j}\right]\\
&+\sum_{j=2}^{n_i-1}\left[\frac{1}{2\iota_{i,j}}\bar{\varepsilon}_{i,j}^2+\frac{\gamma_{i,j}}{2\Gamma_{i,j}}W_{i,j}^{\mathrm{T}}W_{i,j}+\frac{1}{2\rho_{i,j}}\bar{s}_{i,j}^2\right]\\
&-\left(k_{i,n_i}-\frac{\bar{g}_{i,n_i}}{2\nu_{i,n_i}}-\frac{\iota_{i,n_i}}{2}\right)e_{i,n_i}^2+\frac{1}{2\iota_{i,n_i}}\bar{\varepsilon}_{i,n_i}^2\\
&-\frac{\Gamma_{i,n_i}\gamma_{i,n_i}}{2}\tilde{W}_{i,n_i}^{\mathrm{T}}\tilde{W}_{i,n_i}+\frac{\gamma_{i,n_i}}{2\Gamma_{i,n_i}}W_{i,n_i}^{\mathrm{T}}W_{i,n_i}+\frac{\nu_{i,n_i}}{2}m_i^2.
\end{aligned}
$$

Therefore, we can obtain

$$
\begin{aligned}
\dot{V}_i\le&\sum_{j=1}^{n_i}\frac{1}{h_{i,j}}\left[\beta_{i,j}g_{i,j}\left(\bar{x}_{i,j}\right)N_j\left(\xi_{i,j}\right)+1\right]\dot{\xi}_{i,j}\\
&-\sum_{j=1}^{n_i}K_{i,j}e_{i,j}^2-\sum_{j=1}^{n_i}\frac{\Gamma_{i,j}\gamma_{i,j}}{2}\tilde{W}_{i,j}^{\mathrm{T}}\tilde{W}_{i,j}\\
&-\sum_{j=1}^{n_i-1}\left(\frac{1}{\tau_{i,j}}-\frac{\bar{g}_{i,j}^2}{2\chi_{i,j}}-\frac{\rho_{i,j}}{2}\right)\tilde{\alpha}_{i,j}^2+\sum_{j=1}^{n_i-1}\frac{1}{2\rho_{i,j}}\bar{s}_{i,j}^2\\
&+\frac{1}{2}\sum_{j=1}^{n_i}\left(\frac{1}{\iota_{i,j}}\bar{\varepsilon}_{i,j}^2+\frac{\gamma_{i,j}}{\Gamma_{i,j}}W_{i,j}^{\mathrm{T}}W_{i,j}\right)+\frac{\nu_{i,n_i}}{2}m_i^2
\end{aligned}
\tag{7.61}
$$

with β_i and $K_{i,j}$ being $\beta_{i,j}=\begin{cases}\frac{1}{h_{i,j}}\left(d_i+|b_i|\right), & j=1\\ \frac{1}{h_{i,j}}, & j\neq 1\end{cases}$ and

$$K_{i,j}=\begin{cases} k_{i,1}-\frac{\upsilon_{i,1}+\chi_{i,1}}{2}\left(d_i+|b_i|\right)-\frac{\iota_{i,1}}{2},\ j=1\\ k_{i,2}-\frac{\upsilon_{i,2}+\chi_{i,2}+\iota_{i,2}}{2}-\frac{(d_i+|b_i|)}{2\upsilon_{i,1}}\bar{g}_{i,1}^2,\ j=2\\ k_{i,j}-\frac{\upsilon_{i,j}+\chi_{i,j}+\iota_{i,j}}{2}-\frac{\bar{g}_{i,j-1}^2}{2\upsilon_{i,j-1}},\ j=3,\cdots,n_i-1\\ k_{i,n_i}-\frac{\bar{g}_{i,n_i}}{2\nu_{i,n_i}}-\frac{\iota_{i,n_i}}{2}-\frac{\bar{g}_{i,n_i-1}^2}{2\upsilon_{i,n_i-1}},\ j=n_i. \end{cases}$$

Based on the above, (7.61) can illustrate

$$\begin{aligned} V_i\le&\sum_{j=1}^{n_i}\int_0^t\frac{\beta_{i,j}}{h_{i,j}}g_{i,j}(\sigma)N_j\left(\xi_{i,j}(\sigma)\right)\dot{\xi}_{i,j}(\sigma)\mathrm{e}^{-q_i(t-\sigma)}\mathrm{d}\sigma\\ &+\sum_{j=1}^{n_i}\int_0^t\frac{1}{h_{i,j}}\dot{\xi}_{i,j}(\sigma)\mathrm{e}^{-q_i(t-\sigma)}\mathrm{d}\sigma+\frac{p_i}{q_i}, \end{aligned}\tag{7.62}$$

where $q_i=\min\limits_{j\in[1,n_i]}\left\{2K_{i,j},\Gamma_{i,j}\gamma_{i,j},2\left(\frac{1}{\tau_{i,j}}-\frac{\bar{g}_{i,j}^2}{2\chi_{i,j}}-\frac{\rho_{i,j}}{2}\right)\right\}$ and $p_i=\sum\limits_{j=1}^{n_i-1}\frac{1}{2\rho_{i,j}}\bar{s}_{i,j}^2+\frac{1}{2}\sum\limits_{j=1}^{n_i}\left(\frac{1}{\iota_{i,j}}\bar{\varepsilon}_{i,j}^2+\frac{\gamma_{i,j}}{\Gamma_{i,j}}W_{i,j}^{\mathrm{T}}W_{i,j}\right)+\frac{\nu_{i,n_i}}{2}m_i^2$. From Lemma 7.2, we know that the bipartite synchronization error $e_{i,1}$, the tracking errors $e_{i,j}$, the Nussbaum-type gain $\xi_{i,j}$, the adaptive parameters errors $\tilde{W}_{i,j}$, and the filter errors $\tilde{\alpha}_{i,j}$ are bounded, which implies that the MASs could achieve practical bipartite consensus. This completes the proof. □

Remark 7.1 The local output interaction (7.2) plays an important role in the stability analysis for the multiagent systems. However, from the definition of $e_{i,j}$ in (7.2), the local interaction consists of each agent's output as well as the neighbor's output, which is influenced by unknown control directions, as shown in (7.1). Consequently, the issue of unknown control directions of interest in this chapter is more challenging because the multiple control directions are involved in the local interaction, as compared to conventional problems with a single unknown control direction [19, 20].

The continuous dynamical system with event-triggered control input can be viewed as a hybrid system, in which the Zeno phenomenon might occur with infinite triggering over a finite horizon. The control design with Zeno behavior cannot be implemented in engineering applications. Besides the stability analysis for the event-triggered control design, the feasibility analysis is another important issue to be considered. The feasibility of the event-triggered bipartite synchronization protocol design under a fixed threshold strategy is discussed as follows.

Theorem 7.2 *Considering the distributed control input design* (7.54)*, the Zeno phenomenon can be excluded from the event-triggered mechanism under the fixed threshold strategy* (7.55).

Proof To establish the Zeno-free property of the event-triggering condition (7.55), it is sufficient to show that the inter-event time is lower-bounded by a positive constant. Differentiating $|\tilde{v}_i(t)|$ yields

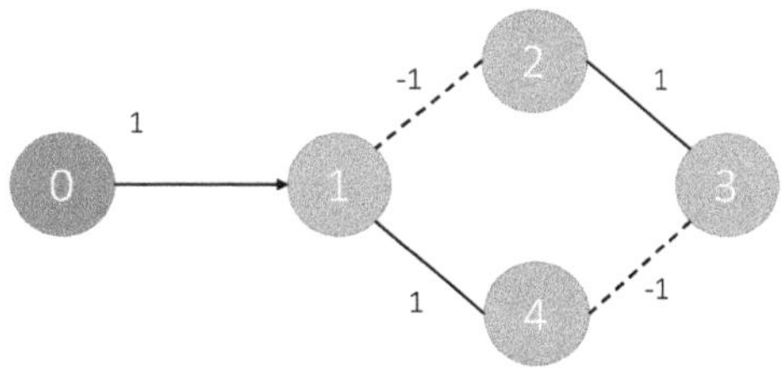

Fig. 7.1 The communication topology with 4 followers

$$\frac{d}{dt}\left|\tilde{\upsilon}_i(t)\right| = \frac{d}{dt}\left\{\tilde{\upsilon}_i(t)\, sign\left[\tilde{\upsilon}_i(t)\right]\right\} \le \left|\dot{\tilde{\upsilon}}_i(t)\right| = \left|\dot{\upsilon}_i(t)\right|. \tag{7.63}$$

Based on (7.53) and (7.54), we have

$$\begin{aligned} \dot{r}_{i,n_i} &= k_{i,n_i}\dot{e}_{i,n_i} + \dot{\hat{W}}_{i,n_i}^{\mathrm{T}}(t)\,\varphi_{i,n_i}\left(\bar{x}_{i,n_i}\right) \\ &\quad + \hat{W}_{i,n_i}^{\mathrm{T}}(t)\,\dot{\varphi}_{i,n_i}\left(\bar{x}_{i,n_i}\right) - \ddot{\alpha}_{i,n_i-1} \\ \dot{\upsilon}_i(t) &= \frac{1}{h_{i,n_i}}\left[N\left(\xi_{i,n_i}\right)\dot{r}_{i,n_i} + \frac{\partial N_{n_i}\left(\xi_{i,n_i}\right)}{\partial \xi_{i,n_i}}\dot{\xi}_{i,n_i} r_{i,n_i}\right]. \end{aligned} \tag{7.64}$$

It is evident that $\dot{\upsilon}_i(t)$ is continuous and bounded on a compact set. Therefore, it can be inferred that there exists a constant $\bar{\upsilon}_i > 0$ such that $\left|\dot{\upsilon}_i(t)\right| \le \bar{\upsilon}_i$ for all t. Based on this fact, we have $\int_{t_{i,k}}^{t_{i,k+1}} \frac{\mathrm{d}\left|\tilde{\upsilon}_i(t)\right|}{\mathrm{d}t} \le \bar{\upsilon}_i\left(t_{i,k+1} - t_{i,k}\right)$. Then, one has $t_{i,k+1} - t_{i,k} \ge \frac{\left|\tilde{\upsilon}_i(t)\right|}{\bar{\upsilon}_i} \ge \frac{m_i}{\bar{\upsilon}_i} > 0$, which implies that the Zeno behavior could be excluded. This completes the proof. □

7.5 Simulation Example

In this section, we use a numerical example to confirm the effectiveness of the proposed approach. The multiagent group consists of a leader designated as 0 and four followers (identified as 1–4). The communication topology is illustrated in Fig. 7.1.

The leader signal is $y_0 = 0.5 * (sin(0.25 * t) + sin(t))$ and the dynamics of the four followers are

$$
\begin{cases}
\dot{x}_{1,1} = \left[0.9 + 0.3\cos\left(x_{1,1}\right)\right] x_{1,2} + 0.2 x_{1,1} \mathrm{e}^{-0.5 x_{1,1}}, \\
\dot{x}_{1,2} = \left[-1.2 + 0.2\sin\left(x_{1,1} \cdot x_{1,2}\right)\right] u_1 + 0.2 x_{1,1} \sin\left(x_{1,2}\right), \\
y_1 = x_{1,1},
\end{cases}
$$

$$
\begin{cases}
\dot{x}_{2,1} = \left[1 + 0.1\sin\left(x_{2,1}^2\right)\right] x_{2,2} + \cos\left(0.8 x_{2,1}\right) \\
\dot{x}_{2,2} = \left[1 + 0.1\cos\left(x_{2,1} x_{2,2}\right)\right] u_2 + 2\sin\left(0.15 x_{2,1} x_{2,2}\right) \\
y_2 = x_{2,1}
\end{cases}
$$

$$
\begin{cases}
\dot{x}_{3,1} = \left[0.8 + 0.2\mathrm{e}^{-0.5 x_{3,1}}\right] x_{1,2} \\
\dot{x}_{3,2} = \left[-0.9 + 0.3\sin\left(x_{3,1} x_{3,2}\right)\right] u_3 \\
y_3 = x_{3,1}
\end{cases}
$$

$$
\begin{cases}
\dot{x}_{4,1} = 1.1 x_{4,2} + 0.1\sin\left(x_{4,1}\right) \\
\dot{x}_{4,2} = 0.9 u_4 + 0.1\sin\left(x_{4,1} \cdot x_{4,2}\right) \\
y_4 = x_{4,1}.
\end{cases}
$$

In the simulation experiments, the FBFs are selected as

$$
\varphi_{i,j}\left(X_{i,j}\right) = \left[\varphi_{i,j,1}\left(X_{i,j}\right) \cdots \varphi_{i,j,m}\left(X_{i,j}\right) \cdots \varphi_{i,j,M_{i,j}}\left(X_{i,j}\right)\right]^{\mathrm{T}}
$$

with $\varphi_{i,j,m}\left(X_{i,j}\right) = \dfrac{\prod_{p=1}^{P_{i,j}} \mu_{F_{i,j,p}^m}\left(X_{i,j,p}\right)}{\sum_{m=1}^{M_{i,j}}\left[\prod_{p=1}^{P_{i,j}} \mu_{F_{i,j,p}^m}\left(X_{i,j,p}\right)\right]}$.

The adaptive Nussbaum-based control for the i-agent with two subsystem is designed as

$$
\begin{cases}
r_{i,1} = -k_{i,1} e_{i,1} - \hat{W}_{i,1}^{\mathrm{T}}(t)\varphi_{i,1}\left(X_{i,1}\right) + b_i \dot{y}_0, \\
\dot{\xi}_{i,1} = h_{i,1} e_{i,1} r_{i,1}, \\
s_{i,1} = \frac{1}{h_{i,1}} N\left(\xi_{i,1}\right) r_{i,1}, \\
\dot{\hat{W}}_{i,1} = e_{i,1}\varphi_{i,1}\left(X_{i,1}\right) - \gamma_{i,1}\hat{W}_{i,1}, \\
\dot{\alpha}_{i,1} = \frac{1}{\tau_{i,1}} s_{i,1} - \frac{1}{\tau_{i,1}}\alpha_{i,1},
\end{cases}
$$

and

$$
\begin{cases}
r_{i,2} = -k_{i,2} e_{i,2} - \hat{W}_{i,2}^{\mathrm{T}}(t)\varphi_{i,2}\left(\bar{x}_{i,2}\right) + \dot{\alpha}_{i,1}, \\
\dot{\xi}_{i,2} = h_{i,2} e_{i,2}, \\
r_{i,2} v_i = \frac{1}{h_{i,2}} N\left(\xi_{i,2}\right) r_{i,2}, \\
\dot{\hat{W}}_{i,2} = e_{i,2}\varphi_{i,2}\left(X_{i,2}\right) - \gamma_{i,2}\hat{W}_{i,2}, \\
u_i(t) = v_i\left(t_{i,k}\right).
\end{cases}
$$

with $i = 1, 2, 3, 4$. The event-triggered mechanism is designed as

$$
t_{i,k+1} = \inf\left\{t > t_{i,k} \middle| \left|\tilde{v}_i(t)\right| \geq m_i\right\}.
$$

The parameters in the above design are selected as follows: $k_{1,1} = 50$, $k_{1,2} = 50$, $h_{1,1} = 5$, $h_{1,2} = 0.1$, $k_{2,1} = 50$, $k_{2,2} = 50$, $h_{2,1} = 5$, $h_{2,2} = 0.1$, $k_{3,1} = 50$, $k_{3,2} = 50$, $h_{3,1} = 5$, $h_{3,2} = 0.05$, $k_{4,1} = 50$, $k_{4,2} = 50$, $h_{4,1} = 10$, $h_{4,2} = 0.1$, $\gamma_{1,1} = \gamma_{1,2} = \gamma_{2,1} = \gamma_{2,2} = \gamma_{3,1} = \gamma_{3,2} = \gamma_{4,1} = \gamma_{4,2} = 100$, $\tau_{1,1} = \tau_{2,1} = \tau_{3,1} = \tau_{4,1} = 0.3$, $m_1 = 10$, $m_2 = 6$, $m_3 = 6$, and $m_4 = 8$.

The simulation results are depicted in Figs. 7.2, 7.3, 7.4, 7.5 and highlight the performance characteristics. The figure depicts a time-variable value chart, where time progresses from left to right along the time axis. The y-axis represents changes in the variable value starting from an initial value. The tracking performance of the multiagent system is displayed in Fig. 7.2, where the output of the followers y_1 and y_4 can track the leader y_0 and the followers' outputs y_2 and y_3 can track $-y_0$. The MASs can achieve a bipartite consensus solution with rapid convergence. Figure 7.3a and b displays the tracking errors of the MASs. The convergence of bipartite consensus error $e_{i,j}$ for each follower verifies the distributed consensus performance with the presented design. In Fig. 7.3c, the Nussbaum-type gains are presented. Here, the boundedness of the Nussbaum-type gains proves the validity of our proposed method. As displayed in Fig. 7.3d, the fuzzy logic system achieves parameter parameter stability and convergence through ongoing parameter estimation and adjustment, which clearly demonstrates the effectiveness of the fuzzy system. The DSC filter output $\alpha_{i,1}$ and the state $x_{i,2}$ of second subsystem for follower i is shown in Fig. 7.4. The state $x_{i,2}$ tracks the DSC filter output $\alpha_{i,1}$, which illustrates the stability of second subsystem within the fuzzy backstepping control design. The control input signals for all the followers are shown in Fig. 7.5. It is evident that the update frequency of the control signal u_i is low when the control signal did not fluctuate. The event-triggered mechanism serves to decrease the update frequency of the control signal and alleviate the communication burden in the context of the bipartite output consensus problem.

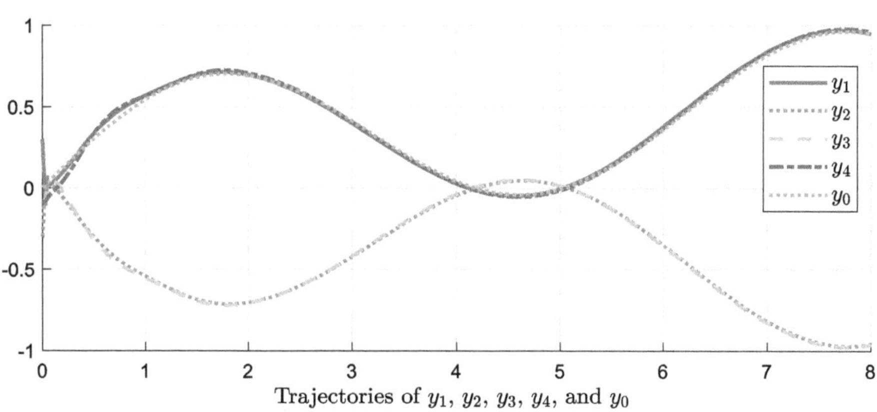

Fig. 7.2 The output of MASs and reference signal

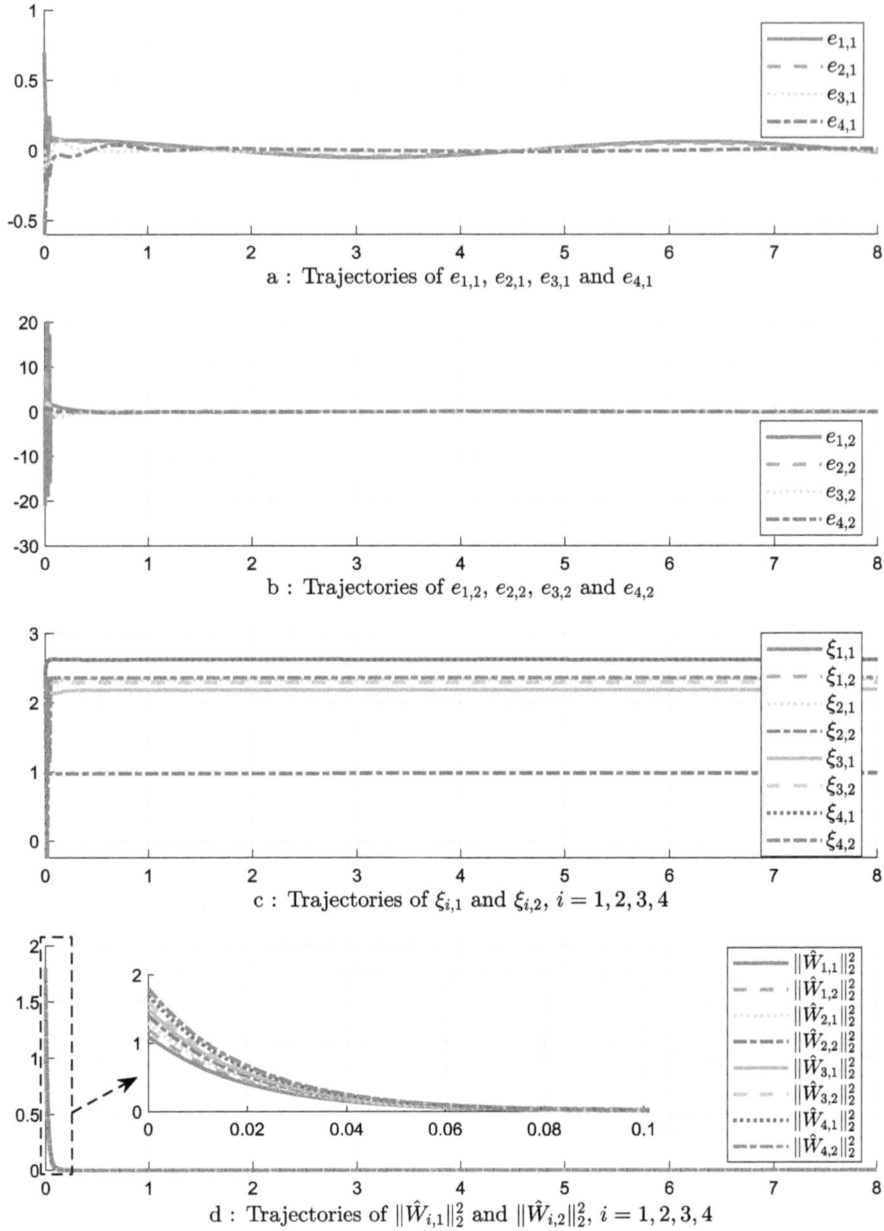

Fig. 7.3 Bipartite consensus errors, tracking errors, Nussbaum gains, and FLS weights

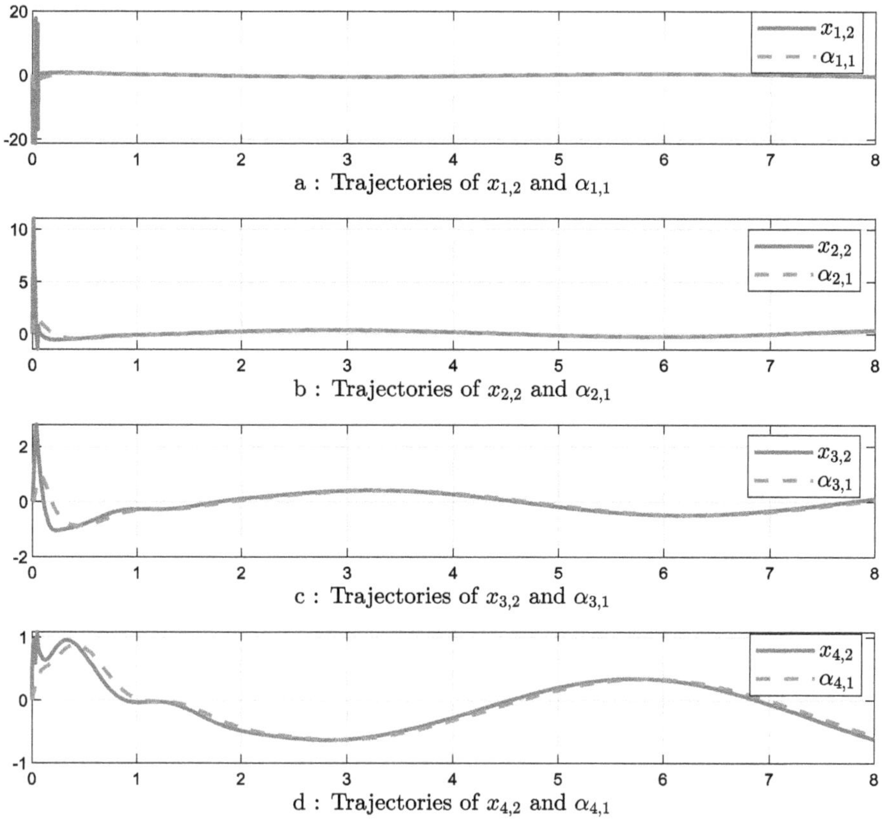

Fig. 7.4 State signals and DSC output

By examining the figures presented in the simulation results above, we demonstrate the efficacy of the distributed control protocol employing the event-triggered mechanism in addressing the bipartite synchronization problem within MASs characterized by multiple unknown control directions.

7.6 Conclusion

This chapter has presented a solution to the challenge of multiagent nonlinear systems characterized by multiple unknown control directions. We have introduced an adaptive fuzzy bipartite tracking control scheme that incorporated an event-triggered mechanism and Nussbaum-type gains. A novel Nussbaum-type function has been also developed to address multiple unknown control directions in a group of heterogeneous MASs. To alleviate the communication and computational load on individual

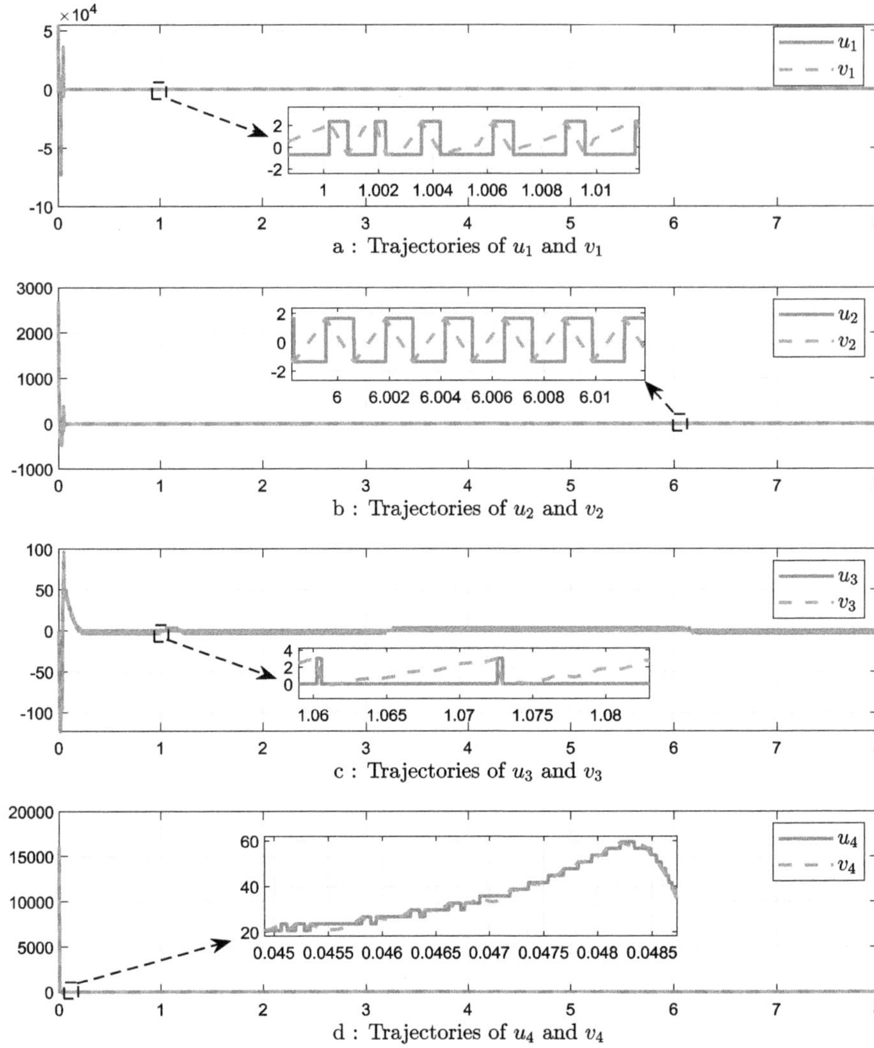

Fig. 7.5 Control inputs with event-triggered mechanism

agents, we have introduced an event-triggered control mechanism. Additionally, the FLSs have offered an effective approach for handling uncertain information in the controller design. The simulation example has validated the feasibility of the proposed design.

References

1. Yang, Y., Modares, H., Wunsch, D.C., Yin, Y.: Leader-follower output synchronization of linear heterogeneous systems with active leader using reinforcement learning. IEEE Trans. Neural Netw. Learn. Syst. **29**(6), 2139–2153 (2018)
2. Yang, Y., Xu, C.Z.: Adaptive fuzzy leader-follower synchronization of constrained heterogeneous multiagent systems. IEEE Trans. Fuzzy Syst. **30**(1), 205–219 (2022)
3. Yang, Y., Modares, H., Vamvoudakis, K.G., Yin, Y., Wunsch, D.C.: Dynamic intermittent feedback design for H_∞ containment control on a directed graph. IEEE Trans. Cybern. **50**(8), 3752–3765 (2020)
4. Zhang, H., Wang, F., Dong, Y., Chiclana, F., Herrera-Viedma, E.: Social trust driven consensus reaching model with a minimum adjustment feedback mechanism considering assessments-modifications willingness. IEEE Trans. Fuzzy Syst. **30**(6), 2019–2031 (2022)
5. Liu, J., Deng, C., Fu, L., Long, H., Gan, X., Wang, X., Chen, G., Xu, J.: Evolving bipartite model reveals the bounded weights in mobile social networks. IEEE Trans. Mob. Comput. **21**(3), 971–985 (2022)
6. Zou, Y., Meng, Z.: Targeted bipartite consensus of opinion dynamics in social networks with credibility intervals. IEEE Trans. Cybern. **52**(1), 372–383 (2022)
7. Wasserman, S., Faust, K.: Social Network Analysis: Methods and Applications. Cambridge University Press, Cambridge (1994)
8. Easley, D., Kleinberg, J.: Networks, Crowds, and Markets: Reasoning about a Highly Connected World. Cambridge University Press, Cambridge (2010)
9. Altafini, C.: Consensus problems on networks with antagonistic interactions. IEEE Trans. Autom. Control **58**(3), 935–946 (2013)
10. Liu, G., Yang, Y., Zhao, X., Ahn, C.K.: Adaptive fuzzy practical bipartite synchronization for multi-agent systems with intermittent feedback under multiple unknown control directions. IEEE Trans. Fuzzy Syst. **32**(5), 3211–3225 (2024)
11. Li, Q., Xia, L., Song, R., Liu, J.: Leader-follower bipartite output synchronization on signed digraphs under adversarial factors via data-based reinforcement learning. IEEE Trans. Neural Netw. Learn. Syst. **31**(10), 4185–4195 (2020)
12. Wang, X., Cao, Y., Niu, B., Song, Y.: A novel bipartite consensus tracking control for multiagent systems under sensor deception attacks. IEEE Trans. Cybern. **53**(9), 5984–5993 (2023)
13. Huang, Y., Liu, Y.: Practical tracking via adaptive event-triggered feedback for uncertain nonlinear systems. IEEE Trans. Autom. Control **64**(9), 3920–3927 (2019)
14. Cao, Z., Niu, Y., Lam, H.K., Zhao, J.: Sliding mode control of Markovian jump fuzzy systems: a dynamic event-triggered method. IEEE Trans. Fuzzy Syst. **29**(10), 2902–2915 (2021)
15. Donkers, M.C., Heemels, W.P.M.H.: Output-based event-triggered control with guaranteed L_∞-gain and improved and decentralized event-triggering. IEEE Trans. Autom. Control **57**(6), 1362–1376 (2012)
16. Zheng, Y., Li, Y.X.: Event-triggered output feedback consensus tracking control for nonlinear multiagent systems with sensor failures. Automatica **148**, 110787 (2023)
17. Xu, C., Xu, H., Guan, Z., Ge, Y.: Observer-based dynamic event-triggered semiglobal bipartite consensus of linear multi-agent systems with input saturation. IEEE Trans. Cybern. **53**(5), 3139–3152 (2023)
18. Cai, Y., Zhang, H., Duan, J., Zhang, J.: Distributed Bipartite Consensus of Linear Multiagent Systems Based on Event-Triggered Output Feedback Control Scheme. IEEE Trans. Syst. Man Cybern.: Syst. **51**(11), 6743–6756 (2021)
19. Wang, L., Wang, M., Meng, W.: System transformation-based event-triggered fuzzy control for state constrained nonlinear systems with unknown control directions. IEEE Trans. Fuzzy Syst. **31**(7), 2331–2344 (2023)
20. Liu, L., Wang, Z., Huang, Z., Zhang, H.: Adaptive predefined performance control for MIMO systems with unknown direction via generalized fuzzy hyperbolic model. IEEE Trans. Fuzzy Syst. **25**(3), 527–542 (2017)

21. Liu, G., Yang, Y., Gao, W., Wunsch, D.C.: Adaptive Nussbaum design for nonholonomic systems with asymptotic stabilization against false data injection. IEEE Trans. Cybern. **55**(11), 5189–5201 (2025)
22. Liang, H., Guo, X., Pan, Y., Huang, T.: Event-triggered fuzzy bipartite tracking control for network systems based on distributed reduced-order observers. IEEE Trans. Fuzzy Syst. **29**(6), 1601–1614 (2021)
23. Yu, J., Ahn, C., Shi, P.: Event-triggered bipartite consensus for fuzzy multiagent systems under Markovian switching signed topology. IEEE Trans. Fuzzy Syst. **30**(7), 2610–2620 (2022)

Chapter 8
Adaptive Resilient Nussbaum Design for Nonholonomic Systems with False Data Injection

8.1 Introduction

Complex engineering applications require the control algorithm to be implemented on remotely located high-performance computers. With the advancement of control, computing, and communication technologies, security has become a critical concern in cyber-physical systems [1, 2], because the signal's transmission is vulnerable to cyber-attacks [3]. It is necessary to develop a secure control strategy with high resilience against cyber-physical threats in safety-critical scenarios.

Common malicious attacks include denial of service (DoS) [4] and false data injection (FDI) [5]. The DoS attack refers to the prevention of the target from providing services [6], which aims to disrupt the information transmission from sensor measurements and control inputs [7]. The FDI-type attack aims to tamper with the transmitted information and affect the data integrity [8]. FDI attacks are usually challenging to handle because only the compromised signals are available for feedback design [9]. Therefore, a secure strategy to effectively counteract the impact of FDI attacks is of vital importance. It is noteworthy that FDI-type attacks could further lead to time-varying unknown control gains. Existing security control designs depend on the adaptive estimation scheme to compensate for unknown attack signals [10]. With the time-varying unknown control gains, such adaptive security schemes might deteriorate the control performance. To tackle the challenge presented by the unknown control direction in controller design, the Nussbaum design has been widely considered [11, 12]. Furthermore, the attack in nonholonomic systems has the inevitable consequence of multiple time-varying control directions. To address this issue, this chapter introduces a novel Nussbaum-type function that provides flexibility to handle multiple unknown control directions.

As an important category of mechanical systems, nonholonomic systems have attracted a lot of attention, such as vehicles [13] and wheeled mobile robots [14], in the last decade. For controllable nonholonomic systems, even continuous pure-state feedback cannot ensure the stability of the closed-loop systems [15]. As for

Y. Yang et al., *Adaptive Nussbaum Design for Nonlinear Dynamical Systems*,
https://doi.org/10.1007/978-981-95-4997-9_8

Brockett conditions, invariant continuous feedback laws for nonholonomic systems do not exist [16]. To stabilize the nonholonomic systems, numerous efforts have been devoted, such as time-varying feedback [17, 18], discontinuous time-invariant feedback [19, 20], and hybrid feedback [21]. However, these methods can achieve rapid transient response at the sacrifice of control input's continuity [22]. By discontinuous state scaling, the nonholonomic system can be equivalently converted into the cascade system, which consists of a scalar subsystem and a lower triangular subsystem [23]. For systems in cascade form, a backstepping scheme can be effectively employed for feedback design [24]. However, the adaptive backstepping design for nonholonomic systems requires that the system operates in a secure condition without malicious attacks. In practice, the signal transmissions are susceptible to attacks, which could disrupt the normal operation of the system and render the well-developed strategies ineffective for nonholonomic systems. Therefore, there is an urgent need to develop novel strategies that can stabilize nonholonomic systems in the presence of security threats. The contributions of this chapter are as follows.

Compared to existing results on nonholonomic systems with ideal state measurement [25, 26], this chapter takes the presence of FDI attacks into account to investigate the impact of corrupted state information on the secure stabilization of nonholonomic systems. As opposed to estimating the unknown control gains caused by FDI attacks [27, 28], this chapter introduces a novel adaptive Nussbaum design incorporating additional online learning schemes to compensate for unknown control gains, effectively countering FDIs in multiple transmission channels.

The remainder of this chapter is organized as follows. Section 8.2 outlines the preliminaries on nonholonomic systems. Section 8.3 introduces the adaptive backstepping design methodology for nonholonomic systems with FDI attacks, presenting the main results of this chapter. Section 8.5 demonstrates the effectiveness of the proposed controller through simulation results using a wheel robot example. Finally, the conclusion is provided in Sect. 8.6.

8.2 Problem Formulation

Consider the nonholonomic systems evolving in continuous time as

$$\begin{aligned} \dot{x}_0 &= u_0 + f_0(x_0) \\ \dot{x}_i &= u_0 x_{i+1} + f_i(u_0, x_0, \bar{x}_i), \quad i = 1, \ldots, n-1 \\ \dot{x}_n &= u_1 + f_n(u_0, x_0, \bar{x}_n) \end{aligned} \tag{8.1}$$

where $\bar{x}_i = [x_1, x_2, \ldots, x_i]^{\mathrm{T}} \in \mathbb{R}^i$ represents the state, $u \in \mathbb{R}$ denotes the control input, $y = x_1 \in \mathbb{R}$ stands for the system output, and $f_i(\cdot) : \mathbb{R}^i \to \mathbb{R}$ denotes unknown smooth nonlinear functions.

For the FDI attack, we define

$$\check{x}_0 = [1 + \beta_0(t)]\, x_0, \quad \check{x}_i = [1 + \beta(t)]\, x_i \tag{8.2}$$

where $\check{x}_0$ and $\check{x}_i$ denote the compromised system states after the attack, and the time-varying coefficients $\beta_0(t)$ and $\beta(t)$ are both unknown.

The following lemmas and assumptions are necessary for the subsequent analysis.

Assumption 8.1 ([29]) The weights β_0 and β satisfy $1+\beta_0(t) \neq 0$ and $1+\beta(t) \neq 0$. It is assumed that $|\beta_0| \leq \bar{\beta}_0$, $|\dot{\beta}_0| \leq \bar{\beta}_{d0}$, $|\beta| \leq \bar{\beta}$, and $|\dot{\beta}| \leq \bar{\beta}_d$, where $\bar{\beta}_0$, $\bar{\beta}_{d0}$, $\bar{\beta}$, and $\bar{\beta}_d$ are unknown positive constants.

Assumption 8.2 ([30]) Define $\rho_0 = \frac{1}{1+\beta_0(t)}$ and $\rho_i = \frac{(1+\beta_0(t))^{n-i}}{1+\beta(t)}$. Since the signals ρ_0 and ρ_i are positive, there exist unknown positive constants $\underline{\rho}_0$, $\bar{\rho}_0$, $\bar{\rho}_{d0}$, $\underline{\rho}_i$, $\bar{\rho}_i$, and $\bar{\rho}_{di}$ such that

$$0 < \underline{\rho}_0 \leq |\rho_0| \leq \bar{\rho}_0,\ |\dot{\rho}_0| \leq \bar{\rho}_{d0},\ 0 < \underline{\rho}_i \leq |\rho_i| \leq \bar{\rho}_i,\ |\dot{\rho}_i| \leq \bar{\rho}_{di}.$$

Assumption 8.3 ([15]) For the function $f_0(\cdot)$, there exists a function $\phi_0(\cdot)$ such that

$$f_0(x_0) = x_0 \phi_0(x_0), \tag{8.3}$$

for all $x_0 \in \Omega_0$.

To overcome the issue of multiple unknown control directions caused by FDI attacks on nonholonomic systems, a novel Nussbaum-type function is considered in this chapter as

$$N(\xi) = \sin(\xi)\left[2\cosh\left(\xi^2\right) + \left(4\xi^2+1\right)\sinh\left(\xi^2\right)\right]. \tag{8.4}$$

Lemma 8.1 *If there exists a positive definite, radially unbounded function $V(t)$ for all $t \in [0, t_f)$ with the Nussbaum-type functions $N(\cdot)$ in (8.4). Then, $V(t)$ and ξ_j are bounded if*

$$V(t) \leq \sum_{j=1}^{M} \int_0^t \left(g_j(\tau)N(\xi_j(\tau)) + 1\right)\dot{\xi}_j(\tau)d\tau + c_0, \tag{8.5}$$

where $a > 0$, $c_0 > 0$, and $\{g_j(\cdot)\}_{j=1}^{M}$ are continuous functions satisfying $|g_j(\cdot)| \in [g_{j,\min}, g_{j,\max}]$ with constants $g_{j,\min} > 0$ and $g_{j,\max} > 0$.

Proof Denote

$$Y(t_n, t_m) = \sum_{i=1}^{N} \int_{t_n}^{t_m} \left(g_i(\tau) N(\xi_i(\tau)) + 1\right)\dot{\xi}_i(\tau)\,\mathrm{d}\tau \tag{8.6}$$

which means $Y(0,t) = V(t) - c_0$. Then, the boundedness of $\{\xi_1, \xi_2, \ldots, \xi_N\}$ is proved by contradiction. For the convenience of discussion, assume that $\xi_1, \ldots, \xi_\Delta$ are unbounded and $\xi_{\Delta+1}, \ldots, \xi_N$ are bounded. In the next, we investigate two cases.

First, consider the scenario where the upper bound of $\xi_i(t)$ does not exist on the interval $\left[0, t_f\right)$ for $1 \leq i \leq \Delta$. Recall the Nussbaum-type function property (8.4), there exist two increasing sequences $\left\{\omega_m^{i,1}\right\}_{m=1}^{\infty}$ and $\left\{\omega_m^{i,2}\right\}_{m=1}^{\infty}$ with

$$\begin{cases} \omega_m^{i,1} = 2m\pi, \omega_m^{i,2} = 2m\pi + \pi, \text{if } g_i(\cdot) > 0 \\ \omega_m^{i,1} = 2m\pi - \pi, \omega_m^{i,2} = 2m\pi, \text{if } g_i(\cdot) < 0 \end{cases} \tag{8.7}$$

Accordingly, define two increasing time sequences $\left\{t_m^{i,1}\right\}_{m=1}^{\infty}$ and $\left\{t_m^{i,2}\right\}_{m=1}^{\infty}$ as $\xi_i\left(t_m^{i,1}\right) = \omega_m^{i,1}$ and $\xi_i\left(t_m^{i,2}\right) = \omega_m^{i,2}$, respectively. In addition, one has $\lim_{m\to\infty} t_m^{i,1} = t_f$, $\lim_{m\to\infty} t_m^{i,2} = t_f$ $\lim_{m\to\infty} \omega_m^{i,1} = \infty$, and $\lim_{m\to\infty} \omega_m^{i,2} = \infty$.

From the definition of $t_m^{i,1}$, there exists a constant $k \in \{1, 2, \ldots, \Delta\}$ satisfying $\xi_k\left(t_{m+1}^{k,1}\right) = \omega_{m+1}^{k,1}$ and $\xi_i\left(t_{m+1}^{k,1}\right) \leq \omega_{m+1}^{k,1} = \omega_{m+1}^{i,1}$ for $i \in \{1, 2, \ldots, \Delta\}$, $i \neq k$ and $t \leq t_{m+1}^{k,1}$. Additionally, observe that $\xi_{\Delta+1}, \ldots, \xi_N$ are bounded. This implies the boundedness of the sum $\sum_{i=\Delta+1}^{N} \int_0^{t_{m+1}^{k,1}} g_i(\tau) N(\xi_i(\tau)) \dot{\xi}_i(\tau) \mathrm{d}\tau$. According to (8.6), one has

$$\begin{aligned} Y\left(0, t_{m+1}^{k,1}\right) \leq & \sum_{i=1}^{\Delta} \int_0^{t_{m+1}^{k,1}} g_i(\tau) N(\xi_i(\tau)) \dot{\xi}_i(\tau) \mathrm{d}\tau \\ & + \sum_{i=1}^{N} \int_0^{t_{m+1}^{k,1}} \dot{\xi}_i(\tau) \mathrm{d}\tau + c_1 \end{aligned} \tag{8.8}$$

with $c_1 = \sum_{i=\Delta+1}^{N} \int_0^{t_{m+1}^{k,1}} g_i(\tau) N(\xi_i(\tau)) \dot{\xi}_i(\tau) \mathrm{d}\tau$ being bounded. Based on time intervals, the term $\sum_{i=1}^{\Delta} \int_0^{t_{m+1}^{k,1}} g_i(\tau) N(\xi_i(\tau)) \dot{\xi}_i(\tau) \mathrm{d}\tau$ can be written as

$$\begin{aligned} & \sum_{i=1}^{\Delta} \int_0^{t_{m+1}^{k,1}} g_i(\tau) N(\xi_i(\tau)) \dot{\xi}_i(\tau) \mathrm{d}\tau \\ = & \sum_{i=1}^{\Delta} \int_0^{t_m^{i,1}} g_i(\tau) N(\xi_i(\tau)) \dot{\xi}_i(\tau) \mathrm{d}\tau \\ & + \sum_{i=1}^{\Delta} \int_{t_m^{i,1}}^{t_m^{i,2}} g_i(\tau) N(\xi_i(\tau)) \dot{\xi}_i(\tau) \mathrm{d}\tau \\ & + \sum_{i=1, i\neq k}^{\Delta} \int_{t_m^{i,2}}^{t_{m+1}^{k,1}} g_i(\tau) N(\xi_i(\tau)) \dot{\xi}_i(\tau) \mathrm{d}\tau \\ & + \int_{t_m^{k,2}}^{t_{m+1}^{k,1}} g_k(\tau) N(\xi_k(\tau)) \dot{\xi}_k(\tau) \mathrm{d}\tau. \end{aligned} \tag{8.9}$$

By Defining $G_{i,\max} = \max\limits_i \left\{g_{i,\max}\right\}$ and $G_{i,\min} = \min\limits_i \left\{g_{i,\min}\right\}$, one has

$$\sum_{i=1}^{\Delta} \int_{t_m^{i,1}}^{t_m^{i,2}} g_i(\tau) N(\xi_i(\tau)) \dot{\xi}_i(\tau) \mathrm{d}\tau \le \sum_{i=1}^{\Delta} G_{i,\max} \int_{\omega_m^{i,1}}^{\omega_m^{i,2}} \operatorname{sign}(g_i) N(\xi_i) \mathrm{d}\xi_i \quad (8.10)$$

for $t \in \left[t_m^{k,1}, t_m^{k,2}\right]$,

$$\sum_{i=1, i\neq k}^{\Delta} \int_{t_m^{i,2}}^{t_{m+1}^{k,1}} g_i(\tau) N(\xi_i(\tau)) \dot{\xi}_i(\tau) \mathrm{d}\tau \le 0 \quad (8.11)$$

for the interval $t \in \left[t_m^{k,2}, t_{m+1}^{k,1}\right]$, and

$$\int_{t_m^{k,2}}^{t_{m+1}^{k,1}} g_k(\tau) N(\xi_k(\tau)) \dot{\xi}_k(\tau) \mathrm{d}\tau \le G_{k,\min} \int_{\omega_m^{k,1}}^{\omega_{m+1}^{k,1}} \operatorname{sign}(g_k) N(\xi_k) \mathrm{d}\xi_k \quad (8.12)$$

for $t \in \left[t_m^{k,2}, t_{m+1}^{k,1}\right]$. By the definition of $N(\cdot)$, $\left\{\omega_m^{i,1}\right\}_{m=1}^{\infty}$ and $\left\{\omega_m^{i,2}\right\}_{m=1}^{\infty}$, one has

$$\begin{aligned}
\int_{\omega_m^{i,1}}^{\omega_m^{i,2}} \operatorname{sign}(g_i) N(\xi_i) \mathrm{d}\xi_i &= \sinh\left(\left(\omega_m^{i,2}\right)^2\right) + \sinh\left(\left(\omega_m^{i,1}\right)^2\right) \\
\int_{\omega_m^{k,2}}^{\omega_{m+1}^{k,1}} \operatorname{sign}(g_k) N(\xi_i) \mathrm{d}\xi_i &= -\sinh\left(\left(\omega_{m+1}^{k,1}\right)^2\right) - \sinh\left(\left(\omega_m^{k,2}\right)^2\right).
\end{aligned}$$

Considering the above facts and $\sum\limits_{i=1}^{N} \int_0^{t_{m+1}^{k,1}} \dot{\xi}_i(\tau) \mathrm{d}\tau \le \sum\limits_{i=1}^{N} \left[\omega_{m+1}^{k,1} - \xi_i(0)\right]$, one can obtain

$$\begin{aligned}
Y\left(0, t_{m+1}^{k,1}\right) \le c_2 &+ \sum_{i=1}^{\Delta} \int_0^{\omega_m^{i,1}} g_i(\tau) N(\xi_i(\tau)) \mathrm{d}\xi_i + N\omega_{m+1}^{k,1} \\
&+ \Delta G_{\max}\left[\sinh\left(\left(\omega_m^{k,2}\right)^2\right) + \sinh\left(\left(\omega_m^{k,1}\right)^2\right)\right] \\
&- G_{k,\min}\left[\sinh\left(\left(\omega_{m+1}^{k,1}\right)^2\right) + \sinh\left(\left(\omega_m^{k,2}\right)^2\right)\right]
\end{aligned}$$

with $c_2 = c_1 - \sum\limits_{i=1}^{N} \xi_i(0)$.

Define $t_{i,j}$ as $\xi_i\left(t_{i,j}\right) = j\pi$, $j \in \{0, 1, \ldots, 2m\}$. Therefore, the second term in $Y\left(0, t_{m+1}^{k,1}\right)$ can be described as

$$
\begin{aligned}
&\sum_{i=1}^{\Delta}\int_{0}^{\omega_m^{i,1}} g_i(\tau)\,N(\xi_i(\tau))\,\dot{\xi}_i(\tau)\,\mathrm{d}\tau \\
\le& \sum_{i=1}^{\Delta}\sum_{j=1}^{m-1}\left[G_{i,\max}S(2j\pi,2j\pi-\pi)-G_{i,\min}S(2j\pi,2j\pi-\pi)\right] \\
&+\sum_{i=1}^{\Delta}s_iG_{i,\max}S(2m\pi-\pi,2(m-1)\pi) \\
&-\sum_{i=1}^{\Delta}G_{i,\min}S(2m\pi,2m\pi-\pi)
\end{aligned}
$$

with $S(\iota_1,\iota_2)=\sinh\left(\iota_1^2\right)+\sinh\left(\iota_2^2\right)$ and $s_i=\frac{1+sign(g_i)}{2}$. Based on the Nussbaum-type function (8.4) with Definition 1.1, one has

$$
\lim_{m\to+\infty}\sum_{i=1}^{\Delta}\int_{0}^{\omega_m^{i,1}} g_i(\tau)\,N(\xi_i(\tau))\,\mathrm{d}\xi_i=-\infty. \tag{8.13}
$$

Furthermore, one can know

$$
\begin{aligned}
&\Delta G_{\max}\left[\sinh\left(\left(\omega_m^{k,2}\right)^2\right)+\sinh\left(\left(\omega_m^{k,1}\right)^2\right)\right] \\
&-G_{k,\min}\left[\sinh\left(\left(\omega_{m+1}^{k,1}\right)^2\right)+\sinh\left(\left(\omega_m^{k,2}\right)^2\right)\right] \\
=& -G_{k,\min}\sinh\left(\left(\omega_{m+1}^{k,1}\right)^2\right)\Theta_{m,k}
\end{aligned} \tag{8.14}
$$

where

$$
\begin{aligned}
\Theta_{m,k}=& -\frac{\Delta G_{\max}}{G_{k,\min}}\frac{\sinh\left(\left(\omega_m^{k,1}\right)^2\right)}{\sinh\left(\left(\omega_{m+1}^{k,1}\right)^2\right)}+\frac{N\omega_{m+1}^{k,1}}{G_{k,\min}\sinh\left(\left(\omega_{m+1}^{k,1}\right)^2\right)} \\
&+\left(1-\frac{\Delta G_{\max}}{\bar{\vartheta}_4}\right)\frac{\sinh\left(\left(\omega_m^{k,2}\right)^2\right)}{\sinh\left(\left(\omega_{m+1}^{k,1}\right)^2\right)}+1.
\end{aligned} \tag{8.15}
$$

It can be easily observed that $\lim_{m\to+\infty}\bar{\vartheta}_1=-\infty$.

Therefore, from (8.13), $Y\left(0,t_{m+1}^{k,1}\right)\to-\infty$ as $m\to+\infty$, i.e., $\lim_{t\to+\infty}V(x(t),t)=-\infty$, which is a contradiction to the fact in Lemma 8.1. Therefore, an upper bound for $\xi_i(t)$ exists on the interval $\left[0,t_f\right)$ for all $1\le i\le\Delta$.

Furthermore, assume that a lower bound for $\xi_i(t)$ does not exist on $[0, t_f)$ for $1 \leq \Delta \leq N$, i.e., $-\xi_i(t)$ does not have an upper bound. Recall (8.6), one has

$$Y(t_n, t_m) = \sum_{i=1}^{N} \int_{t_n}^{t_m} (g_i(\tau) N(\xi_i(\tau)) - 1) \dot{\xi}_i(\tau) \mathrm{d}\tau$$

because $N(\cdot)$ is an odd function. Similar to the previous discussions, one has

$$\begin{aligned} Y\left(0, t_{m+1}^{k,1}\right) \leq c_2 &+ \sum_{i=1}^{\Delta} \int_0^{\omega_m^{i,1}} g_i(\tau) N(\xi_i(\tau)) \mathrm{d}\xi_i - N\omega_{m+1}^{k,1} \\ &+ \Delta G_{\max}\left[\sinh\left(\left(\omega_m^{k,2}\right)^2\right) + \sinh\left(\left(\omega_m^{k,1}\right)^2\right)\right] \\ &- G_{k,\min}\left[\sinh\left(\left(\omega_{m+1}^{k,1}\right)^2\right) + \sinh\left(\left(\omega_m^{k,2}\right)^2\right)\right] \end{aligned}$$

with $c_2 = c_1 + \sum_{i=1}^{N} \xi_i(0)$. It can be easily observed that $Y\left(0, t_{m+1}^{k,1}\right) \to -\infty$ as $m \to +\infty$, i.e., $\lim_{t \to +\infty} V(x(t), t) = -\infty$, which is a contradiction to the fact in Lemma 8.1. Therefore, one can obtain that, on $[0, t_f)$, the lower bound of $\xi_i(t)$ exists. This completes the proof. □

8.3 Adaptive Nussbaum Design

8.3.1 Design of u_0

In order to design the control signal of u_0, we use the following FLS to approximate the smooth function $\phi_0(\cdot)$, which can be written as

$$\phi_0(x_0) = W_0^{\mathrm{T}} \varphi_0(x_0) + \varepsilon_0 \tag{8.16}$$

with $\bar{\varepsilon}_0 > 0$ satisfying

$$\|\varepsilon_0(x_0)\| \leq \bar{\varepsilon}_0 \tag{8.17}$$

and W_1 is the unknown ideal weight.

In the following, denote $\hat{\theta}_0(t)$, $\hat{\omega}_0(t)$, and $\hat{\chi}_0(t)$ as the adaptive learning of the unknown parameters $\theta_0 = \|W_0\|^2$, $\omega_0 = \frac{1}{2}\bar{\varepsilon}_0^2 + \bar{\rho}_0^4$, and $\chi_0 = \bar{\rho}_0$, respectively. Then, the adaptive learning errors can be defined as $\tilde{\theta}_0 = \theta_0 - \hat{\theta}_0$, $\tilde{\omega}_0 = \omega_0 - \hat{\omega}_0$, and $\tilde{\chi}_0 = \chi_0 - \hat{\chi}_0$.

Design u_0 as

$$u_0 = \check{x}_0 g_0\left(\check{X}_0\right) \tag{8.18}$$

where $\check{X}_0 = \left[\check{x}_0, \hat{\theta}_0, \hat{\omega}_0, \hat{\chi}_0\right]^{\mathrm{T}}$ and

$$\begin{aligned} g_0\left(\check{X}_0\right) &= N(\xi_0) r_0 \left(1 + \tanh\left(\frac{N(\xi_0) r_0}{\varsigma}\right)\right) + (\kappa+1)\varsigma \\ r_0 &= k_0 + \frac{1}{2}\hat{\theta}_0\left(\varphi_0^{\mathrm{T}}(\check{x}_0)\varphi_0(\check{x}_0) + 1\right) + \hat{\omega} + \hat{\chi}_0(\kappa+1)\varsigma \check{x}_0^2 \end{aligned} \tag{8.19}$$

with the positive constant ς.

It can be seen that $g_0\left(\check{X}_0\right) \neq 0$ is guaranteed regardless of the values of $\check{X}_0$. Hence, $1/g_0\left(\check{X}_0\right)$ is well defined.

Furthermore, the Nussbaum-type gain can be designed as

$$\dot{\xi}_0 = \check{x}_0^2 r_0. \tag{8.20}$$

To estimate the unknown parameters, the dynamics of the adaptive learning laws are designed as

$$\begin{aligned} \dot{\hat{\theta}}_0 &= \frac{1}{2}\check{x}_0^2\varphi_0^{\mathrm{T}}(\check{x}_0)\varphi_0(\check{x}_0) + \frac{1}{2}\check{x}_0^2 \\ \dot{\hat{\omega}}_0 &= \check{x}_0^2 \\ \dot{\hat{\chi}}_0 &= (\kappa+1)\varsigma\check{x}_0^2 \end{aligned} \tag{8.21}$$

8.3.2 Design of u_1

Denote the discontinuous state scaling transformation [15]

$$z_i = \frac{x_i}{x_0^{n-i}}, \quad i = 1, 2, \ldots, n \tag{8.22}$$

which processes the property of increasing the resolution around a given point. Then, one can obtain the dynamics of z_i as

$$\begin{aligned} \dot{z}_i &= (1+\beta_0(t))\, g_0\left(\check{X}_0\right) z_{i+1} + \frac{f_i(u_0, x_0, \bar{x}_i)}{x_0^{n-i}} \\ &\quad - (n-i)(1+\beta_0(t))\, g_0\left(\check{X}_0\right) z_i - \frac{(n-i) z_i f_0(x_0)}{x_0}, 1 \le i \le n-1 \\ \dot{z}_n &= u_1 + f_n(u_0, x_0, \bar{x}_n) \end{aligned} \tag{8.23}$$

With the FDI attacks, define

$$\check{z}_i = \frac{\check{x}_i}{\check{x}_0^{n-i}} = \frac{1+\beta(t)}{(1+\beta_0(t))^{n-i}} \frac{x_i}{x_0^{n-i}} = \frac{1+\beta(t)}{(1+\beta_0(t))^{n-i}} z_i. \tag{8.24}$$

The design process for u_1 involves n steps, including the development of both the control law and adaptive laws, which depends on coordinate transformation as

$$\begin{aligned} &\check{e}_1 = \check{z}_1, e_1 = z_1 = \rho_1 \check{e}_1 \\ &\check{e}_i = \check{z}_i - \alpha_{i-1}, e_i = z_i - \rho_i \alpha_{i-1} = \rho_i \check{e}_i \end{aligned} \tag{8.25}$$

where α_i is the virtual control signal.

From (8.23) and (8.25), one has

$$\begin{aligned} \dot{e}_1 &= (1+\beta_0(t))\, g_0\left(\check{X}_0\right) \rho_2 \alpha_1 + (1+\beta_0(t))\, g_0\left(\check{X}_0\right) e_2 \\ &\quad + \frac{f_1(u_0, x_0, x_1)}{x_0^{n-1}} - (n-1)(1+\beta_0(t))\, g_0\left(\check{X}_0\right) z_1 - \frac{(n-1)\, z_1 f_0(x_0)}{x_0} \\ \dot{e}_i &= (1+\beta_0(t))\, g_0\left(\check{X}_0\right) \rho_{i+1} \alpha_i + (1+\beta_0(t))\, g_0\left(\check{X}_0\right) e_{i+1} \\ &\quad + \frac{f_i(u_0, x_0, \bar{x}_i)}{x_0^{n-1}} - (n-i)(1+\beta_0(t))\, g_0\left(\check{X}_0\right) z_i \\ &\quad - \frac{(n-i) z_i f_0(x_0)}{x_0} - \dot{\rho}_i \alpha_{i-1} - \rho_i \dot{\alpha}_{i-1} \\ \dot{e}_n &= u_1 + f_n(u_0, x_0, \bar{x}_n) - \dot{\rho}_n \alpha_{n-1} - \rho_n \dot{\alpha}_{n-1}. \end{aligned} \tag{8.26}$$

Step 1

Define $\theta_1 = \|W_1\|^2$ and $\omega_1 = \bar{\rho}_1^2 + \frac{1}{2}\bar{\varepsilon}_1^2$, which are unknown terms. In the following, the adaptive learning laws $\hat{\theta}_1$ and $\hat{\omega}_1$ are designed with the learning errors $\tilde{\theta}_1 = \theta_1 - \hat{\theta}_1$ and $\tilde{\omega}_1 = \omega_1 - \hat{\omega}_1$.

Define $X_1 = [u_0, x_0, x_1]^{\mathrm{T}}$ and $\check{X}_1 = \left[u_0, \check{x}_0, \check{x}_1\right]^{\mathrm{T}}$. Then, the adaptive Nussbaum virtual control is designed as

$$\begin{aligned} \alpha_1 &= \frac{1}{g_0(X_0)} N(\xi_1)\, r_1 \\ \dot{\xi}_1 &= \check{e}_1 r_1 \end{aligned} \tag{8.27}$$

where

$$r_1 = k_1\check{e}_1 + \frac{1}{2}\hat{\theta}_1\varphi_1^{\mathrm{T}}\left(\check{X}_1\right)\varphi_1\left(\check{X}_1\right)\tanh\left(\frac{\check{e}_1\varphi_1^{\mathrm{T}}\left(\check{X}_1\right)\varphi_1\left(\check{X}_1\right)}{\sigma(t)}\right)$$
$$+\frac{1}{2}\hat{\theta}_1\tanh\left(\frac{\check{e}_1}{\sigma(t)}\right) + \hat{\omega}_1\tanh\left(\frac{\check{e}_1}{\sigma(t)}\right) \tag{8.28}$$

with $\sigma(t) = \mathrm{e}^{-rt}$ satisfying $\sigma(t) > 0$ and

$$\lim_{t\to\infty}\int_{t_0}^{t}\sigma(\iota)\,d\iota \le \bar{\sigma} < \infty, \quad \bar{\sigma} > 0. \tag{8.29}$$

To compensate for the terms θ_1 and ω_1, consider

$$\dot{\hat{\theta}}_1 = \frac{1}{2}\check{e}_1\tanh\left(\frac{\check{e}_1}{\sigma(t)}\right) + \frac{1}{2}\check{e}_1\varphi_1^{\mathrm{T}}\left(\check{X}_1\right)\varphi_1\left(\check{X}_1\right)\tanh\left(\frac{\check{e}_1\varphi_1^{\mathrm{T}}\left(\check{X}_1\right)\varphi_1\left(\check{X}_1\right)}{\sigma(t)}\right)$$
$$\dot{\hat{\omega}}_1 = \check{e}_1\tanh\left(\frac{\check{e}_1}{\sigma(t)}\right) \tag{8.30}$$

Step i ($2 \le i \le n-1$)

For Step i, one can also define $\theta_i = \|W_i\|^2$, $\omega_i = \bar{\rho}_i^2 + \frac{1}{2}\bar{\varepsilon}_i^2$, $\chi_i = \bar{\rho}_i$, and $\vartheta_i = 2(n-i)\left(1+\bar{\beta}_0\right)\bar{g}_0\bar{\rho}_i^2 + \bar{\rho}_i\bar{\rho}_{di}$, which are unknown terms. Then, to estimate the unknown terms, the adaptive learning laws $\hat{\theta}_i$, $\hat{\omega}_i$, $\hat{\chi}_i$, and $\hat{\vartheta}_i$ are designed with the learning errors $\tilde{\theta}_i = \theta_i - \hat{\theta}_i$, $\tilde{\omega}_i = \omega_i - \hat{\omega}_i$, $\tilde{\vartheta}_i = \vartheta_i - \hat{\vartheta}_i$, and $\tilde{\chi}_i = \chi_i - \hat{\chi}_i$.

Define $X_i = [u_0, x_0, \bar{x}_i]^{\mathrm{T}}$, $\check{X}_i = \left[u_0, x_0, \check{\bar{x}}_i\right]^{\mathrm{T}}$. Then, the adaptive Nussbaum virtual control α_i is designed as

$$\alpha_i = \frac{1}{g_0(X_0)}N(\xi_i)\,r_i, \quad \dot{\xi}_i = \check{e}_i r_i \tag{8.31}$$

with

$$r_i = k_i\check{e}_i + \hat{\vartheta}_i\check{e}_i\alpha_{i-1}\tanh\left(\frac{\check{e}_i\alpha_{i-1}}{\sigma(t)}\right) + \frac{1}{2}\hat{\theta}_i\check{e}_i\tanh\left(\frac{\check{e}_i}{\sigma(t)}\right)$$
$$+\hat{\chi}_i\check{e}_i\dot{\alpha}_{i-1}\tanh\left(\frac{\check{e}_i\dot{\alpha}_{i-1}}{\sigma(t)}\right) + \hat{\omega}_i\check{e}_i\tanh\left(\frac{\check{e}_i}{\sigma(t)}\right)$$
$$+\frac{1}{2}\hat{\theta}_i\check{e}_i\varphi_i^{\mathrm{T}}\left(\check{X}_i\right)\varphi_i\left(\check{X}_i\right)\tanh\left(\frac{\check{e}_i\varphi_i^{\mathrm{T}}\left(\check{X}_i\right)\varphi_i\left(\check{X}_i\right)}{\sigma(t)}\right). \tag{8.32}$$

Design the adaptive law of the parameter as

$$\begin{aligned}
\dot{\hat{\theta}}_i &= \frac{1}{2}\check{e}_i\varphi_i^{\mathrm{T}}\left(\check{X}_i\right)\varphi_i\left(\check{X}_i\right)\tanh\left(\frac{\check{e}_i\varphi_i^{\mathrm{T}}\left(\check{X}_i\right)\varphi_i\left(\check{X}_i\right)}{\sigma(t)}\right) \\
&\quad+\frac{1}{2}\check{e}_i\tanh\left(\frac{\check{e}_i}{\sigma(t)}\right) \\
\dot{\hat{\vartheta}}_i &= \check{e}_i\alpha_{i-1}\tanh\left(\frac{\check{e}_i\alpha_{i-1}}{\sigma(t)}\right) \\
\dot{\hat{\omega}}_i &= \check{e}_i\tanh\left(\frac{\check{e}_i}{\sigma(t)}\right), \\
\dot{\hat{\chi}}_i &= \check{e}_i\dot{\alpha}_{i-1}\tanh\left(\frac{\check{e}_i\dot{\alpha}_{i-1}}{\sigma(t)}\right).
\end{aligned} \tag{8.33}$$

Step n

In Step n, we introduce the adaptive estimates $\hat{\theta}_n$, $\hat{\omega}_n$, $\hat{\chi}_n$, and $\hat{\vartheta}_n$ to approximate the unknown parameters $\theta_n = \|W_n\|^2$, $\omega_n = \bar{\rho}_n^2 + \frac{1}{2}\bar{\varepsilon}_n^2$, and $\chi_n = \bar{\rho}_n$, $\vartheta_n = \bar{\rho}_n\bar{\rho}_{dn}$ with the learning errors $\tilde{\theta}_n = \theta_n - \hat{\theta}_n$, $\tilde{\omega}_n = \omega_n - \hat{\omega}_n$, $\tilde{\chi}_n = \chi_n - \hat{\chi}_n$, and $\tilde{\vartheta}_n = \vartheta_n - \hat{\vartheta}_n$, respectively.

Based on the definition $X_n = [u_0, x_0, \bar{x}_n]^{\mathrm{T}}$, and $\check{X}_n = \left[u_0, x_0, \check{\bar{x}}_n\right]^{\mathrm{T}}$, the control u_1 and the Nussbaum-type gain ξ_n for step n are designed as

$$u_1 = N(\xi_n)r_n, \quad \dot{\xi}_n = \check{e}_n r_n \tag{8.34}$$

with

$$\begin{aligned}
r_n &= k_n\check{e}_n + \hat{\vartheta}_n\check{e}_n\alpha_{n-1}\tanh\left(\frac{\check{e}_n\alpha_{n-1}}{\sigma(t)}\right) + \hat{\chi}_n\check{e}_n\dot{\alpha}_{n-1}\tanh\left(\frac{\check{e}_n\dot{\alpha}_{n-1}}{\sigma(t)}\right) \\
&\quad+\hat{\omega}_n\check{e}_n\tanh\left(\frac{\check{e}_n}{\sigma(t)}\right) + \frac{1}{2}\hat{\theta}_n\check{e}_n\tanh\left(\frac{\check{e}_n}{\sigma(t)}\right) \\
&\quad+\frac{1}{2}\hat{\theta}_n\check{e}_n\varphi_n^{\mathrm{T}}\left(\check{X}_n\right)\varphi_n\left(\check{X}_n\right)\tanh\left(\frac{\check{e}_n\varphi_n^{\mathrm{T}}\left(\check{X}_n\right)\varphi_n\left(\check{X}_n\right)}{\sigma(t)}\right).
\end{aligned} \tag{8.35}$$

Design the adaptive law of the parameter as

$$\begin{aligned}
\dot{\hat{\theta}}_n &= \frac{1}{2}\check{e}_n\varphi_n^{\mathrm{T}}\left(\check{X}_n\right)\varphi_n\left(\check{X}_n\right)\tanh\left(\frac{\check{e}_n\varphi_n^{\mathrm{T}}\left(\check{X}_n\right)\varphi_n\left(\check{X}_n\right)}{\sigma(t)}\right) \\
&\quad +\frac{1}{2}\check{e}_n\tanh\left(\frac{\check{e}_n}{\sigma(t)}\right) \\
\dot{\hat{\omega}}_n &= \frac{1}{2}\check{e}_n\tanh\left(\frac{\check{e}_n}{\sigma(t)}\right), \\
\dot{\hat{\vartheta}}_n &= \check{e}_n\alpha_{n-1}\tanh\left(\frac{\check{e}_n\alpha_{n-1}}{\sigma(t)}\right) \\
\dot{\hat{\chi}}_n &= \check{e}_n\dot{\alpha}_{n-1}\tanh\left(\frac{\check{e}_n\dot{\alpha}_{n-1}}{\sigma(t)}\right).
\end{aligned} \tag{8.36}$$

8.4 Stability Analysis

8.4.1 Stability Analysis for x_0

With the adaptive design in (8.18)–(8.21), the stability analysis of state x_0 can be discussed as follows.

Theorem 8.1 *Consider the uncertain nonholonomic system* (8.1) *against FDI attacks under Assumptions 8.1–8.3 with the control design u_0. Then, for bounded initial conditions, one can obtain the signals for x_0 are bounded and x_0 converges to zero when $t \to \infty$.*

Proof Under Assumption 8.3 and FLS approximation (8.16), the x_0-subsystem can be rewritten as

$$\begin{aligned}
\dot{x}_0 &= u_0 + f_0(x_0) = u_0 + x_0\phi_0(x_0) \\
&= u_0 + x_0 W_0^{\mathrm{T}}\varphi_0(x_0) + x_0\varepsilon_0.
\end{aligned} \tag{8.37}$$

Consider the Lyapunov candidate as

$$V_0 = \frac{1}{2}x_0^2 + \frac{1}{2}\tilde{\theta}_0^2 + \frac{1}{2}\tilde{\omega}_0^2 + \frac{1}{2}\tilde{\chi}_0^2. \tag{8.38}$$

Differentiating V_0, one has

$$\dot{V}_0 = x_0\left[u_0 + x_0 W_0^{\mathrm{T}}\varphi_0(x_0) + x_0\varepsilon_0\right] - \tilde{\theta}_0\dot{\hat{\theta}}_0 - \tilde{\omega}_0\dot{\hat{\omega}}_0 - \tilde{\chi}_0\dot{\hat{\chi}}_0. \tag{8.39}$$

Based on Young's inequality, one has

$$\begin{aligned} x_0^2 W_0^{\mathrm{T}} \varphi_0 (x_0) &\leq \frac{1}{2}\bar{\rho}^4 \check{x}_0^2 + \frac{1}{2}\check{x}_0^2 \theta_0 \varphi_0^{\mathrm{T}} (x_0) \varphi_0 (x_0) \\ x_0^2 \varepsilon_0 &\leq \frac{1}{2}\bar{\rho}^4 \check{x}_0^2 + \frac{1}{2}\check{x}_0^2 \bar{\varepsilon}_0^2. \end{aligned} \tag{8.40}$$

Then, one has

$$\begin{aligned} \dot{V}_0 \leq{}& \rho_0 \check{x}_0 u_0 + \frac{1}{2}\bar{\rho}_0^4 \check{x}_0^2 + \frac{1}{2}\check{x}_0^2 \theta_0 \varphi_0^{\mathrm{T}} (x_0) \varphi_0 (x_0) \\ &+ \frac{1}{2}\bar{\rho}_0^4 \check{x}_0^2 + \frac{1}{2}\check{x}_0^2 \bar{\varepsilon}_0^2 - \tilde{\theta}_0 \dot{\hat{\theta}}_0 - \tilde{\omega}_0 \dot{\hat{\omega}}_0 - \tilde{\chi}_0 \dot{\hat{\chi}}_0 \\ &+ \frac{1}{2}\check{x}_0^2 \theta_0 \varphi_0^{\mathrm{T}} (x_0) \varphi_0 (x_0) - \frac{1}{2}\check{x}_0^2 \theta_0 \varphi_0^{\mathrm{T}} (\check{x}_0) \varphi_0 (\check{x}_0). \end{aligned} \tag{8.41}$$

From the FLS, one has the fact $0 < \varphi_0^{\mathrm{T}} (\cdot) \varphi_0 (\cdot) \leq 1$, which implies

$$\frac{1}{2}\check{x}_0^2 \theta_0 \varphi_0^{\mathrm{T}} (x_0) \varphi_0 (x_0) - \frac{1}{2}\check{x}_0^2 \theta_0 \varphi_0^{\mathrm{T}} (\check{x}_0) \varphi_0 (\check{x}_0) \leq \frac{1}{2}\check{x}_0^2 \theta_0. \tag{8.42}$$

Inserting (8.42) into (8.41) yields

$$\begin{aligned} \dot{V}_0 \leq{}& \rho_0 \check{x}_0 u_0 + \frac{1}{2}\check{x}_0^2 \hat{\theta}_0 \varphi_0^{\mathrm{T}} (\check{x}_0) \varphi_0 (\check{x}_0) + \frac{1}{2}\check{x}_0^2 \hat{\theta}_0 + \check{x}_0^2 \hat{\omega}_0 \\ &+ \tilde{\theta}_0 \left(\frac{1}{2}\check{x}_0^2 \varphi_0^{\mathrm{T}} (\check{x}_0) \varphi_0 (\check{x}_0) + \frac{1}{2}\check{x}_0^2 - \dot{\hat{\theta}}_0 \right) \\ &+ \tilde{\omega}_0 \left(\check{x}_0^2 - \dot{\hat{\omega}}_0 \right) - \tilde{\chi}_0 \dot{\hat{\chi}}_0. \end{aligned} \tag{8.43}$$

Taking the adaptive design (8.21) into (8.43), one can obtain

$$\dot{V}_0 \leq \rho_0 \check{x}_0 u_0 + \frac{1}{2}\check{x}_0^2 \hat{\theta}_0 \varphi_0^{\mathrm{T}} (\check{x}_0) \varphi_0 (\check{x}_0) + \frac{1}{2}\check{x}_0^2 \hat{\theta}_0 + \check{x}_0^2 \hat{\omega}_0 - \tilde{\chi}_0 \dot{\hat{\chi}}_0. \tag{8.44}$$

Inserting the control design into (8.44) one has

$$\begin{aligned} \dot{V}_0 \leq{}& h_0 N (\xi_0) \dot{\xi}_0 \left(1 + \tanh \left(\frac{N (\xi_0) r_0}{\varsigma} \right) \right) + \dot{\xi}_0 \\ &+ \rho_0 (\kappa + 1) \varsigma \check{x}_0^2 - k_0 \check{x}_0^2 - \hat{\chi}_0 (\kappa + 1) \varsigma \check{x}_0^2 - \tilde{\chi}_0 \dot{\hat{\chi}}_0 \end{aligned} \tag{8.45}$$

with $h_0 = \rho_0 \left[1 + \tanh \left(\frac{N(\xi_0) r_0}{\varsigma} \right) \right]$. Therefore, $\dot{V}_0$ further satisfies

$$\dot{V}_0 \leq [h_0 N (\xi_0) + 1] \dot{\xi}_0 - k_0 \check{x}_0^2 + \tilde{\chi}_0 \left((\kappa + 1) \varsigma \check{x}_0^2 - \dot{\hat{\chi}}_0 \right). \tag{8.46}$$

Taking the adaptive learning design for $\hat{\chi}_0$, one can obtain

$$\begin{aligned}\dot{V}_0 &\leq [h_0 N(\xi_0) + 1]\dot{\xi}_0 - k_0 \check{x}_0^2 \\ &\leq [h_0 N(\xi_0) + 1]\dot{\xi}_0.\end{aligned} \tag{8.47}$$

Therefore, one can get the conclusion that the signals for x_0 are all bounded. Furthermore, $\check{x}_0^2$ is uniformly continuous, and $\dot{\check{x}}_0$ is bounded within the interval $[0, +\infty)$. Then, based on Lemma 1.10, one can easily get $\lim_{t\to\infty} \check{x}_0 = 0$. Based on (8.2) and Assumption 8.1, one has $\lim_{t\to\infty} x_0 = 0$. This completes the proof. □

8.4.2 Stability Analysis for x_i

Step 1

Consider the Lyapunov candidate

$$V_1 = \frac{1}{2}e_1^2 + \frac{1}{2}\tilde{\theta}_1^2 + \frac{1}{2}\tilde{\omega}_1^2. \tag{8.48}$$

Differentiating V_1 yields

$$\begin{aligned}\dot{V}_1 = &(1 + \beta_0(t)) g_0\left(\check{X}_0\right) e_1 e_2 + g_0\left(\check{X}_0\right)\rho_1 e_1 \alpha_1 \\ &- (n-1)(1 + \beta_0(t)) e_1 g_0\left(\check{X}_0\right) z_1 \\ &+ e_1 \frac{f_1(u_0, x_0, x_1)}{x_0^{n-1}} - e_1 \frac{(n-1) z_1 f_0(x_0)}{x_0} \\ &- \tilde{\theta}_1 \dot{\hat{\theta}}_1 - \tilde{\omega}_1 \dot{\hat{\omega}}_1.\end{aligned} \tag{8.49}$$

According to Young's inequality, one has

$$(1 + \beta_0(t)) g_0\left(\check{X}_0\right) e_1 e_2 \leq \frac{1}{2}\bar{\rho}_1^2 \check{e}_1^2 + \frac{1}{2}\left(1 + \bar{\beta}_0\right)^2 \bar{g}_0^2 \bar{\rho}_2^2 \check{e}_2^2. \tag{8.50}$$

For the term $-(n-1)(1 + \beta_0(t)) e_1 g_0\left(\check{X}_0\right) z_1$ in (8.49), one has

$$\begin{aligned}-(n-1)(1 + \beta_0(t)) e_1 g_0\left(\check{X}_0\right) z_1 &\leq \left|(n-1)(1 + \beta_0(t)) g_0\left(\check{X}_0\right) e_1^2\right| \\ &\leq (n-1)\left(1 + \bar{\beta}_0\right) \bar{g}_0 \bar{\rho}_1^2 \check{e}_1^2.\end{aligned} \tag{8.51}$$

From the (8.50) and (8.51), $\dot{V}_1$ satisfies

$$\dot{V}_1 \leq g_0(X_0)\rho_1 e_1 \alpha_1 + \bar{\rho}_1^2\left[\frac{1}{2} + (n-1)\left(1+\bar{\beta}_0\right)\bar{g}_0\right]\check{e}_1^2$$
$$+e_1\left[\frac{f_1(u_0, x_0, x_1)}{x_0^{n-1}} - \frac{(n-1)e_1 f_0(x_0)}{x_0}\right]$$
$$-\tilde{\theta}_1\dot{\hat{\theta}}_1 - \tilde{\omega}_1\dot{\hat{\omega}}_1 + \frac{1}{2}\left(1+\bar{\beta}_0\right)^2\bar{g}_0^2\bar{\rho}_2^2\check{e}_2^2. \tag{8.52}$$

For the unknown function, one can define $F_1(X_1)$ as

$$F_1(X_1) = \frac{f_1(u_0, x_0, x_1)}{x_0^{n-1}} - \frac{(n-1)e_1 f_0(x_0)}{x_0}. \tag{8.53}$$

Using the FLS, the function $F_1(X_1)$ can be expressed as

$$F_1(X_1) = W_1^{\mathrm{T}}\varphi_1(X_1) + \varepsilon_1 \tag{8.54}$$

where $\bar{\varepsilon}_1 > 0$ is a constant such that $|\varepsilon_1| \leq \bar{\varepsilon}_1$ and W_1 is the unknown ideal weight. The adaptive learning $\hat{W}_1$ and $\hat{\varepsilon}_1$ in (8.30) is designed to compensate the unknown terms W_1 and $\bar{\varepsilon}_1$, respectively.

Inserting (8.54) into (8.52) yields

$$\dot{V}_1 \leq g_0(X_0)\rho_1 e_1 \alpha_1 + \bar{\rho}_1^2\left[\frac{1}{2} + (n-1)\left(1+\bar{\beta}_0\right)\bar{g}_0\right]\check{e}_1^2 - \tilde{\theta}_1\dot{\hat{\theta}}_1$$
$$-\tilde{\omega}_1\dot{\hat{\omega}}_1 + \frac{1}{2}\left(1+\bar{\beta}_0\right)^2\bar{g}_0^2\bar{\rho}_2^2\check{e}_2^2 + e_1 W_1^{\mathrm{T}}\varphi_1(X_1) + e_1\varepsilon_1. \tag{8.55}$$

Using Young's inequality to the terms in (8.55), one has

$$e_1 W_1^{\mathrm{T}}\varphi_1(X_1) \leq \frac{1}{2}\left|\check{e}_1\right|\theta_1\varphi_1^{\mathrm{T}}(X_1)\varphi_1(X_1) + \frac{1}{2}\bar{\rho}_1^2\left|\check{e}_1\right|$$
$$e_1\varepsilon_1 \leq \frac{1}{2}\bar{\rho}_1^2\left|\check{e}_1\right| + \frac{1}{2}\bar{\varepsilon}_1^2\left|\check{e}_1\right|. \tag{8.56}$$

Inserting (8.27) into (8.55), $\dot{V}_1$ further satisfies

$$\dot{V}_1 \leq g_0(X_0)\rho_1 e_1 \alpha_1 + \bar{\rho}_1^2\left[\frac{1}{2} + (n-1)\left(1+\bar{\beta}_0\right)\bar{g}_0\right]\check{e}_1^2$$
$$-\tilde{\theta}_1\dot{\hat{\theta}}_1 - \tilde{\omega}_1\dot{\hat{\omega}}_1 + \bar{\rho}_1^2\left|\check{e}_1\right| + \frac{1}{2}\bar{\varepsilon}_1^2\left|\check{e}_1\right|$$
$$+\frac{1}{2}\left(1+\bar{\beta}_0\right)^2\bar{g}_0^2\bar{\rho}_2^2\check{e}_2^2 + \frac{1}{2}\left|\check{e}_1\right|\theta_1\varphi_1^{\mathrm{T}}(X_1)\varphi_1(X_1). \tag{8.57}$$

Adding and subtracting the term $\frac{1}{2}\left|\check{e}_1\right|\theta_1\varphi_1^{\mathrm{T}}\left(\check{X}_1\right)\varphi_1\left(\check{X}_1\right)$ to the above inequality, one can obtain

$$\begin{aligned}\dot{V}_1 \leq\ & g_0\left(X_0\right)\rho_1 e_1\alpha_1+\bar{\rho}_1^2\left[\frac{1}{2}+(n-1)\left(1+\bar{\beta}_0\right)\bar{g}_0\right]\check{e}_1^2\\ & -\tilde{\theta}_1\dot{\hat{\theta}}_1-\tilde{\omega}_1\dot{\hat{\omega}}_1+\bar{\rho}_1^2\left|\check{e}_1\right|+\frac{1}{2}\bar{\varepsilon}_1^2\left|\check{e}_1\right|\\ & +\frac{1}{2}\left(1+\bar{\beta}_0\right)^2\bar{g}_0^2\bar{\rho}_2^2\check{e}_2^2+\frac{1}{2}\left|\check{e}_1\right|\theta_1\varphi_1^{\mathrm{T}}\left(\check{X}_1\right)\varphi_1\left(\check{X}_1\right)\\ & +\frac{1}{2}\left|\check{e}_1\right|\theta_1\varphi_1^{\mathrm{T}}\left(X_1\right)\varphi_1\left(X_1\right)-\frac{1}{2}\left|\check{e}_1\right|\theta_1\varphi_1^{\mathrm{T}}\left(\check{X}_1\right)\varphi_1\left(\check{X}_1\right).\end{aligned}\tag{8.58}$$

Based on the fact $0<\varphi_1^{\mathrm{T}}(\cdot)\varphi_1(\cdot)\leq 1$, one has

$$\frac{1}{2}\left|\check{e}_1\right|\theta_1\varphi_1^{\mathrm{T}}\left(X_1\right)\varphi_1\left(X_1\right)-\frac{1}{2}\left|\check{e}_1\right|\theta_1\varphi_1^{\mathrm{T}}\left(\check{X}_1\right)\varphi_1\left(\check{X}_1\right)\leq\frac{1}{2}\left|\check{e}_1\right|\theta_1.\tag{8.59}$$

Then, $\dot{V}_1$ can be expressed as

$$\begin{aligned}\dot{V}_1 \leq\ & g_0\left(X_0\right)\rho_1 e_1\alpha_1+\bar{\rho}_1^2\left[\frac{1}{2}+(n-1)\left(1+\bar{\beta}_0\right)\bar{g}_0\right]\check{e}_1^2\\ & -\tilde{\theta}_1\dot{\hat{\theta}}_1-\tilde{\omega}_1\dot{\hat{\omega}}_1+\frac{1}{2}\left|\check{e}_1\right|\theta_1+\omega_1\left|\check{e}_1\right|\\ & +\frac{1}{2}\left(1+\bar{\beta}_0\right)^2\bar{g}_0^2\bar{\rho}_2^2\check{e}_2^2+\frac{1}{2}\left|\check{e}_1\right|\theta_1\varphi_1^{\mathrm{T}}\left(\check{X}_1\right)\varphi_1\left(\check{X}_1\right).\end{aligned}\tag{8.60}$$

Based on Lemma 1.8, one has

$$\begin{aligned}\left|\check{e}_1\right|\varphi_1^{\mathrm{T}}\left(\check{X}_1\right)\varphi_1\left(\check{X}_1\right) & \leq\kappa\sigma(t)+\check{e}_1\varphi_1^{\mathrm{T}}\left(\check{X}_1\right)\varphi_1\left(\check{X}_1\right)\tanh\left(\frac{\check{e}_1\varphi_1^{\mathrm{T}}\left(\check{X}_1\right)\varphi_1\left(\check{X}_1\right)}{\sigma(t)}\right)\\ \left|\check{e}_1\right| & \leq\kappa\sigma(t)+\check{e}_1\tanh\left(\frac{\check{e}_1}{\sigma(t)}\right).\end{aligned}\tag{8.61}$$

Taking (8.30) and (8.61) into (8.60), $\dot{V}_1$ further satisfies

$$\begin{aligned}\dot{V}_1 \leq\ & g_0\left(X_0\right)\rho_1^2\check{e}_1\alpha_1+\bar{\rho}_1^2\left[\frac{1}{2}+(n-1)\left(1+\bar{\beta}_0\right)\bar{g}_0\right]\check{e}_1^2+\frac{1}{2}\left(1+\bar{\beta}_0\right)^2\bar{g}_0^2\bar{\rho}_2^2\check{e}_2^2\\ & +\frac{1}{2}\hat{\theta}_1\check{e}_1\varphi_1^{\mathrm{T}}\left(\check{X}_1\right)\varphi_1\left(\check{X}_1\right)\tanh\left(\frac{\check{e}_1\varphi_1^{\mathrm{T}}\left(\check{X}_1\right)\varphi_1\left(\check{X}_1\right)}{\sigma(t)}\right)\\ & +\left(\theta_1+\omega_1\right)\kappa\sigma(t)+\hat{\omega}_1\check{e}_1\tanh\left(\frac{\check{e}_1}{\sigma(t)}\right)+\frac{1}{2}\hat{\theta}_1\check{e}_1\tanh\left(\frac{\check{e}_1}{\sigma(t)}\right).\end{aligned}\tag{8.62}$$

Therefore, one has

$$\dot{V}_1 \leq \left[\rho_1^2 N(\xi_1) + 1\right]\dot{\xi}_1 - a_1\check{e}_1^2 + \frac{1}{2}\left(1+\bar{\beta}_0\right)^2\bar{g}_0^2\bar{\rho}_2^2\check{e}_2^2 + b_1\sigma(t) \tag{8.63}$$

with $a_1 = k_1 - \bar{\rho}_1^2\left[\frac{1}{2} + (n-1)\left(1+\bar{\beta}_0\right)\bar{g}_0\right]$ and $b_1 = (\theta_1 + \omega_1)\kappa$.

Step i ($2 \leq i \leq n-1$)

Consider the Lyapunov candidate as

$$V_i = \frac{1}{2}e_i^2 + \frac{1}{2}\tilde{\theta}_i^2 + \frac{1}{2}\tilde{\vartheta}_i^2 + \frac{1}{2}\tilde{\omega}_i^2 + \frac{1}{2}\tilde{\chi}_i^2. \tag{8.64}$$

Then, one can get

$$\begin{aligned}\dot{V}_i = {} & g_0\left(\check{X}_0\right)\rho_i^2\check{e}_i\alpha_i + (1+\beta_0(t))\,g_0\left(\check{X}_0\right)e_ie_{i+1} - e_i\dot{\rho}_i\alpha_{i-1}\\ & -(n-i)(1+\beta_0(t))\,e_ig_0\left(\check{X}_0\right)z_i\\ & +e_i\frac{f_i(u_0,x_0,\bar{x}_i)}{x_0^{n-i}} - e_i\frac{(n-i)z_if_0(x_0)}{x_0}\\ & -e_i\rho_i\dot{\alpha}_{i-1} - \tilde{\theta}_i\dot{\hat{\theta}}_i - \tilde{\vartheta}_i\dot{\hat{\vartheta}}_i - \tilde{\omega}_i\dot{\hat{\omega}}_i - \tilde{\chi}_i\dot{\hat{\chi}}_i.\end{aligned} \tag{8.65}$$

From Young's inequality, one has

$$(1+\beta_0(t))\,g_0(X_0)\,e_ie_{i+1} \leq \frac{1}{2}\bar{\rho}_i^2\check{e}_i^2 + \frac{1}{2}\left(1+\bar{\beta}_0\right)^2\bar{g}_0^2\bar{\rho}_{i+1}^2\check{e}_{i+1}^2. \tag{8.66}$$

For the term $-(n-i)(1+\beta_0(t))\,e_ig_0\left(\check{X}_0\right)z_i$ in (8.65), one has

$$\begin{aligned}-(n-i)(1+\beta_0(t))\,e_ig_0\left(\check{X}_0\right)z_i & \leq (n-i)\left(1+\bar{\beta}_0\right)\bar{g}_0\bar{\rho}_i^2\left|\check{e}_i^2 + \check{e}_i\alpha_{i-1}\right|\\ & \leq 2(n-i)\left(1+\bar{\beta}_0\right)\bar{g}_0\bar{\rho}_i^2\check{e}_i^2\\ & \quad +2(n-i)\left(1+\bar{\beta}_0\right)\bar{g}_0\bar{\rho}_i^2\left|\check{e}_i\alpha_{i-1}\right|\end{aligned} \tag{8.67}$$

Substituting (8.66) and (8.67) into inequality (8.65), one can obtain

$$\begin{aligned}\dot{V}_i \le\ & g_0\left(\check{X}_0\right)\rho_i^2\check{e}_i\alpha_i + \frac{1}{2}\bar{\rho}_i^2\check{e}_i^2 + \frac{1}{2}\left(1+\bar{\beta}_0\right)^2\bar{g}_0^2\bar{\rho}_{i+1}^2\check{e}_{i+1}^2 \\ & + e_i\left[\frac{f_i\left(u_0, x_0, \bar{x}_i\right)}{x_0^{n-i}} - \frac{(n-i)z_i f_0\left(x_0\right)}{x_0}\right] \\ & + 2\left(n-i\right)\left(1+\bar{\beta}_0\right)\bar{g}_0\bar{\rho}_i^2\check{e}_i^2 + 2\left(n-i\right)\left(1+\bar{\beta}_0\right)\bar{g}_0\bar{\rho}_i^2\left|\check{e}_i\alpha_{i-1}\right| \\ & - e_i\dot{\rho}_i\alpha_{i-1} - e_i\rho_i\dot{\alpha}_{i-1} - \tilde{\theta}_i\dot{\hat{\theta}}_i - \tilde{\vartheta}_i\dot{\hat{\vartheta}}_i - \tilde{\omega}_i\dot{\hat{\omega}}_i - \tilde{\chi}_i\dot{\hat{\chi}}_i. \end{aligned} \tag{8.68}$$

Furthermore, the unknown function can be expressed as

$$F_i\left(X_i\right) = \frac{f_i\left(u_0, x_0, \bar{x}_i\right)}{x_0^{n-i}} - \frac{(n-i)z_i f_0\left(x_0\right)}{x_0} = W_i^{\mathrm{T}}\varphi_i\left(X_i\right) + \varepsilon_i \tag{8.69}$$

where $\bar{\varepsilon}_i > 0$ is a constant such that $|\varepsilon_i| \le \bar{\varepsilon}_i$ and W_i is the unknown ideal weight. The adaptive learning $\hat{W}_i$ and $\hat{\varepsilon}_i$ in (8.30) is designed to compensate the unknown terms W_i and $\bar{\varepsilon}_i$, respectively. Therefore, one has

$$\begin{aligned}\dot{V}_i \le\ & g_0\left(X_0\right)\rho_i^2\check{e}_i\alpha_i + \left[\frac{1}{2} + 2\left(n-i\right)\left(1+\bar{\beta}_0\right)\bar{g}_0\right]\bar{\rho}_i^2\check{e}_i^2 \\ & + \frac{1}{2}\left(1+\bar{\beta}_0\right)^2\bar{g}_0^2\bar{\rho}_{i+1}^2\check{e}_{i+1}^2 + e_i W_i^{\mathrm{T}}\varphi_i\left(X_i\right) + e_i\varepsilon_i \\ & + 2\left(n-i\right)\left(1+\bar{\beta}_0\right)\bar{g}_0\bar{\rho}_i^2\left|\check{e}_i\alpha_{i-1}\right| - e_i\dot{\rho}_i\alpha_{i-1} \\ & - e_i\rho_i\dot{\alpha}_{i-1} - \tilde{\theta}_i\dot{\hat{\theta}}_i - \tilde{\vartheta}_i\dot{\hat{\vartheta}}_i - \tilde{\omega}_i\dot{\hat{\omega}}_i - \tilde{\chi}_i\dot{\hat{\chi}}_i. \end{aligned} \tag{8.70}$$

Consider the following facts:

$$\begin{aligned} - e_i\dot{\rho}_i\alpha_{i-1} &\le \left|\check{e}_i\alpha_{i-1}\right|\bar{\rho}_i\bar{\rho}_{di} \\ - e_i\rho_i\dot{\alpha}_{i-1} &\le \left|\check{e}_i\dot{\alpha}_{i-1}\right|\bar{\rho}_i \\ e_i W_i^{\mathrm{T}}\varphi_i\left(X_i\right) &\le \frac{1}{2}\left|\check{e}_i\right|\theta_i\varphi_i^{\mathrm{T}}\left(X_i\right)\varphi_i\left(X_i\right) + \frac{1}{2}\left|\check{e}_i\right|\bar{\rho}_i^2 \\ e_i\varepsilon_i &\le \frac{1}{2}\bar{\rho}_i^2\left|\check{e}_i\right| + \frac{1}{2}\bar{\varepsilon}_i^2\left|\check{e}_i\right|. \end{aligned} \tag{8.71}$$

Then, $\dot{V}_i$ satisfies

$$\begin{aligned}\dot{V}_i \le\ & g_0\left(X_0\right)\rho_i^2\check{e}_i\alpha_i + \left[\frac{1}{2} + 2\left(n-i\right)\left(1+\bar{\beta}_0\right)\bar{g}_0\right]\bar{\rho}_i^2\check{e}_i^2 \\ & + \frac{1}{2}\left(1+\bar{\beta}_0\right)^2\bar{g}_0^2\bar{\rho}_{i+1}^2\check{e}_{i+1}^2 + \vartheta_i\left|\check{e}_i\alpha_{i-1}\right| + \frac{1}{2}\left|\check{e}_i\right|\theta_i\varphi_i^{\mathrm{T}}\left(X_i\right)\varphi_i\left(X_i\right) \\ & + \omega_i\left|\check{e}_i\right| + \chi_i\left|\check{e}_i\dot{\alpha}_{i-1}\right| - \tilde{\theta}_i\dot{\hat{\theta}}_i - \tilde{\vartheta}_i\dot{\hat{\vartheta}}_i - \tilde{\omega}_i\dot{\hat{\omega}}_i - \tilde{\chi}_i\dot{\hat{\chi}}_i. \end{aligned} \tag{8.72}$$

Adding and subtracting the term $\frac{1}{2}\left|\check{e}_i\right|\theta_i\varphi_i^{\mathrm{T}}\left(\check{X}_i\right)\varphi_i\left(\check{X}_i\right)$ to above inequality, one can obtain

$$\begin{aligned}\dot{V}_i \le\ & g_0\left(X_0\right)\rho_i^2\check{e}_i\alpha_i+\left[\frac{1}{2}+2\left(n-i\right)\left(1+\bar{\beta}_0\right)\bar{g}_0\right]\bar{\rho}_i^2\check{e}_i^2\\&+\frac{1}{2}\left(1+\bar{\beta}_0\right)^2\bar{g}_0^2\bar{\rho}_{i+1}^2\check{e}_{i+1}^2+\vartheta_i\left|\check{e}_i\alpha_{i-1}\right|+\omega_i\left|\check{e}_i\right|\\&+\chi_i\left|\check{e}_i\dot{\alpha}_{i-1}\right|+\frac{1}{2}\left|\check{e}_i\right|\theta_i\varphi_i^{\mathrm{T}}\left(X_i\right)\varphi_i\left(X_i\right)\\&+\frac{1}{2}\left|\check{e}_i\right|\theta_i\varphi_i^{\mathrm{T}}\left(\check{X}_i\right)\varphi_i\left(\check{X}_i\right)-\frac{1}{2}\left|\check{e}_i\right|\theta_i\varphi_i^{\mathrm{T}}\left(\check{X}_i\right)\varphi_i\left(\check{X}_i\right)\\&-\tilde{\theta}_i\dot{\hat{\theta}}_i-\tilde{\vartheta}_i\dot{\hat{\vartheta}}_i-\tilde{\omega}_i\dot{\hat{\omega}}_i-\tilde{\chi}_i\dot{\hat{\chi}}_i.\end{aligned}\tag{8.73}$$

Note that the FLS basis function satisfies $0<\varphi_i^{\mathrm{T}}\left(\cdot\right)\varphi_i\left(\cdot\right)\le 1$, which implies

$$\frac{1}{2}\left|\check{e}_i\right|\theta_i\varphi_i^{\mathrm{T}}\left(X_i\right)\varphi_i\left(X_i\right)-\frac{1}{2}\left|\check{e}_i\right|\theta_i\varphi_i^{\mathrm{T}}\left(\check{X}_i\right)\varphi_i\left(\check{X}_i\right)\le\frac{1}{2}\left|\check{e}_i\right|\theta_i.\tag{8.74}$$

Then, $\dot{V}_i$ further satisfies

$$\begin{aligned}\dot{V}_i \le\ & g_0\left(X_0\right)\rho_i^2\check{e}_i\alpha_i+\left[\frac{1}{2}+2\left(n-i\right)\left(1+\bar{\beta}_0\right)\bar{g}_0\right]\bar{\rho}_i^2\check{e}_i^2\\&+\frac{1}{2}\left(1+\bar{\beta}_0\right)^2\bar{g}_0^2\bar{\rho}_{i+1}^2\check{e}_{i+1}^2+\vartheta_i\left|\check{e}_i\alpha_{i-1}\right|+\omega_i\left|\check{e}_i\right|\\&+\frac{1}{2}\left|\check{e}_i\right|\theta_i\varphi_i^{\mathrm{T}}\left(\check{X}_i\right)\varphi_i\left(\check{X}_i\right)+\frac{1}{2}\left|\check{e}_i\right|\theta_i\\&+\chi_i\left|\check{e}_i\dot{\alpha}_{i-1}\right|-\tilde{\theta}_i\dot{\hat{\theta}}_i-\tilde{\vartheta}_i\dot{\hat{\vartheta}}_i-\tilde{\omega}_i\dot{\hat{\omega}}_i-\tilde{\chi}_i\dot{\hat{\chi}}_i.\end{aligned}\tag{8.75}$$

Based on Lemma 1.8, one has

$$\begin{aligned}\left|\check{e}_i\right|\varphi_i^{\mathrm{T}}\left(\check{X}_i\right)\varphi_i\left(\check{X}_i\right)&\le\kappa\sigma\left(t\right)+\check{e}_i\varphi_i^{\mathrm{T}}\left(\check{X}_i\right)\varphi_i\left(\check{X}_i\right)\tanh\left(\frac{\check{e}_i\varphi_i^{\mathrm{T}}\left(\check{X}_i\right)\varphi_i\left(\check{X}_i\right)}{\sigma\left(t\right)}\right)\\\left|\check{e}_i\right|&\le\kappa\sigma\left(t\right)+\check{e}_i\tanh\left(\frac{\check{e}_i}{\sigma\left(t\right)}\right)\\\left|\check{e}_i\alpha_{i-1}\right|&\le\kappa\sigma\left(t\right)+\check{e}_i\alpha_{i-1}\tanh\left(\frac{\check{e}_i\alpha_{i-1}}{\sigma\left(t\right)}\right)\\\left|\check{e}_i\dot{\alpha}_{i-1}\right|&\le\kappa\sigma\left(t\right)+\check{e}_i\dot{\alpha}_{i-1}\tanh\left(\frac{\check{e}_i\dot{\alpha}_{i-1}}{\sigma\left(t\right)}\right).\end{aligned}\tag{8.76}$$

Considering the design in (8.33), one has

$$\begin{aligned}\dot{V}_i \leq\ & g_0(X_0)\rho_i^2\check{e}_i\alpha_i + \left[\frac{1}{2} + 2(n-i)\left(1+\bar{\beta}_0\right)\bar{g}_0\right]\bar{\rho}_i^2\check{e}_i^2 \\ & + \frac{1}{2}\left(1+\bar{\beta}_0\right)^2\bar{g}_0^2\bar{\rho}_{i+1}^2\check{e}_{i+1}^2 + \hat{\vartheta}_i\check{e}_i\alpha_{i-1}\tanh\left(\frac{\check{e}_i\alpha_{i-1}}{\sigma(t)}\right) \\ & + \hat{\chi}_i\check{e}_i\dot{\alpha}_{i-1}\tanh\left(\frac{\check{e}_i\dot{\alpha}_{i-1}}{\sigma(t)}\right) + \hat{\omega}_i\check{e}_i\tanh\left(\frac{\check{e}_i}{\sigma(t)}\right) \\ & + \frac{1}{2}\hat{\theta}_i\check{e}_i\varphi_i^{\mathrm{T}}\left(\check{X}_i\right)\varphi_i\left(\check{X}_i\right)\tanh\left(\frac{\check{e}_i\varphi_i^{\mathrm{T}}\left(\check{X}_i\right)\varphi_i\left(\check{X}_i\right)}{\sigma(t)}\right) \\ & + \frac{1}{2}\hat{\theta}_i\check{e}_i\tanh\left(\frac{\check{e}_i}{\sigma(t)}\right) + (\theta_i + \vartheta_i + \omega_i + \chi_i)\kappa\sigma(t).\end{aligned} \tag{8.77}$$

Based on the above, one get the results as

$$\begin{aligned}\dot{V}_i \leq\ & \left[\rho_i^2 N(\xi_i) + 1\right]\dot{\xi}_i + \frac{1}{2}\left(1+\bar{\beta}_0\right)^2\bar{g}_0^2\bar{\rho}_{i+1}^2\check{e}_{i+1}^2 + b_i\sigma \\ & - \left\{k_i - \bar{\rho}_i^2\left[\frac{1}{2} + 2(n-i)\left(1+\bar{\beta}_0\right)\bar{g}_0\right]\right\}\check{e}_i^2\end{aligned} \tag{8.78}$$

where $b_i = (\theta_i + \vartheta_i + \omega_i + \chi_i)\kappa$.

Step n

The Lyapunov function is considered as

$$V_n = \frac{1}{2}e_n^2 + \frac{1}{2}\tilde{\theta}_n^2 + \frac{1}{2}\tilde{\vartheta}_n^2 + \frac{1}{2}\omega_n^2 + \frac{1}{2}\tilde{\chi}_n^2. \tag{8.79}$$

The expression of the Lyapunov function V_n can be written as

$$\begin{aligned}\dot{V}_n =\ & e_n u_1 + e_n f_n(u_0, x_0, \bar{x}_n) - e_n\dot{\rho}_n\alpha_{n-1} \\ & - e_n\rho_n\dot{\alpha}_{n-1} - \tilde{\theta}_n\dot{\hat{\theta}}_n - \tilde{\omega}_n\dot{\hat{\omega}}_n - \tilde{\vartheta}_n\dot{\hat{\vartheta}}_n - \tilde{\chi}_n\dot{\hat{\chi}}_n.\end{aligned} \tag{8.80}$$

For the unknown function $f_n(u_0, x_0, \bar{x}_n)$, one can define

$$\begin{aligned}& F_n(X_n) = f_n(u_0, x_0, \bar{x}_n) = W_n^{\mathrm{T}}\varphi_n(X_n) + \varepsilon_n, \\ & |\varepsilon_n| \leq \bar{\varepsilon}_n\end{aligned} \tag{8.81}$$

with $\bar{\varepsilon}_i > 0$ being a constant such that $|\varepsilon_i| \leq \bar{\varepsilon}_i$, W_i being the unknown ideal weight. Taking $F_n(X_n)$ into (8.80), one has

$$\begin{aligned}\dot{V}_n &= e_n u_1 + e_n W_n^{\mathrm{T}} \varphi_n(X_n) + e_n \varepsilon_n - e_n \dot{\rho}_n \alpha_{n-1} \\ &\quad - e_n \rho_n \dot{\alpha}_{n-1} - \tilde{\theta}_n \dot{\hat{\theta}}_n - \tilde{\omega}_n \dot{\hat{\omega}}_n - \tilde{\vartheta}_n \dot{\hat{\vartheta}}_n - \tilde{\chi}_n \dot{\hat{\chi}}_n. \end{aligned} \tag{8.82}$$

Furthermore, the fact is satisfied as

$$\begin{aligned} e_n W_n^{\mathrm{T}} \varphi_n(X_n) &\leq \frac{1}{2} \left|\breve{e}_n\right| \theta_n \varphi_n^{\mathrm{T}}(X_n) \varphi_n(X_n) + \frac{1}{2} \left|\breve{e}_n\right| \bar{\rho}_n^2 \\ e_n \varepsilon_n &\leq \frac{1}{2} \left|\breve{e}_n\right| \bar{\rho}_n^2 + \frac{1}{2} \left|\breve{e}_n\right| \bar{\varepsilon}_n^2 \\ -e_n \dot{\rho}_n \alpha_{n-1} &\leq \left|\breve{e}_n \alpha_{n-1}\right| \vartheta_n \\ -e_n \rho_n \dot{\alpha}_{n-1} &\leq \left|\breve{e}_n \dot{\alpha}_{n-1}\right| \chi_n. \end{aligned} \tag{8.83}$$

Similar to the design process in Step i, the time derivative of V_n is expressed as

$$\begin{aligned}\dot{V}_n &\leq e_n u_1 + \frac{1}{2} \left|\breve{e}_n\right| \theta_n \varphi_n^{\mathrm{T}}\left(\breve{X}_n\right) \varphi_n\left(\breve{X}_n\right) + \frac{1}{2} \left|\breve{e}_n\right| \theta_n \varphi_n^{\mathrm{T}}(X_n) \varphi_n(X_n) \\ &\quad - \frac{1}{2} \left|\breve{e}_n\right| \theta_n \varphi_n^{\mathrm{T}}\left(\breve{X}_n\right) \varphi_n\left(\breve{X}_n\right) + \left|\breve{e}_n\right| \omega_n + \left|\breve{e}_n \alpha_{n-1}\right| \vartheta_n + \left|\breve{e}_n \dot{\alpha}_{n-1}\right| \chi_n \\ &\quad - \tilde{\theta}_n \dot{\hat{\theta}}_n - \tilde{\omega}_n \dot{\hat{\omega}}_n - \tilde{\vartheta}_n \dot{\hat{\vartheta}}_n - \tilde{\chi}_n \dot{\hat{\chi}}_n. \end{aligned} \tag{8.84}$$

From the fact $0 < \varphi_n^{\mathrm{T}}(\cdot) \varphi_n(\cdot) \leq 1$, one can note that

$$\frac{1}{2} \left|\breve{e}_n\right| \theta_n \left[\varphi_n^{\mathrm{T}}(X_n) \varphi_n(X_n) - \varphi_n^{\mathrm{T}}\left(\breve{X}_n\right) \varphi_n\left(\breve{X}_n\right)\right] \leq \frac{1}{2} \left|\breve{e}_n\right| \theta_n. \tag{8.85}$$

Therefore, one can describe $\dot{V}_n$ as

$$\begin{aligned}\dot{V}_n &\leq e_n u_1 + \frac{1}{2} \left|\breve{e}_n\right| \theta_n \varphi_n^{\mathrm{T}}\left(\breve{X}_n\right) \varphi_n\left(\breve{X}_n\right) + \frac{1}{2} \left|\breve{e}_n\right| \theta_n \\ &\quad + \left|\breve{e}_n\right| \omega_n + \left|\breve{e}_n \alpha_{n-1}\right| \vartheta_n + \left|\breve{e}_n \dot{\alpha}_{n-1}\right| \chi_n \\ &\quad - \tilde{\theta}_n \dot{\hat{\theta}}_n - \tilde{\omega}_n \dot{\hat{\omega}}_n - \tilde{\vartheta}_n \dot{\hat{\vartheta}}_n - \tilde{\chi}_n \dot{\hat{\chi}}_n. \end{aligned} \tag{8.86}$$

From Lemma 1.8, one has

$$\begin{aligned} \left|\breve{e}_n\right| \varphi_n^{\mathrm{T}}\left(\breve{X}_n\right) \varphi_n\left(\breve{X}_n\right) &\leq \kappa\sigma + \breve{e}_n \varphi_n^{\mathrm{T}}\left(\breve{X}_n\right) \varphi_n\left(\breve{X}_n\right) \tanh\left[\frac{\breve{e}_n \varphi_n^{\mathrm{T}}\left(\breve{X}_n\right) \varphi_n\left(\breve{X}_n\right)}{\sigma(t)}\right] \\ \left|\breve{e}_n\right| &\leq \kappa\sigma + \breve{e}_n \tanh\left(\frac{\breve{e}_n}{\sigma(t)}\right) \end{aligned}$$

$$\begin{aligned}
\left|\check{e}_n\alpha_{n-1}\right| &\leq \kappa\sigma + \check{e}_n\alpha_{n-1}\tanh\left(\frac{\check{e}_n\alpha_{n-1}}{\sigma(t)}\right) \\
\left|\check{e}_n\dot{\alpha}_{n-1}\right| &\leq \kappa\sigma + \check{e}_n\dot{\alpha}_{n-1}\tanh\left(\frac{\check{e}_n\dot{\alpha}_{n-1}}{\sigma(t)}\right).
\end{aligned} \tag{8.87}$$

From the design of (8.34) and (8.36), one has

$$\begin{aligned}
\dot{V}_n \leq\ & e_n u_1 + \hat{\vartheta}_n\check{e}_n\alpha_{n-1}\tanh\left(\frac{\check{e}_n\alpha_{n-1}}{\sigma(t)}\right) + \hat{\chi}_n\check{e}_n\alpha_{n-1}\tanh\left(\frac{\check{e}_n\alpha_{n-1}}{\sigma(t)}\right) \\
& + \hat{\omega}_n\check{e}_n\tanh\left(\frac{\check{e}_n}{\sigma(t)}\right) + \hat{\omega}_n\check{e}_n\tanh\left(\frac{\check{e}_n}{\sigma(t)}\right) + \frac{1}{2}\hat{\theta}_n\check{e}_n\tanh\left(\frac{\check{e}_n}{\sigma(t)}\right) \\
& + \frac{1}{2}\hat{\theta}_n\check{e}_n\varphi_n^{\mathrm{T}}\left(\check{X}_n\right)\varphi_n\left(\check{X}_n\right)\tanh\left(\frac{\check{e}_n\varphi_n^{\mathrm{T}}\left(\check{X}_n\right)\varphi_n\left(\check{X}_n\right)}{\sigma(t)}\right) \\
& + \left(\theta_n + \omega_n + \vartheta_n + \chi_n\right)\kappa\sigma(t).
\end{aligned} \tag{8.88}$$

Based on the above analysis, one has

$$\dot{V}_n \leq \left[\rho_n N\left(\xi_n\right) + 1\right]\dot{\xi}_n - k_n\check{e}_n^2 + b_n\sigma(t) \tag{8.89}$$

where $b_n = \left(\theta_n + \omega_n + \vartheta_n + \chi_n\right)\kappa$. Collecting the discussions in Steps $1 - n$ above, the stability analysis for state $x = \left[x_1 \cdots x_n\right]$ can be summarized as follows.

Theorem 8.2 *Consider the uncertain nonholonomic system* (8.1) *against FDI attacks under Assumptions 8.1–8.3 with the control designs u_0 and u_1. Then, for bounded initial conditions, one has that all signals in the closed-loop system remain bounded and all the system states x_i for $i = 0, 1, 2, \ldots, n$ converge to zero when $t \to \infty$.*

Proof Define the Lyapunov function as

$$V = \frac{1}{2}\left(\sum_{i=1}^{n} e_i^2 + \sum_{i=1}^{n}\tilde{\theta}_i^2 + \sum_{i=1}^{n}\tilde{\omega}_i^2 + \sum_{i=2}^{n}\tilde{\vartheta}_i^2 + \sum_{i=2}^{n}\tilde{\chi}_i^2\right). \tag{8.90}$$

Based on (8.63)–(8.89), the expression of V can be written as

$$\dot{V} \leq \sum_{i=1}^{n}\left[h_i N\left(\xi_i\right) + 1\right]\dot{\xi}_i - \sum_{i=1}^{n} a_i\check{e}_i^2 + \sum_{i=1}^{n} b_i\sigma(t) \tag{8.91}$$

with $h_i = \rho_i^2$ for $i = 1, 2, \ldots, n$, $h_n = \rho_n$, $a_i = k_i - \bar{\rho}_i^2\left[\frac{1}{2} + 2(n-i)\left(1 + \bar{\beta}_0\right)\bar{g}_0\right] - \frac{1}{2}\left(1 + \bar{\beta}_0\right)^2\bar{g}_0^2\bar{\rho}_i^2$ for $i = 2, 3, \ldots, n$, and $a_n = k_n - \frac{1}{2}\left(1 + \bar{\beta}_0\right)^2\bar{g}_0^2\bar{\rho}_n^2$. Then, one has

$$V \leq \sum_{i=1}^{n} \int_{0}^{t} \left[h_i N\left(\xi_i\left(\tau\right)\right) + 1\right] \dot{\xi}_i\left(\tau\right) d\tau$$
$$- \sum_{i=1}^{n} a_i \int_{0}^{t} \check{e}_i^2\left(\tau\right) d\tau + \sum_{i=1}^{n} b_i \bar{\sigma}$$
$$\leq \sum_{i=1}^{n} \int_{0}^{t} \left[h_i N\left(\xi_i\left(\tau\right)\right) + 1\right] \dot{\xi}_i\left(\tau\right) d\tau + \sum_{i=1}^{n} b_i \bar{\sigma}. \tag{8.92}$$

From Lemma 8.1, it can be concluded from (8.92) that all the signals for the uncertain nonholonomic system remain bounded. Furthermore, $\check{e}_i^2$ is uniformly continuous, and $\dot{\check{e}}_i$ is bounded within the interval $[0, +\infty)$. Then, Lemma 1.10 implies that $\lim_{t\to\infty} \check{e}_i = 0$. From the design in (8.27), one can observe that $\lim_{t\to\infty} \alpha_1 = 0$ and $\lim_{t\to\infty} \check{z}_1 = 0$. Then, based on the definition of $\check{e}_i$, one can ensure that $\lim_{t\to\infty} \check{z}_2 = \lim_{t\to\infty} \check{e}_2 + \lim_{t\to\infty} \alpha_1 = 0$. Similarly, one has $\lim_{t\to\infty} z_i = 0$ for $i = 1, 2, \ldots, n$. Furthermore, according to the definition of (8.22), one can ascertain that $\lim_{t\to\infty} \check{x}_i = 0$ for $i = 1, 2, \ldots, n$ and $\check{x}_i$ and $\check{x}_0$ are equivalent infinitesimal. Hence, one can ensure that $\lim_{t\to\infty} \check{x}_i = 0$. Based on (8.2) and Assumption 8.1, one has $\lim_{t\to\infty} x_i = 0$. This completes the proof. □

8.5 Simulation Example

In this section, a wheeled mobile robot [31] is considered as

$$\begin{aligned} \dot{x}_c &= \varpi \cos\theta \\ \dot{y}_c &= \varpi \sin\theta \\ \dot{\theta} &= \omega \end{aligned} \tag{8.93}$$

where ϖ and ω are the linear velocity and the angular velocity, θ is the orientation, and (x_c, y_c) is the Cartesian position of the wheeled mobile robot center. To transform the wheeled mobile robot dynamics into the nonholonomic form as in (8.1), consider the variables transformation

$$\begin{aligned} x_0 &= \theta, \\ x_1 &= x_c \cos\theta - y_c \sin\theta, \\ x_2 &= x_c \cos\theta + y_c \sin\theta, \\ u_0 &= \omega, u_1 = \varpi. \end{aligned}$$

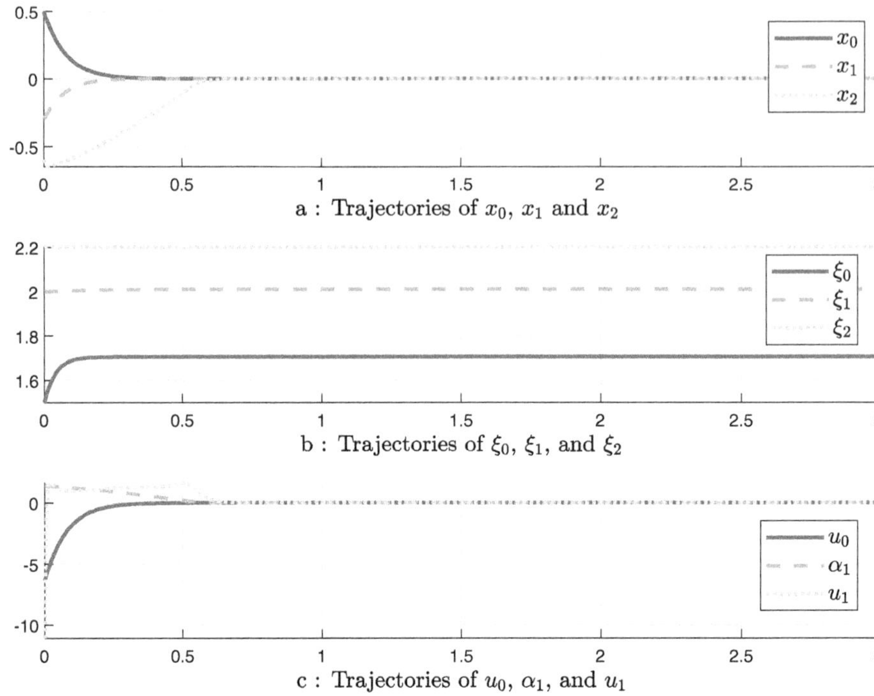

Fig. 8.1 System states, Nussbaum-type gains, virtual control signal, and control signals

Then, (8.93) can be written as

$$\begin{aligned}\dot{x}_0 &= u_0\\ \dot{x}_1 &= x_2 u_0\\ \dot{x}_2 &= u_1 - x_1 u_0\end{aligned}$$

The FDI attack signals in the simulation are $\beta_0 = -2 - 0.2\sin(t)$ and $\beta = -3 - 0.1\cos(t)$. Based on the design in Sects. 8.3.1 and 8.3.2 with the parameters being $k_0 = 20$, $k_1 = 3.3$, $k_2 = 8$, $\varsigma = 10$, and $\sigma = \mathrm{e}^{-0.1\mathrm{t}}$, the simulation results are as follows.

The simulation results for uncertain nonholonomic systems are illustrated in Figs. 8.1 and 8.2. Figure 8.1a shows the evolution of the system states x_0, x_1, and x_2, respectively. All the system states can converge to zero, which validates the asymptotic stability of the presented secure control strategy. In Fig. 8.1b, the evolution of ξ_0, ξ_1, and ξ_2 is shown, which verifies the boundedness of Nussbaum-type gains designed in this chapter. The boundedness of the Nussbaum-type gains explains the discussions presented in Lemma 8.1. The virtual control signal α_1 and the control signals u_0 and u_1 are shown in Fig. 8.1c. According to the discussion in Theorem 8.2, the virtual control signal can converge to zero, which becomes a reference for x_2 to

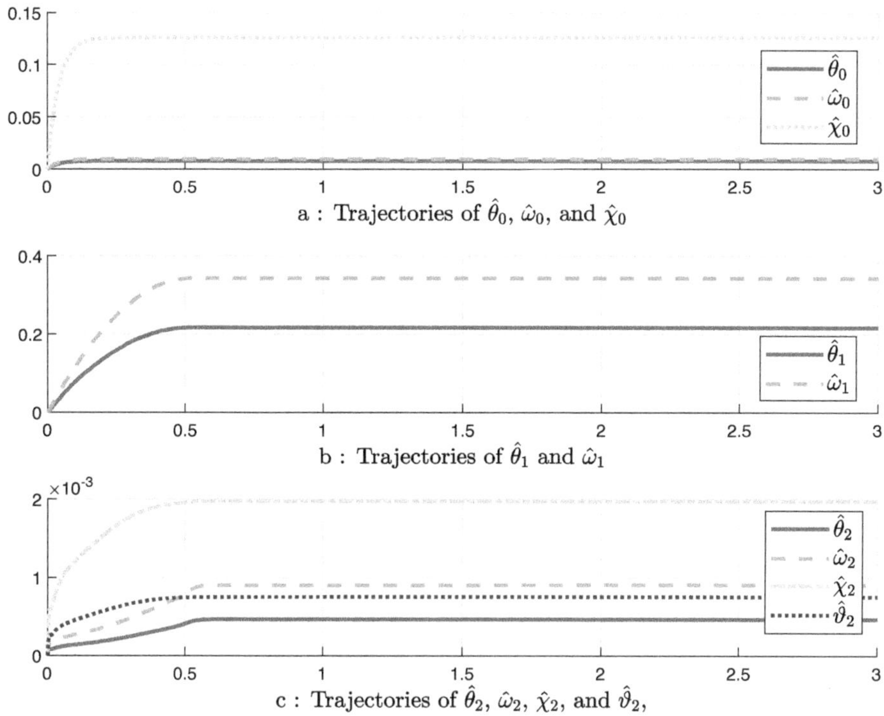

Fig. 8.2 Adaptive learning laws

converge to zero. As shown in Fig. 8.2, subfigures (a), (b), and (c) represent the evolution of adaptive parameters in each subsystem of the nonholonomic system. According to the presented adaptive Nussbaum design, each subsystem has distinct learning rate, which ultimately ensures the boundedness of the closed-loop signals.

8.6 Conclusion

In this chapter, we develop an adaptive Nussbaum design to asymptotically stabilize the uncertain nonholonomic systems subject to unknown FDI attacks. Considering the characteristics of nonholonomic systems and the challenges posed by FDI attacks, a secure control strategy is developed to achieve the asymptotic stability in the presence of attacks. Simulation studies are implemented to validate the effectiveness of the proposed secure control scheme for a mobile robot. The experimental results demonstrate that our control strategy can ensure stable operation of the system even in the presence of FDI attacks.

References

1. Barboni, A., Rezaee, H., Boem, F., Parisini, T.: Detection of Covert Cyber-Attacks in Interconnected Systems: A Distributed Model-Based Approach. IEEE Trans. Autom. Control **65**(9), 3728–3741 (2020)
2. Rezaee, H., Zhang, K., Parisini, T., Polycarpou, M.M.: Cooperative adaptive cruise control in the presence of communication and radar stochastic data loss. IEEE Trans. Intell. Transp. Syst. **25**(6), 4964–4976 (2024)
3. Zhang, K., Keliris, C., Parisini, T., Jiang, B., Polycarpou, M.M.: Passive attack detection for a class of stealthy intermittent integrity attacks. IEEE/CAA J. Autom. Sinica. **10**(4), 898–915 (2023)
4. Rezaee, H., Parisini, T., Polycarpou, M.M.: Almost sure resilient consensus under stochastic interaction: links failure and noisy channels. IEEE Trans. Autom. Control **66**(12), 5727–5741 (2021)
5. Zhang, K., Keliris, C., Parisini, T., Polycarpou, M.M.: Stealthy integrity attacks for a class of nonlinear cyber-physical systems. IEEE Trans. Autom. Control **67**(12), 6723–6730 (2022)
6. An, L., Yang, G.H.: Decentralized adaptive fuzzy secure control for nonlinear uncertain interconnected systems against intermittent DoS attacks. IEEE Trans. Cybern. **49**(3), 827–838 (2019)
7. Lu, A.Y., Yang, G.H.: Observer-based control for cyber-physical systems under denial-of-service with a decentralized event-triggered scheme. IEEE Trans. Cybern. **50**(12), 4886–4895 (2020)
8. Liu, G., Yang, Y., Gao, W., Wunsch, D.C.: Adaptive Nussbaum design for nonholonomic systems with asymptotic stabilization against false data injection. IEEE Trans. Cybern. **55**(11), 5189–5201 (2025)
9. Song, S., Park, J.H., Zhang, B., Song, X.: Adaptive NN finite-time resilient control for nonlinear time-delay systems with unknown false data injection and actuator faults. IEEE Trans. Neural Netw. Learn. Syst. **33**(10), 5416–5428 (2022)
10. Zhao, J., Yang, G.H.: Fuzzy adaptive secure tracking control against unknown false data injection attacks for uncertain nonlinear systems with input quantization. Appl. Math. Comput. **437**, 127551 (2023)
11. Wen, C., Zhou, J., Liu, Z., Su, H.: Robust adaptive control of uncertain nonlinear systems in the presence of input saturation and external disturbance. IEEE Trans. Autom. Control **56**(7), 1672–1678 (2011)
12. Sun, W., Wang, L., Wu, Y.: Adaptive dynamic surface fuzzy control for state constrained time-delay nonlinear nonstrict feedback systems with unknown control directions. IEEE Trans. Syst. Man Cybern.: Syst. **51**(12), 7423–7434 (2021)
13. Wang, Y., Miao, Z., Zhong, H., Pan, Q.: Simultaneous stabilization and tracking of nonholonomic mobile robots: a Lyapunov-based approach. IEEE Trans. Control Syst. Technol. **23**(4), 1440–1450 (2015)
14. Sun, W., Su, S.F., Xia, J., Wu, Y.: Adaptive tracking control of wheeled inverted pendulums with periodic disturbances. IEEE Trans. Cybern. **50**(5), 1867–1876 (2020)
15. Ge, S.S., Wang, Z., Lee, T.H.: Adaptive stabilization of uncertain nonholonomic systems by state and output feedback. Automatica **39**(8), 1451–1460 (2003)
16. Brockett, R.W.: Asymptotic Stability and Feedback Stabilization. Defense Technical Information Center (1982)
17. Samson, C.: Control of chained systems application to path following and time-varying point-stabilization of mobile robots. IEEE Trans. Autom. Control **40**(1), 64–77 (1995)
18. Tian, Y.P., Li, S.: Exponential stabilization of nonholonomic dynamic systems by smooth time-varying control. Automatica **38**(7), 1139–1146 (2002)
19. M'Closkey, R.T., Murray, R.M.: Exponential stabilization of driftless nonlinear control systems using homogeneous feedback. IEEE Trans. Autom. Control **42**(5), 614–628 (1997)
20. Astolfi, A.: Discontinuous control of nonholonomic systems. Syst. & Control Lett. **27**(1), 37–45 (1996)

21. Kolmanovsky, I., McClamroch, N.H.: Hybrid feedback laws for a class of cascade nonlinear control systems. IEEE Trans. Autom. Control **41**(9), 1271–1282 (1996)
22. Jiang, Z.P., Nijmeijer, H.: A recursive technique for tracking control of nonholonomic systems in chained form. IEEE Trans. Autom. Control **44**(2), 265–279 (1999)
23. Wang, W., Huang, J., Wen, C., Fan, H.: Distributed adaptive control for consensus tracking with application to formation control of nonholonomic mobile robots. Automatica **50**(4), 1254–1263 (2014)
24. Huang, J., Wen, C., Wang, W., Jiang, Z.P.: Adaptive output feedback tracking control of a nonholonomic mobile robot. Automatica **50**(3), 821–831 (2014)
25. Li, Q., Yang, H., Xia, Y., Zhao, H.: Switched model predictive control for nonholonomic mobile robots under adaptive dwell time. IEEE Trans. Cybern. **54**(6), 3444–3453 (2024)
26. Ju, S., Wang, J., Dou, L.: MPC-based cooperative enclosing for nonholonomic mobile agents under input constraint and unknown disturbance. IEEE Trans. Cybern. **53**(2), 845–858 (2023)
27. Zhao, J., Yang, G.H.: Fuzzy adaptive finite-time resilient control against unknown false data injection attacks for MIMO Nonlinear switched systems with unknown dead zone. IEEE Trans. Cybern. **54**(1), 586–598 (2024)
28. Li, X.M., Zhou, Q., Li, P., Li, H., Lu, R.: Event-triggered consensus control for multi-agent systems against false data-injection attacks. IEEE Trans. Cybern. **50**(5), 1856–1866 (2020)
29. Song, S., Park, J.H., Zhang, B., Song, X.: Event-based adaptive fuzzy fixed-time secure control for nonlinear CPSs against unknown false data injection and backlash-like hysteresis. IEEE Trans. Fuzzy Syst. **30**(6), 1939–1951 (2022)
30. Zhang, Q., He, D.: Adaptive neural control of nonlinear cyber-physical systems against randomly occurring false data injection attacks. IEEE Trans. Syst. Man Cybern.: Syst. **53**(4), 2444–2455 (2023)
31. Zhang, P., Liu, T., Jiang, Z.P.: Systematic design of robust event-triggered state and output feedback controllers for uncertain nonholonomic systems. IEEE Trans. Autom. Control **66**(1), 213–228 (2021)

www.ingramcontent.com/pod-product-compliance
Ingram Content Group UK Ltd.
Pitfield, Milton Keynes, MK11 3LW, UK
UKHW020634260926
475829UK00003BA/82
* 9 7 8 9 8 1 9 5 4 9 9 6 2 *